SCIENCE
AS IT
COULD HAVE
BEEN

SCIENCE AS IT COULD HAVE BEEN

Discussing the **CONTINGENCY/ INEVITABILITY** *Problem*

EDITED BY
Léna Soler, Emiliano Trizio,
and Andrew Pickering

UNIVERSITY OF PITTSBURGH PRESS

Published by the University of Pittsburgh Press, Pittsburgh, Pa., 15260
Manufactured in the United States of America
Printed on acid-free paper
10 9 8 7 6 5 4 3 2 1

ISBN 13: 978-0-8229-4445-4
ISBN 10: 0-8229-4445-6

Cataloging-in-Publication data is available from the Library of Congress.

To Cathy Dufour, our very dear friend and colleague who suddenly and prematurely passed away in March 2011.

Cathy was a bright, dynamic, and generous person, and a pivotal contributor to the PratiScienS research devoted to contingency in science. She always engaged enthusiastically in the various demanding collective activities of the PratiScienS team. In particular, she actively participated in organizing the international conference that gave the first impulse to this book. There she gave a fascinating talk on contingency in the history of quantum physics, which would have been part of this book had she still been with us later to finalize a written version.

We miss her.

CONTENTS

ACKNOWLEDGMENTS

We offer our greatest thanks to many people and institutions for their indispensable roles in realizing this volume.

First of all, let us express our great appreciation to all the authors: it is their investment of time and effort that brought the present project to a successful end. We are grateful for their important contributions to the still-underdeveloped issue of contingentism/inevitabilism in relation to science.

Second, we acknowledge French institutions for their support of activities related to this research program, namely, the Agence Nationale de la Recherche (ANR), the Maison des Sciences de l'Homme Lorraine (MSHL), the Région Lorraine, the Laboratoire d'Histoire des Sciences et de Philosophie (LHSP)–Archives Henri Poincaré (UMR 7117 of the CNRS), and the Université de Lorraine. In particular, the support of these institutions along with the Fondation des Treilles (Provence, France) enabled us to organize, in September 2009, an international conference on the contingentist/inevitabilist issue, in which many of this book's contributors participated.

We also thank the two reviewers solicited by the University of Pittsburgh Press. We greatly appreciated their helpful recommendations and were very gratified to read their supportive comments. Many thanks as well to the multiple people who, at an earlier stage, reviewed each chapter of this book in parallel with the editors. Their insightful suggestions highly contributed to the quality of this volume.

Finally, special thanks to the staff of the University of Pittsburgh Press for their collaboration in the publishing process. We are specifically grateful to Abby Collier for her warm support and invaluable advice, as well as for her unexampled expertise and efficiency all through the process.

INTRODUCTION

The Contingentist/Inevitabilist Debate

Current State of Play, Paradigmatic Forms of Problems
and Arguments, Connections to More Familiar
Philosophical Themes

LÉNA SOLER

This book aims to explore a crucial but neglected philosophical
issue: the issue of contingency versus inevitability within science.
I will refer to it below, borrowing a vocabulary introduced by Ian
Hacking (1999), as the "contingentist"/"inevitabilist" debate.[1] On first approach, the issue points to questions such as those that follow.

Could science have been otherwise?[2] Could human science, or some
aspects of it, have been dramatically different from science as we know it?
Or is there something inevitable in any sound scientific enterprise, and if
so, what, exactly? Was what we currently identify with our most reliable
scientific knowledge inevitable, that is, necessary under some conditions?
Were quarks, genes, continental drift, the standard quantum theory, or
mathematical theorems inevitable components of any human science?
Or could we have developed an alternative successful science based on
essentially different notions, conceptions, and results? Could our taken-as-
established scientific conclusions, theories, experimental data, ontological
commitments, and other scientific "results" (in a broad sense of result),
have been significantly different? Widening the scope from scientific results to the means of scientific inquiry, could all or part of what we take
as reliable scientific methods, procedures, reasoning, explanations, techniques, material and immaterial instruments have been significantly dif-

ferent—contingent in this sense? Can aspects of our science be detached from the accidental details of history and be granted the status of inevitable elements of any possible successful science?

I will not endeavor here to specify the question further, for two reasons. First, to clarify its meaning and to frame it in interesting ways are parts of the philosophical work that remains to be accomplished. As Joseph Rouse notes in chapter 13—and as many authors of this volume express in different forms—in the present state of the contingentist/inevitabilist issue, "the problem is not how to answer a well-defined question in a given context. We need to find a context and define the question so as to tell us something important about science as it has been and will be." In other words, and for reasons that will become clearer throughout this book, an important part of the philosophical task is to clarify the meaning of the terms involved, to design ways of formulating the question that prove most illuminating, tractable, or productive for some specified purpose, to distinguish multiple versions and levels of the problems involved, and so forth. Second, although one core aim of almost all contributions to this collection is precisely to take some steps in this direction, the privileged strategies and the solutions taken as more fertile prove to be diversified.

In this introduction, I provide landmarks that situate the contingentist/inevitabilist issue in the science studies literature,[3] offer insights about its philosophical significance and interest, and specify the structure and contents of this book. All along the path, a transversal aim is to reveal important themes and difficulties that prove essentially associated with the debate on contingency in science and are repeatedly involved in the available discussions of the topic.

The Contingentist/Inevitabilist Issue and Related Matters in Science Studies: The Current State of Play

A Neglected Issue

The so-called turn to practice within the field of science studies, which can retrospectively be taken to have begun in the 1970s, but was more self-consciously advocated and then culminated in the 1990s,[4] has fostered a sensitivity to context-dependent, sometimes very localized variations within scientific practices, diachronically as well as synchronically, and has revealed significant differences, both from one scientific group to another and among contemporary individual practitioners of science. This recorded variability in time and space resulted in a tendency to emphasize contingent aspects of science.

Such a tendency took up different forms.

1. It became manifest mainly through numerous scattered declarations, especially in the constructivists' writings. As an example among innumerable possible ones, let me mention some writings of Barry Barnes. Barnes (1991) states that "Reality will tolerate alternative descriptions without protest" (§ 6.1).[5] He then suggests, against the adoption of an "idealist metaphysics," to maintain "the traditional usage of 'world' and 'reality' without denying the referential aspects of speech," and simply to "say that there is one world, and any number of different descriptions of it" (§ 6.8). Such a "multiplicity thesis" (Trizio 2008) about scientific accounts of one supposedly unique reality is what I have called the "realistically framed" formulation of the contingentist issue: to conceive the issue as a relation between "how the world is" and "what science says about the world" (see Soler 2008a, 222, point A). The realistically framed formulation seems to correspond to the most "natural"—or in any case the more widespread—way of understanding contingency in relation to science. Contingentist declarations of this type have sometimes prompted criticisms, specifically by realist-inclined thinkers.

2. More rarely, the tendency to emphasize contingentist aspects of science took the form of full-blown contingentist theses purporting to hold for specific episodes of the history of science. Two cases have become exemplary in this respect and are overwhelmingly mobilized in the diverse contributions of this book: Andrew Pickering's reading of the history of particle physics in the course of the twentieth century; and James Cushing's account of the development of quantum mechanics in the first half of the twentieth century. Pickering famously argued that the "old paradigm" of particle physics in the 1960s could have successfully continued to develop instead of the "new" quarky paradigm that actually came to supersede it in the 1980s, so that we could have been left with a very different but no less sound nonquarky physics (see Pickering 1984a, 1995a). Cushing endeavored to back up what he takes to be a historically plausible counterfactual history of quantum physics, according to

which physicists would have ended with an ontology incompatible with the one associated today with the dominant standard quantum theory (see Cushing 1992, 1994).

3. Still more rarely, contingentist trends took up the form of general arguments intended to hold independently of the peculiarities of historical cases, especially in relation to the social dimension of science. In this vein we can mention among the earliest attempts a paper by Harry Collins (1981), in which the author aims both to substantiate the claim that scientific method is irreducibly contingent and to elucidate why this kind of claim is so difficult to vindicate.[6] Pickering's *The Mangle of Practice* (1995a) is another case in point, albeit of a different type. The book explicitly defends a contingentist thesis of wide scope, based on general features of science as it is practiced, such as "opportunism in context" and the "mangle" of an indefinite multiplicity of material, conceptual, and social ingredients, from which robust but nevertheless contingent holistic fits, called "scientific symbioses," sometimes emerge.[7]

Yet, beyond an undeniably increased sensitivity to contingent aspects of science in science studies, there have been few attempts to provide systematic philosophical analyses of the contingentist/inevitabilist issue.[8] In this respect, some circumscribed works of Hacking have played a seminal and decisive role. Hacking first isolated the contingentism/inevitabilism issue as an autonomous question in his famous book *The Social Construction of What?* (1999). Here, he pictured the dichotomy as one "sticking point," distinct from the realist/antirealist dichotomy, about which thinkers can endlessly disagree. Then, in a lesser-known article published in 2000, tellingly titled "How Inevitable Are the Results of Successful Science?," he articulated a more detailed and complete characterization of the issue. The crucial question that needs to be answered is, according to him: "If the results R of a scientific investigation are correct, would any investigation of roughly the same subject matter, if successful, at least implicitly contain or imply the same results?" (Hacking 2000a, 61). Having proposed a well-defined formulation of the general problem, the paper delineates and discusses some intrinsic conceptual difficulties associated with any similar problem formulation, thereby offering a global theoretical framework to address the issue.[9]

It is often under the influence of these writings from Hacking that philosophers of science have been led to mention the contingentism/inevitabilism issue, and sometimes to discuss it further (Franklin 1998; Giere 2006; Radick 2005a). My own inspiration first came from Pickering (1995a) and Cushing (1994), but received a decisive impetus from Hacking's later analyses. These analyses helped me both to grasp the distinctive philosophical import of the issue and to realize that the issue was not recognized as an autonomous philosophical problem of major significance in science studies—unlike, say, the debate on scientific realism. This motivated me to contribute to the stimulation of further interest in the topic through conferences and publications,[10] and to institute the contingentist/inevitabilist issue as one core theme of the research project "PratiScienS" that I launched in 2007 in Nancy.[11] In this context, an international conference was accepted by the Fondation des Treilles and took place in September 2009.[12] Most of the authors of this book attended the conference. Additional contributions to this volume have been solicited for the sake of thematic richness and diversity of viewpoints. In the volume as it stands, various approaches, suggestions, and theses about the contingentist/inevitabilist issue are represented. This reflects the diversity of the contributors' geographic origins, scientific backgrounds, and philosophical orientations.

From the beginning of the 2000s onward, the situation of the contingentist/inevitabilist problem has slowly evolved. A growing number of scholars have shown interest in the problem and have explored it along diverse lines. In the next two sections, I provide information about the main recent contributions I am aware of, most specifically on very recent and still little-known works from the philosophical young guard. Simultaneously, I endeavor to introduce the reader to key substantial and methodological issues pervasively at stake in the contingentist/inevitabilist problem—without any pretense of being exhaustive. Regarding *substantial* issues (see 13–19), I spotlight some connections between the so far still neglected contingentist/inevitabilist problem and other, often more familiar philosophical themes, such as scientific realism or scientific pluralism. Regarding *methodological* issues (see 6–13), I put forward some important problems that are repeatedly raised in the contingentist/inevitabilist debate, such as whether we are dealing with an answerable question or whether a counterfactual history of science can play the role of a warranting resource.

Recurrent Methodological Concerns

A "How-Much" and/or a "What" Question?

Are contingentism and inevitabilism susceptible to degrees? Are there more or less strong forms of contingentism and inevitabilism? Is the force of a contingentist/inevitabilist thesis dependent on what is claimed to be contingent or inevitable within science? Such questions are often more or less implicitly at stake in discussions of the contingentist/inevitabilist issue and call for a more systematic explicit examination.

I raised the question of degrees of contingentism and inevitabilism some years ago, noting that "it suffices to endorse the possibility of two incompatible alternative sciences to be a contingentist." I then asked: "But will the one who endorses the possibility of many more supplementary options be a stronger contingentist? Or should we introduce degrees of contingentism depending on the kind of contingent factors that are supposed to play a role?" (Soler 2008a, 223, point B). These brief remarks on the interrogative mode motivated further research from Joseph D. Martin.[13] Martin starts to "answer Soler's question in the emphatic affirmative," to the extent that in his opinion, "we should distinguish types of contingentist/inevitabilist positions according to what is claimed contingent/inevitable" (Martin 2013, 919). On this basis, he then argues that the answer to the question that is the title of his paper, namely, "Is the Contingentist/Inevitabilist Debate a Matter of Degrees?" is "a resounding no" (919). Rather than a "how much" question, the debate conveys, according to him, a "what" question.

Such issues and related ones are addressed in numerous passages of the present volume. For example, Catherine Allamel-Raffin and Jean-Luc Gangloff wonder, in chapter 2 (102–3), whether we ought to frame the contingentist problem as a dichotomy or as a spectrum. They favor the latter option, arguing that "it is possible to be more or less contingentist, more or less inevitabilist." They also attempt to define the two extremes of the spectrum, starting from (what I have called in Soler 2008a) "benign contingentism," to (what they call) "very radical inevitabilism." The latter corresponds to the idea of a history of science governed by an "inflexible necessity" at all levels, which, they state, "can be seen ironically as an occurrence of the '*fatum mahometanum*.'"

In concert with Martin's suggestion that the contingentist debate should be primarily treated as a "what question," several contributions to this book insist that we should distinguish contingentist/inevitabilist doctrines according to their target, that is, according to what, exactly, they claim to be contingent or inevitable within science. An indefinitely wide variety of targets at different scales and in different scientific fields can be the object of contingentist/inevitabilist claims. The possible targets include not only multiple types of scientific results (a theory, a local theoretical established statement, an experimental fact, etc.) but also many aspects of what can be categorized as scientific method in a broad sense (experimentation, empirical adequacy, anything that is taken as a scientific virtue, etc.). Arguments and positions for and against contingency or inevitability may need differentiation according to the target. In chapter 8, Yves Gingras puts a strong emphasis on the idea that a differentiation of this type should absolutely be introduced. As he writes, "the degree of contingency of science can vary according to the kind of elements we are analyzing: phenomena, entities, laws, mathematical formulations of theories, and so on. In order to make sense, any discussion of inevitability of 'results' must always specify the element being discussed."

When differentiating along such lines, it becomes clear that some forms of contingentism are compatible with some forms of inevitabilism, at least because one can be contingentist about a certain kind of element but inevitabilist about another kind. This is emphasized and illustrated in many essays in this volume: in Léna Soler's chapter 1 (71–73); Catherine Allamel-Raffin and Jean-Luc Gangloff's chapter 2 (102–3); Emiliano Trizio's chapter 4 (137–45); Mieke Boon's chapter 5; Ronald N. Giere's chapter 7 (194–200); Yves Gingras's chapter 8 (209–18); and Joseph Rouse's chapter 13. To worry about differentiation also fosters the concern that the contingentist/inevitabilist issue could be field dependent. This concern is shared by Allamel-Raffin and Gangloff in chapter 2 (104–7); Trizio in chapter 4 (149); Giere in chapter 7 (200–201); Jean Paul Van Bendegem in chapter 9 (223–25), and Jean-Michel Salanskis in chapter 10 (241–55) (both about the contrast between the case of mathematics and the case of the empirical sciences); and Rouse in chapter 13 (319–25). Many of the authors just mentioned—Allamel-Raffin and Gangloff, Trizio, Giere, and Rouse—suggest that the case of physics could be significantly different from the case of biology and the earth sciences, and some of them urge us to pay more attention to the latter.[14]

Concerning the "what" of contingentist/inevitabilist claims, highly diversified cases are discussed in the book. Just to give an idea of this diversity in the

form of an inventory à la Prévert, the book considers the contingency/inevitability of ingredients of the following type: experimental facts; observational and experimental data; scientific laws; mathematical theorems; theoretical hypotheses or whole theories in mathematics, physics, cosmology, biology, geology, the biomedical sciences, the engineering sciences, interdisciplinary inquiries, and in fields contested as genuinely scientific such as parapsychology; empirically equivalent but incompatible physical theories; mathematical formalisms and linguistic formulations of scientific claims in vernacular or specialized languages; scientific questions; human sense data and human modalities of experience; instrumental devices developed and used in the empirical sciences; technical tests, experimental recipes, and other reproducible connections between well-defined initial and final empirical conditions; mathematical proofs; methods in psychology; identity and boundaries of a whole discipline like mathematics; scientific frameworks, scientific paradigms, robust fits, worldviews, or similar multidimensional integrated units. . . .
In addition, the book involves various periods of the history of science, and these periods are studied on different temporal scales according to the cases: examination of the details of scientific episodes in the short run; or adoption of more panoramic perspectives concerned by scientific trajectories considered in the long run. Such diversity will hopefully help readers to make their own judgment about the need to differentiate arguments and conclusions according to the kind of scientific element, the kind of structural scientific configuration, the scientific field or subfield, and the temporal scale.

Finally, note that to treat the contingentist/inevitabilist issue primarily as a "what" question does not necessarily imply a stop to talking in terms of degrees. It is indeed possible to characterize more or less strong forms of contingentism and inevitabilism *depending on their target*, as the book diversely illustrates. For example, to claim that the same scientific result could have been historically reached along a different path, that is, to be contingentist about the paths but inevitabilist about the result, can be viewed as a weaker version of contingentism than to contend that an incompatible scientific result could have been established (this intuitive option is, for example, taken by Arabatzis [2008]). In this option, "how much" judgments are a function of "what" judgments: the former are indexed on the latter according to some criteria. Alternatively, it is also possible to dissociate the two types of judgments. One can define more or less hard versions of contingentism or inevitabilism according to characteristics other than the target of contingentist/inevitabilist claims, whereas in parallel still being keen to carefully distinguish types of contingentist and inevitabilist conceptions according to their target (for an example among other possible ones, see the way in which Trizio distinguishes weak and strong contingentism in this volume, chapter 4).

Uses and Value of Counterfactual Thinking

Counterfactual history of science and "what if" thinking, which sometimes include audacious science fiction or talk about very abstractly formulated possibilities, form an important part of the substance of arguments involved in the contingentist/inevitabilist debate. So far, however, counterfactual thinking *about science* remains an underdeveloped activity.[15] Although in need of more systematic epistemological clarification, today, given the currency of naturalistic and empirically inclined methodological preferences (see the next section), this activity often tends to be devalued without further ado as devoid of any empirical import and hence unable to warrant anything.[16] As Steve Shapin once put it, counterfactuals suffer from a "credibility handicap" (Shapin 2007). Such a handicap seems to apply, not just within the field of history *of science*, but more generally in the field of history tout court. Giere notes that in the latter field, "counterfactual history has many more detractors than defenders" (this volume, chapter 7).

   With respect to epistemological clarifications concerning the use of counterfactuals about science, an important step has been the publication, in 2008 as a focus section of *Isis*, of an illuminating symposium directed by Gregory Radick, "Counterfactuals and the Historian of Science,"[17] with contributions from John Henry (2008), Peter Bowler (2008), Steven French (2008), and Steve Fuller (2008). As the title indicates, the constituent papers approach the question of contingency in science from the angle of a counterfactual history of science. One central aim is to "help clarify the epistemology of history of science counterfactuals: what we can know by the light of the best ones, what makes the best ones so good, and how to tell the best from the rest" (Radick 2008a, 550). Through varied case studies in different scientific fields (biology, physics, etc.), the symposium provides instructive and stimulating analyses of wider scope about the powers and limits of counterfactual reasoning. This is an invaluable contribution to the contingentist/inevitabilist debate, especially given that, as emphasized by Radick, "The theory and practice of counterfactual history of science are just beginning" (Radick 2008a, 551).

Counterfactual thinking is pervasively at stake in the present book (see below, 000). Multiple contributions develop particular counterfactual scenarios, and several provide meta-reflections about the demonstrative power of counterfactual arguments. The latter meta-reflections generally insist on the difference between counterfactuals involving "mere" *logical* possibilities and counterfactuals involving *historical* possibilities. It is often claimed that fictitious scenarios of the first type are not able to show anything significant about our science, and that only *historically plausible* alternative stories have a chance to do the job.

But how are we going to recognize *genuine* historical *possibilities?* What makes a counterfactual scenario *historically plausible?* This is a tricky issue, considered in several essays in this book.[18] Even the boundary between a "purely" logical and a historical possibility is not so sharp. We can conceive various degrees and various kinds of possibilities, not so easy to characterize, to compare, and to situate, when trying to make sense of a scale that would start with the actual history of science and would end with "purely" logical and highly abstract possibilities, including, in between, scenarios that would only slightly differ from our actual history of science, and more creative science fiction (for instance involving twin-earth-like planets or alien beings as the subjects of science).[19] If there seems to be a general agreement about the idea that what confers plausibility to a counterfactual scientific narrative is its "close connection" to the actual history of science, the idea of "close connection" is admittedly vague. Different thinkers can diverge, and de facto often do diverge, about plausibility judgments of this type.

In such an underdetermined situation, some prefer to discuss the contingentist/inevitabilist issue without making any use of counterfactuals. This is the radical stance taken by Harry Collins in chapter 6, at the price of a severe narrowing of the problem formulation. Others, without totally rejecting the use of counterfactuals, privilege the recourse to actual history of science as far as possible. This is the position favored and implemented by Gingras in chapter 8 (206–9). Still others are not afraid to make extensive use of audacious counterfactuals and endeavor to show that the corresponding thought experiments are able to reveal important things about science (see in particular Rouse's chapter 13 and Jean-Marc Lévy-Leblond's chapter 14). A number of contributors also attempt to characterize the features of some counterfactual strategies that are amenable to carrying a high convincing potential (see Soler's chapter 1, 66–70; Giere's chapter 7, 192–94; Rouse's chapter 13, 317–25; Van Bendegem's chapter 9, 223–25, 238–39).

A *Decidable or Undecidable Issue?*

Many thinkers who have struggled with the contingentist/inevitabilist issue are led, at one point or another, to worry, complain, or in some cases
conclude that the issue is undecidable—a judgment that usually conveys
a negative connotation. Declarations in this vein can be found in many
chapters of this book. Concern about decidability occupies a specifically
prominent place in chapters 1 (Soler), 6 (Collins), 7 (Giere), 8 (Gingras), 10
(Salanskis), and 15 (Chang). Assessments of decidability, however, obviously depend on options of what "decidable" should mean and what criteria of
decidability should be used in this context. These options are more or less
explicit and not always consensual. Very often, "decidable" points to the possibility of getting sufficiently decisive *empirical* evidence for or against some
well-defined contingentist or inevitabilist position. This should not come as
a surprise given the naturalist or empirically inclined contemporary mood
that prevails in all fields aiming at knowledge, philosophy of science included, in contrast to more aprioristic or logically inclined previously dominant
tendencies. Yet, even if we are committed without any reservation to this
sense of "philosophical decidability"—which is not universally the case, see
in particular Salanskis's chapter 10 in this volume, and below (34–35)—it is
far from being the end of the story. We still have to specify what is going to
count as *empirical* evidence and as genuine *evidence* of the empirical type.
But with such questions, we open a Pandora's box of the relations between
philosophy of science on the one hand and history and sociology of science
on the other hand—and perhaps more generally of the distinctive nature of
philosophy and its specific mode of demonstration.

Can we consider that historical and sociological accounts of scientific
episodes provide some independent empirical evidence against which the
credibility of philosophical positions could be assessed? What are we going
to do when several conflicting historical-social narratives of the same scientific episode exist? More profoundly, can we really draw an unproblematic
boundary between philosophical and historical-social readings of science?
Lingering questions of this kind can be raised in relation to any philosophical issue about science. But they appear especially problematic when applied to the contingentist/inevitabilist issue, among other reasons because
counterfactual history of science almost always plays a role in addition to
recourse to *actual* history of science (see the previous section).

Most contributors to this volume discuss the contingentist/inevitabilist issue in close relation with (what they take to be) lessons of the history and sociology of science.[20] A number of chapters furthermore identify versions of the above-mentioned difficulties and attempt to cope with some of them according to diverse strategies (see Soler's chapter 1, 58–61, 65–66, and 82–89; Trizio's chapter 4, 149–50; Collins's chapter 6; Giere's chapter 7; and Hasok Chang's chapter 15).[21]

The issue of whether the contingentist/inevitabilist debate is decidable or not and, as a corollary, the issue of the evidential support historical-social case studies are able to provide in this respect are at the heart of the work of a promising young philosopher of science I am keen to mention here, namely, Katherina Kinzel. Kinzel specifically dedicated her PhD dissertation, directed by Martin Kusch at the University of Vienna (Austria), to address the contingency issue in a systematic perspective.[22] The defense of her PhD took place in July 2014. This dissertation, titled "Could the Results of Our Science Have Been Different? Contingency and Inevitability in the Philosophy and Historiography of Science," is certainly the most systematic and complete work presently available on the topic of contingency in science. It offers an overall thorough and rigorous analysis of the contingentist/inevitabilist issue.

To cope with the problem of undecidability, Kinzel's strategy is to distinguish several levels of formulation, in particular the general-abstract and the local-concrete formulations. She then argues that the situation of each problem formulation sharply differs in terms of conceptual clarity, tractability, and decidability. When the question is framed at the general-abstract level, its central concepts remain "vague" and "deeply problematic," and "it is unclear how there could ever be empirical evidence for contingency and inevitability claims." But when the question is framed at the local-concrete level, it "can be made relatively clear sense of," and empirical evidence can be found from historiographical reconstructions of scientific episodes, against which contingentism and inevitabilism can be comparatively assessed.[23] We can note that Kinzel's methodological strategy and her position about the possibility of empirical support at the local-concrete level are very much in line with the options favored and implemented by Harry Collins in chapter 6 of this volume (see below, 30).

Another young philosopher whom I would like to mention in relation to considerations of decidability of the contingentist/inevitabilist issue, is Ian James Kidd. Kidd's research program is, in large part, specifically dedicated to the contingentist/inevitabilist issue, and he has provided recent stimulating contributions to the topic. One important root of Kidd's work on contingency is David E. Cooper's ideas on "hubris."[24] These ideas deserve mention here qua a possible source of fertile suggestions about the contingentist debate—not only about its decidability but also about its relation to scientific pluralism and scientific realism (see in particular Cooper's 2002 book, *The Measure of Things: Humanism, Humility, and Mystery*, specifically chapter 8). "Hubris" refers to criticized positions that are guilty of lack of humility. These positions include the "absolute" realist stance associated with modern science. Instead, Cooper favors humility, and he argues that humility requires an awareness of the contingency of beliefs, and more generally of forms of life. Accordingly, hubristic doctrines are those that deny, ignore, or fail to properly reflect on the consequences of their contingency. Kidd has endeavored to turn Cooper's criticism of hubristic positions against inevitabilism.[25] He has developed several interesting arguments in this spirit, circling around the idea that inevitabilists are guilty of "epistemic hubris." Kidd's central point is that because inevitabilists act as if they possess epistemic powers that they do not possess, their position always lacks warrant. More generally, Kidd insists that thinkers too often assume that they can, in fact, discover the truth about whether or not a given scientific target is contingent or inevitable. Such an assumption is, according to Kidd, hubristic, because in most cases, it is unclear that we could ever settle the matter.

To conclude these brief reflections on decidability, I want to make one last point. Suppose that the contingentist/inevitabilist issue, or some version of it, turns out to be definitely undecidable—whatever "decidable" is assumed to mean, including the current sense of empirical decidability. This would not necessarily imply that the issue has to be rejected as lacking any philosophical interest and fertility. The section titled "About the Philosophical Significance and Imports of the Contingentist/Inevitabilist Issue" below can be taken as a possible substantiation of this point. The suggestions I offer there about the philosophical significance and the possible benefits of exploring the issue are largely independent of the empirical decidability of the latter.

Prominent Philosophical Themes Connected with the Contingentist/Inevitabilist Issue

Let us now turn to three familiar philosophical themes that enjoy substantial conceptual connections with the contingentist/inevitabilist issue: first, scientific realism; second, scientific pluralism; and third, theses such as the underdetermination of theories by observational data, the incommensurability of scientific paradigms, and related matters. The latter theses, although frankly different from one another in important respects, are nonetheless gathered here because they share one common point, which proves crucial in the context of our problem: the conceptions of the determinants of scientific development the theses incorporate imply, or at least strongly suggest, that the collective "choices" actually made in the history of science between competing theories or paradigms could perfectly have been strongly different, and *legitimately* (rationally, methodologically) different.

Because of the substantial connections between the above-mentioned themes and our problem, it is expected, and often recorded, that discussions of the contingentist/inevitabilist problem involve, at one point or another, some of the philosophical themes in question. Conversely, available philosophical analyses of these themes are privileged loci to find hints related to contingency in science. However, what seems to have happened so far, considering science studies globally, is that allusions to contingency surface only as a sort of epiphenomenon in discussions about scientific realism, scientific pluralism, underdetermination, and incommensurability. The ultimate aim remains to discuss the latter issues, and contingentist claims are commonly associated, if not assimilated, to antirealism, constructivism, instrumentalism, conventionalism, relativism, and the like, without any concern about possible disentanglements. Overall, the contingentist/inevitabilist opposition is not isolated as a distinct well-identified interesting question, and the problems and stakes intrinsically related to it are not disengaged.

Connections with the Debate on Scientific Realism

Except in the rare contexts in which the aim is specifically to clarify the relations between the two issues of contingentism/inevitabilism and realism/antirealism, the second issue clearly monopolizes philosophical attention, thereby eclipsing the contingentist/inevitabilist conflict or hiding its specific interest—when the two issues are not simply conflated.

This is to be regretted—for the two issues, in spite of numerous relations at the logical and the psychological levels, do not coincide. This has been coherently shown along different lines by the (admittedly few) scholars who have seriously examined the connections between the two debates so far (see Hacking 1999, e.g., 79–80; Radick 2005a, 20–25; Soler 2008b, 231–32; Sankey 2008). And this is further argued and illustrated in several contributions to the present book, in particular Trizio's chapter 4 and Boon's chapter 5. All these works agree that associations between scientific realism and inevitabilism on the one hand, and antirealism and contingentism on the other hand, correspond to the conceptually most comfortable and empirically most frequent pairs (for representatives of realist-inevitabilism and antirealist-contingentism, see Soler 2008a). However, other combinations are nevertheless possible—as all the above-mentioned works also attest—albeit remaining "for most tastes, decidedly exotic" (Radick 2005a, 25) and more "difficult to substantiate and defend" (Trizio's chapter 4, this volume). As Gregory Radick concludes: "realism goes with inevitabilism, and antirealism with contingentism, as a matter not of logic but of psychology" (2005a, 47).

Another reason that it is beneficial to untangle the two debates on contingentism and realism is that the contingentist/inevitabilist issue, when considered in its own right, has specific imports that turn out to be of great epistemological significance and can even shed new light on the available argumentative network that constitutes the scientific realist/antirealist debate. Both Trizio and Boon substantiate this point and indicate different specific imports in chapters 4 and 5 (see also Soler 2008b, sec. 2, last paragraph). Radick (2005a) makes a similar point at the end of his paper. Investigating the contingency issue through a counterfactual history of science could, according to him, provide "distinctively philosophical returns" regarding our psychologically almost inescapable realist commitments. More precisely, he suggests that "A decent counterfactual historiography of science could be just the therapy needed, to enable either the surrendering of stubborn realist intuitions or, as it could well turn out, a surrendering to them" (Radick 2005a, 47).

Exploring the complex relations of the inevitabilist/contingentist and the realist/constructivist issues helps us to better grasp the contents and implications of each, and to enrich the conceptual space of our philosophical views about science.

Connections with the Debate on Scientific Pluralism

Substantial relations bind contingentism to pluralism and inevitabilism to monism. Any contingentist claim conveys a "multiplicity thesis"—as Trizio

(2008) puts it. If you affirm that X is contingent, you are committed to the claim that X', X", and so forth could have been the case instead of X. Conversely, if you profess that X was inevitable, you are committed to a uniqueness claim about X. In that way, contingentism and pluralism on the one hand, and inevitabilism and monism on the other hand, are conceptually and logically related.

Beyond that, various forms of each of the four positions involved in the quartet,[26] and thus diverse versions of contingentist-pluralist and inevitabilist-monist conceptions, are possible, and can be found. In particular, pluralism and monism can be advocated according to different modes. They can be advocated in a *descriptive* mode: as a claim about actual features of science as we know it (see, e.g., Lévy-Leblond's chapter 14 for a descriptive pluralist thesis applied to physics). Or they can be advocated in a *normative* mode: as a claim about how we should practice and conceive science, sometimes associated with some suggested policies designed to come closer to the preferred ideal (see Hasok Chang's chapter 15 for a defense of normative pluralism). Finally, they can also be endorsed simultaneously in both modes. Chang offers a recent illustration: although in his contribution to the present book, he focuses mainly on *normative* pluralism, he more generally champions a pluralist stance that is both normative *and* descriptive (see, e.g., Chang 2012). Another antecedent emblematic illustration of a pluralist position endorsed in both descriptive and normative modes is Paul Feyerabend's philosophy (see in particular Feyerabend 1965, 1993). And this philosophy is itself not without interesting relations to contingency. Although usually not explicitly framed in terms of a "contingency question" or a "contingentist position," it is obviously sensitive to the historical contingency of scientific method and more generally of reason. This has been explicated by Ian Kidd in his PhD dissertation. There, Kidd discussed what can be reconstructed as Feyerabend's views on contingency (about Ian Kidd, see above, 12–13).

Examples of contingentist-pluralist and inevitabilist-monist associations are provided in the book. The relationships between the two tenets of each pair are a special object of concern at the end of chapter 1 (Soler) and in chapter 15 (Chang). The authors discuss what would plausibly happen to the currently prevailing intuitions and commitments about inevitabilism and contingentism, if a more pluralist regime of science came to be socially valued and supported.

Finally, let us turn to the relations between the contingentist/inevitabilist issue and conceptions of scientific bifurcations in the framework of theory underdetermination, incommensurability of paradigms, robust-fit schemes of scientific development, and associated matters.

Connections to Underdetermination, Incommensurability, and Symbiotic
Schemes of Scientific Development

Discussions of the contingentist/inevitabilist issue intersect with more classical discussions related to the underdetermination of theories by observations, the appraisal of the relative merits of empirically equivalent theories, the comparative evaluation of methodologically and semantically incommensurable frameworks, and the like. Many of the argumentative resources developed in these philosophically familiar contexts can be, and often are, adapted and reworked by contingentists for their own purpose.

Let me sketch the principle of some typical strategies along these lines. Consider first empirically equivalent theories. If several empirically equivalent but descriptively and ontologically incompatible theories are indeed sustainably possible, this can be turned into an argument according to which the theories actually retained in the history of science could have been otherwise. Because these theories have been retained on the basis of nonevidential criteria, and because the application of nonevidential criteria can vary from one individual and one group to another, other competing empirically equivalent theories could plausibly have been accepted. Consider next incommensurable paradigms. Concede that scientific revolutions indeed institute new scientific regimes characterized by partially disparate and largely incommensurable scientific values, norms, beliefs, relevant and significant problems, types of explanations and arguments, and accordingly, truly different scientific practices. Concede, moreover, that there is no absolute Archimedean platform from which to universally dictate one set of values and practices as *the* best one. If you concede this, then, plausibly, the paradigms that have actually prevailed in the history of science are contingent. In particular, it becomes plausible that our current conceptions of sound scientific method, scientific success, and scientific progress could have been significantly different.

The same strategy can actually be applied to any epistemological position that involves a multiplicity thesis, whatever the terms in which the

target of such a thesis is characterized—multiple scientific beliefs, representations, theories, ontologies, tacit commitments, values, norms, methods, habits, practices, or anything else. That being said, when surveying
contingentist strategies of this type, we quickly notice that two versions
of multiplicity theses are repeatedly at stake, and are often distinguished
regarding the nature of their contingentist implications: (1) the underdetermination scenario and (2) the robust-fit scenario, which almost always ends
in claims of incommensurability.

> **1.** The first version surfs on the classical idea of underdetermination
> of theories by observations: it involves multiple incompatible (con-
> temporary or successive) scientific theories of the same data. This
> version points to the possibility of the contingency *of theories*, while
> possibly maintaining the inevitability *of observations and observa-
> tional statements*—a possibility that, incidentally, instantiates one
> obvious way of being compatibilist by differentiating contingentist
> and inevitabilist claims according to the target (see above, 6–9). The
> abstract possibility of theory-underdetermination is frequently illus-
> trated by the current situation of quantum mechanics: by the fact
> that two incompatible but empirically equivalent theories, the stan-
> dard and the Bohmian quantum theories, actually coexist. Whether
> this fact can be turned into a contingentist thesis about the presently
> dominant standard quantum theory is frequently examined relying
> on Cushing (1994) (see above, 3–4).[27]
>
> **2.** The second version corresponds to multiple scientific robust fits
> among various heterogeneous elements—conceptual, material,
> practical, social, and so on. In the long run, it points to the pos-
> sibility of incommensurable scientific "traditions," "paradigms,"
> "practices," "worlds," or similar integrative units. Such units could
> have been otherwise (are contingent), to the extent that many of the
> elements that constitute a robust fit at a given time, as well as the
> copresence of the "initial" antecedent elements from which a "final"
> robust fit came to emerge in a subsequent stage, could perfectly and
> unquestionably not have been instantiated in the historical reality.
> The case of incommensurable large integrative units leaves much
> less room, if any, for the straightforward identification of an invariant
> shared stratum between the two compared units, including at the
> observational level. Accordingly it is much harder, for inevitabilists,

to identify candidates that could be the residual target of an inevitabilist claim (whereas in the theory-underdetermination version, an obvious candidate is available, namely, the stratum of observational data). This second version is overwhelmingly discussed in reference to two works: Andrew Pickering's writings, especially *Constructing Quarks* (1984a) and *The Mangle of Practice* (1995a); and one article from Hacking, "The Self-Vindication of the Laboratory Sciences" (1992).[28] In substance, these works argue as follows. Multiple elements of scientific practices, including the instruments, techniques of data analysis, and so on, are tailored to fit with one another in the course of the history of science. As a result, they enjoy relations of mutual support. Overall, we are left with "self-vindicated" scientific wholes described as "closed systems" (Hacking) or "scientific symbioses" (Pickering). Attempts to ground contingentist arguments in such a robust-fit or symbiotic scheme usually point to the multiple "plays" or "degrees of freedom" of the adjustments at each stage (Pickering talks of "plasticity"),[29] and to the path dependency of the moves from one stage to another.[30] As a consequence, different incompatible or incommensurable self-vindicating scientific symbioses could have plausibly emerged and could have come to dominate in the history of science.

Both the theory-underdetermination version and the robust-fit version of the multiplicity thesis convey holistic schemes of scientific development. Yet the corresponding holistic schemes do not embody the same ingredients, and the ingredients embodied in the robust-fit case are much more diversified. These two configurations can be equated with two varieties of Duhem–Quine-like theses, with the robust-fit variety exemplifying a more radical case than underdetermination of theories by observational data.

The two configurations are pervasively involved, both in general in writings, which deal with the contingentist/inevitabilist issue (see, e.g., French 2008; Soler 2008a, 2008b; Trizio 2008; Kinzel 2014), and in particular in the present book. A comparative analysis of the two cases in the background of the contingentist/inevitabilist issue is attempted in Soler's chapter 1 and Trizio's chapter 4. Some relations between the contingentist/inevitabilist issue and Kuhn's incommensurability thesis are discussed in several places (see, more specifically, Pickering's chapter 3).

About the Philosophical Significance and Imports of the Contingentist/Inevitabilist Issue

A Still-Marginal Issue Today

The recent works mentioned in the sections above show that more and more researchers, especially younger ones, in science studies have been interested in the debate on contingency in the past decades, and have provided valuable contributions on the subject matter. Despite these encouraging advances, however, the contingentist/inevitabilist debate continues to occupy a relatively marginal position in philosophical, sociological, and historical works interested in science. In mid-2014, Katherina Kinzel was still in a position to express a regret similar to the one I expressed six years before in the introduction to the 2008 symposium (Soler and Sankey 2008), about the little attention devoted to the issue as an autonomous question. In Kinzel's terms: "Strikingly, until now there has been relatively little discussion regarding the contingency of scientific knowledge and practice. . . . Only a small amount of systematic philosophical work addresses this problem as an independent topic. When the issue is raised the question of contingency in science is usually discussed as a consequence of other philosophical views and doctrines, with the related notions of contingency and inevitability remaining vague and intuitive" (Kinzel 2014).

It is hoped that the present book will further boost interest in the topic and remedy the lacuna in the present situation. In any case, this is one of its motivations. In this perspective, let me indicate what I identify as the potential philosophical benefits of investigating the contingentist/inevitabilist issue as an autonomous question. I shall stick with benefits *of general scope*, that is, benefits that simultaneously (1) transcend possible returns concerning *particular* other philosophical themes such as realism and the like, (2) are valuable independently of the specific contingentist or inevitabilist stance one may favor "at the start," and (3) are largely independent of the decidability of the controversy (or some version of it).

Some Philosophical Benefits of Investigating the Contingentist/ Inevitabilist Issue

To my eyes, the contingentist/inevitabilist issue is philosophically stimulating, attractive, and fertile, specifically because beyond the particular position each will ultimately favor, a systematic exploration of the issue helps us to identify and assess some very pivotal commitments of our present form of life, of our ways of approaching historical realities, and of our conception of what science and knowledge are and could be.

To clarify how this can happen, I find it appropriate to start from inevitabilism. When engaging in discussions about the contingency and inevitability of scientific accomplishments, when attempting to deploy the space of the arguments potentially available on each side, we quickly suspect or experience, in many interlocutors and possibly in ourselves as well, something like an inevitabilist instinct about science. The inevitabilist instinct seems deeply active in most of us in one form or another, even if depending on the individual sensibilities, different kinds of scientific targets might work as the most powerful attractors of inevitability. Several thinkers have insisted on such a situation (see, e.g., Hacking 1999, 79; Radick 2005a, 25–26, 45; Soler 2006a; Kidd (forthcoming); in the present volume, this situation is a central concern of my chapter 1, Pickering's chapter 3, and Chang's chapter 15. The suspicion or experience of an inevitabilist instinct encourages an examination of what lies behind it. And to examine what lies behind the inevitabilist instinct is instructive along several lines. Below, I consider three lines.

1. Under examination, behind the instinct, it proves hard to find any argument strong enough to support a true inevitabilism. Albeit instinctive, inevitabilism might well be, as Ian Kidd has contended, a "hubristic" position (see above, 13). And although counterintuitive, a nontrivial form of contingentism might appear more plausible. Note that I say "might." The latter remarks are not intended to *positively assert* that contingentism is more credible than inevitabilism or to presume anything about the outcome of the inquiry. The point is that inevitabilism should not just be taken for granted in the philosophy of science, because a sound philosophy of science should not treat any position as the default position without serious discussion. However, very often inevitabilist stances are more or less implicitly treated this way.

This treatment has two possible outcomes. Either no genuine problem is recognized, and the whole issue is then completely ignored; or the burden of proof is entirely placed on the side of contingentism. In the second case, only contingentists are supposed to have some work to accomplish, whereas inevitabilists would have nothing to do and could feel secure until no strong argument has been supplied in support of contingentism. This is not a philosophically satisfying situation, as several thinkers have stressed (see, in particular, in this volume, Soler's chapter 1; Pickering's chapter 3; Chang's chapter 15; and in sources outside of this volume, Radick 2005a; Kidd forthcoming). To recognize and investigate the contingentist/inevitabilist controversy as a worthy issue and as an autonomous question fosters a more acute awareness of a philosophically unsatisfying situation; it also provides a strong incentive to scrutinize the roots, grounds, and implications of the inevitabilist commitment that is entrenched in ordinary and philosophical ways of thinking about science. From a philosophical point of view, it cannot but be beneficial, whatever conclusions are finally drawn, to refrain from taking inevitabilism as the default position, to treat the two poles of the dichotomy on the same footing, and to go into a thorough examination of the inevitabilist arguments we can hope to articulate, in addition to the much more often attempted examination of contingentist arguments.

To testify that the inevitabilist instinct can be tamed, and to be clear about my own sympathies, I can record that when I started to read the contingentist writings of Andrew Pickering and Harry Collins, I was rather skeptical. But after a long and, I must say, often toilsome trajectory, I am inclined to conclude that, at least in a comparative perspective, contingentism—in a sense that has to be spelled out—is more arguable and plausible than inevitabilism. We are not forced, however, to make a radical binary choice between the two poles of the dichotomy. As emphasized above (see 6–9), and as the contributions to this book diversely illustrate, some combinations and compatibilist stances are possible. Or we can also, as Salanskis does in chapter 10, reject the dichotomy as a legitimate yes/no philosophical question.

2. When struggling to make sense of inevitabilism, we quickly understand that what is at stake can only be a *conditional* inevitability; and when trying to specify the conditions under which something could be considered inevitable, we are forced to elucidate how historical reconstructions of scientific development have been or could be built. In other words, we are led to make explicit the kind of intuitions, assumptions, criteria, and so on, according to which

a given narrator of a given episode in the history of science has drawn the boundary between anecdotal or irrelevant circumstances, conditions of possibility, and factors under which what has indeed happened had inevitably to happen. In that way, the exploration of the contingency issue contributes to clarifying the relations between the history and philosophy of science. A brilliant clarification of this kind can be found in Fuller (2008).

3. Last but not least, while analyzing what lies behind inevitabilist intuitions, we find very fundamental and pivotal commitments that seem inherent in the very idea of what we value as science—and perhaps more generally inherent in the idea of any cognitive activity that goes with a *descriptive* pretension. I have more specifically in mind the commitment to uniqueness that goes with the idea of genuine knowledge, at least as a regulative ideal sometimes expressed in reference to a hypothetical "end of research." This commitment is in turn fueled by another one, realist in spirit, according to which knowledge is knowledge *of one unique world* that is what it is once and for all (on the uniqueness commitment and its relation to realism, see in particular, in this volume, Soler's chapter 1, 84–86; Pickering's chapter 3; and Chang's chapter 15, 359–62, 377–81). The contingentist/inevitabilist issue can thus be viewed as a means to reveal such entrenched commitments, to examine the work they accomplish in our lives, and to consider what might happen if they were shifted, relaxed, or transformed. It is an incentive to reflect on what could be alternative enterprises of knowledge and correlated alternative forms of life, and on that basis, to assess the desirability of our present condition.

Structure of the Book and Cursory Overview of Its Contributions

In the final part of this introduction, I explain the logic that has oriented the composition of this volume and offer a brief overview of its diverse contributions.

Global Survey of the Problem Situation

The volume opens with two contributions that, beyond the personal positions defended by their authors on particular points, introduce the reader to the main types of questions, demands, arguments, strategies, and difficulties involved in the contingentist/inevitabilist debate, thereby offering a first global survey of the problem situation.

The heart of my chapter 1 is the "put-up-or-shut-up" demand addressed by inevitabilists to contingentists, that is, the demand, formulated in catchy words by Hacking (2000a), to "put up" an actual alternative science or to "shut up." But the discussion of this demand leads me to considerations of wider scope. The systematic reconstruction of the dialogue between the two camps reveals and assesses a large network of paradigmatic strategies, arguments, and replies that prove to be pervasively involved, not only in exchanges restrictively centered on the put-up-or-shut-up demand but also in various other disputes connected to the contingentist/inevitabilist issue.

I argue that the put-up-or-shut-up demand cannot be satisfied, for two reasons. The first is that inevitabilists and contingentists are committed to different readings of the history of science. The second corresponds to the monist regime of our science. Science, as we conceive and practice it, is governed by a monist regulative ideal that goes hand in hand with a uniqueness commitment so that from a psychological point of view, scientists feel compelled to choose *the* best theory/hypothesis among competing candidates and, from a sociological point of view, the development of a multiplicity of alternatives is not encouraged and not supported, ideologically or financially. I argue that under such a regime, it is quasi-impossible to find anything in our actual history of science that could inhibit the inevitabilist instinct. The presently monist regime of our science is designed to eliminate what the put-up-or-shut-up demand asks for. Moreover, the monist regime inculcates and cultivates intuitions and entrenched commitments that too easily lead us to dismiss any instance contingentists "put up," whatever its specific features. Thus, insofar as a monist regime holds, the contingentist failure to answer the put-up-or-shut-up demand—either taken as an actual failure or as an illusion induced by inevitabilist unwarranted commitments—cannot be taken as empirical evidence against contingentism. Moreover, contingentists can stress that the monist regime is not itself inevitable: more pluralist scientific regimes could be instituted, in which scientific alternatives would be valued and cultivated, so that the situation of actual alternatives in the history of science would be very different.

In fine, my contention is that, in a monist regime, (1) contingentists should refrain from wasting their time trying to answer the put-up-or-shut-up demand; (2) the so-called arguments against contingentism and for inevitabilism, based on the contingentist failure to put up the requested alternative science, are flawed; and (3) inevitabilism should not be considered as a default position supposed to be secured inasmuch as contingentists have not met the put-up-or-shut-up challenge.

Chapter 2 takes a "panoramic" analytical perspective on the contingentist/inevitabilist debate. Catherine Allamel-Raffin and Jean-Luc Gangloff's aim is to provide a survey of some pivotal issues that should be investigated and of some hard difficulties that should be faced in relation to the debate. In particular, they call our attention to the following key questions. Is the contingentist/inevitabilist issue strictly epistemological or is it also ontological? (In chapter 3, Pickering argues that it is also ontological.) Should we conceive the issue as a dichotomy between two mutually exclusive positions or as a spectrum involving different (sometimes possibly compatible) forms and degrees of contingentism and inevitabilism? At which scale should we situate the discussion? How do we define success? Should we define it in a realist perspective? Or in terms of empirical adequacy? Or in the robust-fit framework?

Whereas the first two chapters provide an overview of the problem situation and a general framework, the next three chapters focus on a more particular but still wide set of issues: they explore the relations between the inevitabilist/contingentist debate on the one hand, and ontological issues and the realist/constructivist dichotomy on the other hand.

Contingency, Ontology, and Realism

In chapter 3, Andrew Pickering explores one of the key questions raised by Allamel-Raffin and Gangloff in chapter 2: the relations between the contingentist/inevitabilist issue and ontology. He does so through a discussion of "the sort of place the world is," with the explicit intention, perfectly congruent with his earlier writings, of promoting "a contingentist vision," not only at the epistemological but, moreover, at the ontological level. He starts by elucidating the reasons why our inevitabilist intuitions about science are so strong and deeply entrenched. Then, he intends to introduce "an ontological antidote to inevitabilism." According to this vision, the world—nonhuman and human—is not fixed but is a "place of endlessly emergent performativity." "Contingency and chance are an integral part" of the dynamic open-ended process, tellingly pictured as a "dance of agency," through which all sorts of heterogeneous ingredients—material, practical, conceptual, social, and so forth—are "mangled" in scientific practices, and through which some items sometimes acquire the status of a firm scientific result. However, Pickering adds qualifications that in some sense manage to make some room for inevitabilist intuitions—and this leads to novel contributions with respect to the conceptions articulated in his previous writings. In the scientific enterprise as we know it, the dance of agency is driven by a determined "telos." Scientists aim at "interactive stabilizations" in which "the human and the nonhuman are split apart." And it so happens that they sometimes succeed, thus being left with "islands of dualist purity" and "of stability." The fact that this dualist separation between the human and the nonhuman is de facto possible is considered by Pickering as an ontological discovery, and as one source—perhaps the most fundamental?—of our inevitabilist intuitions. Pickering concedes that there is "something objective and noncontingent about these islands." Yet to concede the existence of "islands of dualist stability" in an ocean of endless emergence does not cancel the contingency of our science. This is because what counts as an island of dualist stability in scientific practice still depends on all the contingencies of the paths through which we are led to reach such islands and through which we subsequently exploit them. Thus overall, our science remains a "genuinely historical" reality.

Pickering's alternative ontology conveys the promise of a welcome radical change of perspective and inversion of values—at this level, we note that Pickering's analysis is very much in line with Chang's in chapter 15. In the novel ontology, contingency is no more "something to be feared or regretted." It is the unavoidable "counterpart of the endless emergence of the performativity" of the world. We are left with a more open, lively, and creative idea of the scientific enterprise than the one associated with the traditional copyist's ideal of mirroring nature. We should furthermore be aware, Pickering insists, that the dualist telos is itself not inevitable. In other words, science as we know it (and "enframing" as the mark of modern science, and the form of life that goes with it) is contingent. This should encourage us to refrain from taking their desirability for granted. We should at least consider other possible ways of being in the world—Pickering mentions some.

Through these suggestions, chapter 3 shows how a reflection on contingency in science can unveil pivotal commitments in our ways of practicing science, of conceiving knowledge, and more generally in our ways of conceiving forms of life in which science occupies a pervasive place. Chapter 3 simultaneously testifies that this reflection on our present situation is easily conducive to an active philosophy struggling to suggest and promote, at a normative level, desirable evolutions in science studies and beyond. Hasok Chang will provide another determined exemplification of this active stance in chapter 15.

The next two chapters, by Emiliano Trizio and by Mieke Boon, explore the complex pattern of relations between the contingency/inevitability issue and the more traditional debate on scientific realism. Both chapters suggest that by taking into consideration the contingency debate, it becomes possible to enrich the space of our philosophical views about science and to benefit from specific contributions that the contingency debate adds to classical arguments about realism.

In chapter 4, Trizio offers a relatively systematic analysis of the relations between the two issues of inevitabilism/contingentism and realism/antirealism. An important and original part of the essay discusses what happens to the relations among the four positions when realism is understood as "preservative realism," and more specifically as "structural realism." For the sake of the discussion, Trizio introduces several useful clarifying conceptual distinctions and specifications. Central to them is the "multiplicity thesis" assumed by any form of contingentism and denied by inevitabilists—a thesis already repeatedly used above. In this framework, Trizio substantiates the following conclusions. (1) About realism and inevitabilism: Although realism often surreptitiously fuels inevitabilist interpretations of the history of science, some forms of inevitabilism nevertheless remain logically compatible with scientific antirealism, even if admittedly, "antirealist inevitabilism" appears less comfortable, more difficult to defend, and more rarely instantiated in practice. (2) About contingentism and scientific realism: These tenets are not reconcilable if they are predicated about one and the same piece of scientific knowledge. However, more nuanced conclusions are drawn when considering specific forms of contingentism and scientific realism—namely, when identifying scientific realism with *structural* realism, and when understanding the multiplicity thesis built into contingentism either as a case of theory-underdetermination, or as a case of robust fit. In particular, in the robust-fit version of the multiplicity thesis, an essential and challenging conflict is revealed. If the multiple scientific units involved in the contingency thesis equate to incommensurable-like robust fits embedding largely disjoint phenomena, then "the whole idea of looking for historically invariant components of theoretical knowledge that are responsible for the predictive success and its retention through theory change becomes problematic." The very strategy of preservative realism, whether structural or not, ceases to be applicable and even ceases to make sense.

Trizio, moreover, wonders about the specific contribution of contingentist arguments to the debate on scientific realism. Contingentist arguments are found to be instructive on their own, because contingentist historical reconstructions of scientific development have a special modal status, in between the purely logical and the actual one (see above 9–10), that strengthens the plausibility of alternative scientific histories and increases the credibility of the multiplicity thesis. Thereby, they provide a novel way of epitomizing scientific realism, including the preservative-selective-structural versions. Here lies, according to Trizio, the "sui generis" import of contingentism with respect to the realist debate. Preservative realists, however, will only feel threatened provided they are ready to lend some plausibility to contingentist reconstructions of the actual history of science and, based on the latter, to contingentist proposals of counterfactual alternative histories that are taken as credible. But the problem, in this respect, is that the two camps do not see the same things when looking at the same targeted historical realities. This leads Trizio to characterize the opposition between contingentist-antirealists and preservative realists, from a methodological point of view, as "a clash of empirical inferences resting on evidence mainly deriving from the history of science."

In chapter 5, Mieke Boon argues, in line with Trizio at this general level, that the two debates, contingentism/inevitabilism and constructivism/realism, do not coincide and should be distinguished but are interrelated in ways that are instructive and remain to be clarified. More specifically, her aim is to show that the position one endorses with respect to the realist/constructivist controversy has important repercussions for the way one frames the definitions of inevitabilism and contingentism; and reciprocally, that an examination of the meanings of inevitabilism and contingentism suggest a "more viable philosophical view" of realism and constructivism regarding the exact power of science. Such an examination enables us, according to Boon, to find a way between the Charybdis of "naive forms of scientific realism," which convey "overly high expectations," and the Scylla of "strong forms of social constructivism," which carry "overly low confidence" with respect to "what science can do and what it cannot do."

Boon intends to think about these issues in close connection with, and in ways that could be beneficial to, scientific activities specifically concerned *with practical applications*, such as the engineering sciences—that is, she does not want to restrict herself to established scientific *theories* and *propositional* knowledge. Her strategy is to start with, and critically discuss, conceptions articulated in previous writings by two contributors to this book, namely, Hacking's seminal proposals on inevitabilism and Ronald Giere's contingentist perspectivism. The discussion leads Boon to denounce several versions of problematic dualisms involved in the positions under scrutiny. As a remedy, she suggests understanding scientific instruments, apparatuses, human motor and perceptual systems, cognitive faculties, and theories, not as perspectives *on* an isolable or independently reachable something but as "interfaces." Interfaces, as Boon understands them, transform aspects of the world that are unperceivable and not accessible in isolation—aspects she identifies with "inputs"—into perceived, experienced, and conceivable "outputs" (such as graphs, numbers, models, etc.). By articulating this suggestion, she defends a middle position in which our scientific knowledge has both a contingent and an inevitable dimension. Boon's essay thereby offers illustrations of possible ways to be both contingentist and inevitabilist, which complete those given in preceding chapters, and she further substantiates Allamel-Raffin and Gangloff's contention that compatibilism is an option.

The issue of scientific realism is also addressed in Ronald Giere's chapter 7, as indicated in its title, "Contingency, Conditional Realism, and the Evolution of the Sciences," but contrary to Trizio's and Boon's contributions, realism is not Giere's primary focus. One characteristic feature of Giere's chapter 7, as well as of Harry Collins's chapter 6 and Yves Gingras's chapter 8, is their common will to find a way to transform the contingentist/inevitabilist conflict *into an empirically tractable question* (a concern also shared by Chang as we shall see), *and* their common decision, in response, to privilege *concrete* formulations and discussions informed by detailed examination of historical case studies.

In Search of a Concrete and Empirically Tractable Way of Framing the Contingentist/Inevitabilist Issue

Both Giere and Collins complain that the question of whether or not some piece of our present scientific knowledge was inevitable or contingent is unanswerable if posed at a general and abstract level. Both, moreover, appeal to the same strategy to cope with the situation, namely, to reframe the question in a concretely tractable way, so as to be in a position, when discussing a given scientific configuration, to gain *empirical* evidence that is *able to support a well-defined yes/no answer.* Such a strategy is what Collins's title (borrowed from Peter Medawar) refers to as "the art of the soluble."

Collins's will to restrict himself to empirically—or "quasi-empirically" —decidable questions leads him, in chapter 6, to design a "modest" (my term) definition of contingency that makes contingency empirically attestable in a relatively easy way: contingency is understood as actually coexistent scientific incompatibilities. More precisely, a scientific option—say a proposition *p* and the network of concepts, actions, experiments, data (or in brief the more or less local "form of life") in which *p* is embedded—is contingent if "the scientific community" simultaneously endorses, during the same period, some incompatible option—say not-*p* and its correlate alternative form of life. Insofar as both *p* and not-*p* are held at the same time, *p* (and not-*p* as well) cannot be considered as inevitable in the corresponding stage of scientific development. Whether or not several incompatible options are actually maintained during a certain period of time by one and the same scientific community is an empirically decidable question. By contrast, to ask whether or not such a situation will continue to hold "indefinitely" in the "long run" or to ask whether things *could* have been otherwise are not soluble questions. Consequently, long-run forms of the issue and what-if conjectures are deliberately excluded from Collins's inquiry. In brief, Collins's methodological option is to set aside projections into a faraway unforeseeable future as well as any appeal to counterfactuals, and to stick with the *actual* history of science in the *short* term.

Collins's answer to his designed-to-be-soluble contingency question is an emphatic "yes." He takes "short term contingency" to have been "empirically demonstrated" by the sociology of science. In addition to having been empirically demonstrated, short-term contingency as the actual coexistence of incompatible scientific positions is viewed as a "theoretically understood" phenomenon. The theoretical explanation lies in tacit knowledge, the "experimenter's regress," and forms of Wittgenstein's "rules do not contain the rules for their application." These features of actual scientific practices prevent the universal imposition in the short run of one unique scientific option that could be considered as inevitable.

Taking short-term contingency for granted, Collins then discusses some problems raised by such a situation of scientific uncertainty for policymakers who have to decide and act in the short term. Relying on his wellknown "table of expertise," he argues against "technological populism" in situations in which political-social choices have to be made despite short-term contingency.

In chapter 7, Giere explains that "The question of whether some conclusions of the sciences are inevitable or remain forever contingent is unanswerable in the abstract," because contingency and inevitability are relative notions (contingent on . . . /inevitable on . . .), and because so many diversified kinds of things can be involved after the "on." Giere's attempt to specify the issue in a way that makes it empirically answerable is reminiscent of Collins's attempt, but significantly enough, Giere's starting point is a definition of *inevitability* rather than of contingency. A scientific option is defined as inevitable when there are "no remaining alternative avenues of investigation that the relevant scientific community as a whole will take seriously."

This "modest" definition of "inevitability of p" as "consensus about p" (my formulations) confines inevitabilist verdicts, if not to the "short" term as in Collins's framework at least to finite and determined temporal durations, and moreover relativizes these verdicts to a given scientific collective subject ("the relevant scientific community"). Something is not inevitable tout court but inevitable in reference to a given group of scientists during a given period of time. Inevitability thus becomes a local, and a possibly transient and reversible, property: "inevitability is not forever."

Giere then attempts to reconcile such a conditional inevitability with both a form of historical contingentism and what he calls "conditional (or perspectival) realism." In substance, once a lot of contingent events of human history that could perfectly well not have occurred are fixed, some scientific conclusions can sometimes become inevitable in the above-defined restricted sense and can then be interpreted realistically, that is, as revealing, within the corresponding perspective, real aspects of the world. Giere illustrates this scheme of contingently conditioned inevitability and realism in the case of the theory of continental drifts. The latter is presented as inevitable, at a point in history, under a multiplicity of heterogeneous contingent antecedent events.

Despite his preference for empirically tractable particular reformulations of the contingentist/inevitabilist issue, Giere nevertheless also looks for general tools to help conceptualize the topic. For this purpose, he invites us to look in two interesting directions: the evolutionary framework as developed in the life sciences,[31] and counterfactual history as practiced and assessed by professional historians not specifically interested *in the sciences* (on such issues, see also Radick 2008b).

In chapter 8 Yves Gingras, like Collins and Giere, insists that just "asking whether or not 'science is contingent' is . . . much too vague." One central aim of his contribution is to clarify what it may mean to say that some aspect of science is contingent. Like Giere, although inspired by different authors and relying on a different philosophical background, Gingras puts at the heart of his reflection the idea of a *conditional* inevitability—the two essays thereby provide welcome specifications and ways of fleshing out one of the conceptual points introduced at an abstract level in chapter 1. Gingras insists that in the history of science, inevitability cannot but be something "contingently necessary": it corresponds to "cases where a *contingent* decision (to study something or not) entails *necessary* consequences." Consequently, inevitabilist theses should specify the antecedents in each case, and contingentists should refrain from claiming that science could have been otherwise without being precise about what, exactly, is supposed to be the target of the claim. Moreover, Gingras urges us not to talk of contingency in a uniform way but, rather, to distinguish kinds of contingency, according to the type of objects under scrutiny, and depending on the "different *mode of existence*" of these objects—for example, "a historical *event* involving humans," "a particular scientific *concept*," a "formal *theory*," "an *entity*" like the electron, or "an *effect*" like the Zeeman effect. In brief, any cogent analysis of the contingency question should, according to Gingras, scrupulously specify the scale at which the question is posed as well as its precise target. According to the scale and target, the same scientific episode can appear contingent or inevitable.

Gingras substantiates this point through a detailed analysis of a case study in physics. Although not completely rejecting counterfactual history (see above, 9–10), and pointing to ways of practicing it that appear specifically fertile to him, he feels that, in regard to the episode under scrutiny, the actual history of science already offers materials that are sufficiently rich to dispense with inventing imagined configurations. Investigating some moments of the discovery of the wave properties of the electron in the mid-1920s, he identifies their contingent and inevitable aspects—thereby offering additional illustrations of "compatibilist" scenarios—and specifies the different meanings of contingency and inevitability in each occurrence.

Contingency and Mathematics

The subsequent three chapters take mathematics as a central object. Compared to the empirical sciences, mathematics is often considered a very special case. Prima facie, if there is a land of inevitability, or perhaps even of unconditioned necessity, it is mathematics.[32] At least, the field of mathematics is expected to be much less subject to contingency than are empirical terrains. Given that, it seems interesting to focus on this supposedly special case, and to examine whether, and in what respects, contingentist and inevitabilist questions and answers significantly differ in the formal sciences compared to the empirical ones.

Jean Paul Van Bendegem's chapter 9 deals with the contingency/inevitability of mathematical theories and proofs. According to him, the idea that mathematics is contingent rather than inevitable is much more plausible than is widely assumed. His aim is to provide some evidence in favor of this claim. Toward this purpose, he discusses some seminal proposals, introduced by David Bloor, of alternative mathematics borrowed from the history of mathematics (Bloor [1976] 1991), and he moreover develops two alternative mathematics of his own, which correspond to imagined, counterfactual cases: an alternative mathematical theory of complex numbers and an arithmetic that dispenses with the notion of formal proof.

Along this path, Van Bendegem is led to discuss some general difficulties that contingentist-inclined philosophers of mathematics are bound to meet when fighting against the inevitabilist instinct that sleeps in each of us and is even much stronger when it comes to mathematics. Among such general difficulties, he mentions a paradigmatic inevitabilist strategy to defuse contingentist claims about some mathematical target, which he calls the "rescue by definition strategy"—namely, to dismiss any alternative mathematics that contingentists would exhibit, by adapting the very definition of mathematics so as to exclude it.[33] Here we meet the issue, also dealt with by Hacking in chapter 11 in a different perspective, of what we are ready to count as mathematics. More generally, any proposed alternative mathematical piece can be rejected by inevitabilists, either as too different to be counted as a *genuinely mathematical* piece or as too similar to be recognized as an *alternative* mathematics. In Van Bendegem's terms, contingentists have a hard route to find between the "Scylla of too great a difference to make comparison possible and the Charybdis of all too easy comparability, making any difference disappear." It can be noted that a structurally similar difficulty applies as well when dealing with the natural sciences. I emphasize this difficulty at a general level in chapter 1 (49–50).

Another general difficulty of the contingentist/inevitabilist debate that Van Bendegem considers in chapter 9 and applies to the case of mathematics concerns the modal status of contingentist scenarios (see above, 10). Contingentist scenarios can involve alternative mathematics that correspond to either "purely logical possibilities," or real historical cases (as provided by ethnomathematics or past history of mathematics). Van Bendegem compares the assets and drawbacks of the contingentist recourse to each mode. Some paradigmatic inevitabilist attempts to dismiss each mode of contingentist candidate are examined along the way. Van Bendegem's own proposals of alternative mathematics are taken to be situated somewhere between "mere" logical possibilities and actual episodes in the history of mathematics (in the Western world and elsewhere). With this position, Van Bendegem joins Trizio, who, in chapter 4, insists on the "intermediate" modal status of some contingentist scenarios, and presents this status as the reason that contingentist arguments based on alternative scenarios that enjoy intermediate status introduce something new into the already available network of antirealist arguments.

Jean-Michel Salanskis's chapter 10 is not exclusively centered on the case of mathematics, but it makes a large place for the latter by exhibiting and discussing interesting and little-known inevitabilist and contingentist declarations of famous mathematicians, and by considering the case of physics as a mathematized field that makes constitutive use of mathematical frameworks. This being said, the scope and conclusions of chapter 10 extend far beyond the question of the contingency/inevitability *of mathematics*. They concern the very status of the philosophical question itself. Ultimately, Salanskis concludes that it is not a legitimate philosophical question because it is undecidable. But the reasons he puts forward for this undecidability are of a different kind from the ones involved in most of the other chapters of the book: not undecidable because we lack, and cannot hope to get, sufficiently decisive empirical evidence for or against one of the two opposed claims, but undecidable from a transcendental perspective inspired by Kant, which puts the "framework function" at the center.

In a large part of chapter 10, Salanskis plays the game of attempting to take seriously the question of contingency or inevitability as a yes/no philosophical question. In particular, he examines what can be said about the contingency or inevitability *of frameworks*. In so doing, he introduces and evaluates multiple ways of formulating the contingentist/inevitabilist question, and multiple existing strategies to support one of the two opposed tenets. But after playing this game for a while, in the end, Salanskis dismisses the contingency/inevitability issue as a legitimate question from the perspective of a transcendental "good modest philosophy." To formulate the issue as a yes/no question manifests, he argues, a "naturalist fallacy" that ignores Kantian lessons about the nature and limits of human knowledge: it is to act as if an all-comprising, absolute perspective and knowledge were possible.

In chapter 11, Ian Hacking's focus is the contingency/inevitability, not primarily of what we take to be mathematical truths, theorems, and proofs, as in Van Bendegem's and Salanskis's contributions, but of what "counts as" mathematics today. In other words, Hacking's concern is about disciplinary identity and disciplinary boundaries applied to the case of mathematics.[34] The question is: Were the identity and boundaries of mathematics as we view them today inevitable, for example, imposed by some "intrinsic character" of some targeted object (as Platonist realists would claim) or by human neurological equipment and capacities (as neurobiological naturalists would claim)? Or are they contingent, that is, dependent on historical events that could well not have occurred or been different? (On these issues, see also Hacking [2014].)

Hacking's aim is to lend plausibility to the insight that what is recognized today as a *mathematical* topic, subdiscipline, piece of reasoning, hypothesis, and so on could well have been otherwise. To achieve this aim, he shows that far from referring to one uniform activity and topic, "mathematics" encompasses a miscellany of practices and objects that show significant differences in contents and value according to the historical period and the society under consideration. Had other historical paths been followed, our very idea of mathematics could have been different. In particular, what we take to be the hallmarks of mathematics today, what looks so impressive and is so highly valued in the Western world today—namely, the a priori, necessary and apodictic character, the richness of contents, and the "unreasonable effectiveness" of mathematics in the natural sciences—could well not be associated with and not be dignified as mathematics. To take this possibility seriously would, according to Hacking, enrich or reconfigure what counts as a significant question in the *philosophy of* mathematics. Instead of being exclusively obsessed by a priori knowledge, necessity, and realist issues, philosophers of mathematics would become open to other kinds of interesting problems. In particular, Hacking contends "that the more difficult but perhaps more answerable question should now become: how have the platonic and neurobiological constraints jointly interacted with the contingent history of mathematics from 'Thales' to now?"

Through their attention to scientific frameworks and disciplinary identity, Salanskis's and Hacking's essays remind us that it can be instructive to shift the interest from a traditional focus on (the contingency/inevitability of) scientific *propositions* to (the contingency/inevitability of) *other dimensions* of science. The remaining chapters of the book largely take this shift. Albeit some are, at one point or another, led to consider the case of propositions, all spotlight and make more room for *other targets* of contingentist/inevitabilist claims than allegedly true/reliable scientific statements, theorems, or the like. For example, the core target of chapter 12 is scientific method, and chapter 13 considers multiple other, rarely discussed or even "exotic" targets, such as the relations instituted between science and religion or science and technology. Chapter 12 furthermore stretches the scope of contingentist/ inevitabilist targets in that it is applied not to the natural sciences as in the other contributions of this book, but to one of the human sciences: psychology. Chapters 14 and 15 also discuss contingentist/inevitabilist targets other than scientific propositions, for example, relevant scientific questions in chapter 14, or the monist regime under which our science is practiced in chapter 15. But since their core concern is the relation between contingency *and scientific pluralism*, these two chapters have been grouped apart under a distinct rubric that explicitly refers to pluralism (see 38–42). Overall, beyond their own specificities, chapters 12 to 14 together invite us to widen the scope of contingentist/inevitability targets, draw our attention to diverse candidates, and encourage us not to ignore these candidates as unimportant.

Widening the Scope of Contingentist/Inevitabilist Targets: Scientific Practices and the Methodological, Material, Tacit, and Social Dimensions of Science

Chapter 12 concerns introspection as a psychological procedure in order to investigate the human mind. According to a standard historical account, introspection has been rejected as a genuinely scientific method during a large part of the twentieth century but has been revived and revalued as a sound procedure in recent decades. Starting from this account, Michel Bitbol and Claire Petitmengin ask: Was the historical dismissal and quasi-elimination of introspection in the psychology of most of the twentieth century inevitable? To address the latter question, the authors provide a thorough analysis of the ways in which introspection has been conceived, practiced, criticized, and differently evaluated, by scientists and philosophers, in the course of the twentieth century. Their inquiry shows that the official historical account is incomplete and distorted. In particular, despite virulent explicit criticism and negative value judgments against introspection, the authors argue that introspection has never completely ceased to be used. They see essential reasons for this. These reasons point to what could be called a "quasi-transcendental" variety of inevitability (although the authors do not employ this vocabulary): introspection would be inevitable as a condition of possibility of a meaningful/interesting psychological inquiry for the kinds of human beings we are.

Provided that introspection is inevitable in this sense, the historical discrediting of introspection can be seen as a regrettable fact that could perhaps have been avoided. But could that have been? Was the largely negative attitude toward, and relative eclipse of, introspection in the science of mind inevitable? Another sense of "inevitable," historical rather than transcendental, is involved here. Regarding the empirical side of their question, Bitbol and Petitmengin identify a constellation of conditions that have contributed to the fact that we went astray as we did. Among these conditions are widely entrenched but unproductive epistemological commitments of the twentieth century such as the representationalist view of knowledge and the correspondence theory of truth. If one takes such contingent commitments, in conjunction with other more specific contingent features of the psychological practices of the twentieth century, as fixed initial historical conditions, then, according to the authors, the "misapprehension" of introspection was "virtually inescapable." Here we have one more illustration of a conditional inevitability dependent on contingent conditions, which, as stressed by Gingras in chapter 8, can be framed either in terms of a contingentist claim or in terms of an inevitabilist thesis, according to the scale on which one considers the situation.

In chapter 13, Joseph Rouse aims to approach the contingentist/inevitabilist issue in a way that draws lessons from the turn to practice to which he has himself contributed so much. Rouse regrets that most existing treatments of the contingency issue do not take such a shift into account. They still put the contingency/inevitability *of theoretical and ontological claims* at the heart of the discussion and largely ignore other dimensions of science—such as material, technical, skillful, social, and prospective ones—as if they were secondary or counted for nothing. But according to Rouse, what matters most is *not* the contingency or inevitability of *theoretical-ontological claims* associated with a given scientific stage. To substantiate this position and to show how and why other dimensions matter,

Rouse's strategy is to design various imaginative thought experiments in which the theoretical-ontological commitments are kept fixed and identified with those of our science, whereas other features of the counterfactual science nevertheless vary with respect to science as we know it. For instance, he argues that if the instruments and the human skills required for their correct use as well as the models of data analysis differed from ours, all of scientific life would be very different, even assuming for the sake of argument an identity at the level of scientific theories. Or he contends that science would be a fundamentally different enterprise, if, contrary to what prevails in our world, science were taken as a primary source of religious insights. Even if the empirical plausibility of (at least some of) Rouse's original thought experiments could be contested, since it is dubious that the ontological-theoretical component could remain invariant under the imagined changes, the corresponding thought experiments help us to grasp the specific contribution of the different dimensions under variation in each scenario. On this basis, they more generally invite us to reflect on and to better understand the very identity of what we call "science."

Another neglected direction in which Rouse encourages us to look concerns the relation between the contingency/inevitability issue on the one hand and natural laws and nomological necessity on the other hand. Relying on Marc Lange's analysis of the prospective role of laws in scientific practices and his understanding of nomological necessity in inferential terms, Rouse revisits the contingency/inevitability issue and offers new interesting ways of reframing it. For example, he points to one specific form of inevitability as conditional necessity when assuming Lange's conception of laws: counterfactual histories that involve different laws should inevitably be committed to different subnomic claims.

Contingency and Scientific Pluralism

Chapters 11 to 13 indirectly suggest that science as we know it is more pluralist or diversified than is commonly thought, diachronically as well as synchronically, and in diverse respects. Jean-Marc Lévy-Leblond's chapter 14 turns the suggestion into an explicit thesis and endeavors to back it up. Chapter 14 is moreover very much in phase with Rouse's chapter 13 in several other important respects. In particular, both insist that science and the rest of the society are highly integrated and cannot be considered as two separate or independent spheres when making use of counterfactual reasoning. Furthermore, neither of them is afraid to build on imaginative and adventurous counterfactual scenarios. But whereas Rouse is more attached to investigating and transforming the formulation of the question, Lévy-Leblond is clearly in favor of contingentism.

Chapter 14 attempts to support contingentism in science relying on the case of physics. This is done by deploying three parallel strategies in a lively and stimulating manner. The first strategy is based on actual scientific case studies. Lévy-Leblond shows that when looking to the history of physics we find several sustainable alternative "viewpoints" rather than one unique, monolithic consensual one, and he argues that this actual pluralism vindicates contingentism. The second strategy relies on taken-as-credible counterfactual histories of science. The latter, among which is "an Einsteinless history of physics," are intended to show that some seemingly anecdotal historical variations—such as the fact that Einstein could have not existed—would have been able to induce genuinely different physical conceptions. The third strategy uses science fiction. Levy-Leblond examines the amusing and insightful idea developed by Jerome Rothstein (1962) of a "wiggleworm physics" carried by intelligent marine creatures. He identifies Rothstein's analysis of the situation—according to which the science of such creatures could be matched with our science—as "a nontrivial inevitabilist argument" and offers an alternative analysis that leaves room for strong incommensurabilities and supports contingentism. Altogether, these strategies intend to substantiate "a pluralistic vision of physics" that goes hand in hand with a contingentist thesis about physics.

This enterprise leads Lévy-Leblond to meet the question of what should count as the *same* (and thus as *one*) or as *different* (and thus as a *plurality* of) scientific viewpoints or units. He argues that aspects often viewed as secondary or even "external" to science can introduce differences that make a difference. In this respect, he puts a lot of weight on scientific *formulations*. To the question of whether we should consider that the use of different scientific vocabularies, or forms of language (e.g., alphabetic versus ideogram writings), or mathematical formulations, make the corresponding sciences different, Lévy-Leblond answers with an emphatic "yes" and specifies the kinds of differences involved.

Beyond the issue of the contingency of multiple formulations of physical *results* or *answers*, Lévy-Leblond is led to discuss the contingency of what is taken as a relevant/interesting/genuinely physical *question*, and more generally the contingency of the boundaries of scientific disciplines—

intersecting at this level with Hacking's concern about the boundaries of mathematics, but applying the reflection to the case of physics. Concerning physics, it is argued that scientific questions and the demarcation between science and nonscience (e.g., religion) can strongly evolve with the historical and cultural context and should thus be considered as contingent upon this context rather than fixed. The point is illustrated through the fascinating case of the location of Hell, which was considered a respectable and important scientific question by physicists and astronomers of the seventeenth and eighteenth centuries.

In chapter 15, Hasok Chang invites us to reconsider in a normative perspective the relations between the two issues of contingency versus inevitability and pluralism versus monism. He begins by noting that the contingency/inevitability issue as usually understood is "unanswerable." Like several contributors to this book, he feels the need for a renewed approach that would make the issue more tractable and empirically decidable, but the solution he favors is of a different type. The contingency/inevitability issue is usually understood as a descriptive question about science as we know it. But here Chang suggests turning the issue into a prescriptive exhortation to cultivate contingency, so as to inspect very concretely "what we are able to do." The prescription to cultivate contingency refers to a pluralist maxim of the type: when a scientific result is taken as established, do your best to sustain an alternative result with "equally strong justification." If we try hard but fail, we would be in a position to support the claim that the result under scrutiny was inevitable *in the sense of practically unavoidable*. This would give inevitability the status of a "negative doctrine" vindicated on a practical basis. More generally, Chang's proposal would provide a "practical means of testing the contingency issue."

Such a proposal conveys and values a form of scientific pluralism that contrasts with the presently dominant monist view of science and is qualified as "normative" and "active." Chang's pluralism is *normative* because it prescribes pluralism as a better regime than monism for science, whatever aims of science are privileged—"Truth with a capital T," "empirical adequacy," and so forth. The discussion leads Chang to distinguish several forms of pluralist and monist scientific regimes and to meet some widespread objections to scientific pluralism. Note that one of these objections intersects with a central concern of Collins in chapter 6: on what basis are policymakers going to decide about the most appropriate actions involving technoscience, in situations where several conflicting knowledge options coexist? Moreover, Chang's pluralism is *active* because it designs strategies and performs concrete actions with the deliberate intention of transforming science as it is currently conceived and practiced (that is, as an activity driven by a monistic regulative ideal). Chapter 15 offers multiple suggestions about what philosophers and historians of science could do—and what Chang has himself extensively done—in order to foster a more pluralist practice of science and to convince policymakers and scientists that it is preferable to let "many things go" in science.

Chang's chapter 15 is among those that most strikingly show how an investigation of the contingency/inevitability issue helps to reveal and possibly to challenge the relevance and fertility of some deeply entrenched and apparently natural intuitions and commitments. Reflecting on the roots of the current strong inevitabilist instinct, Chang points to monism and insists on the relations of mutual support that bind inevitabilism and the monist regime of our science. He correlatively argues that an active pluralism would "change the entire spirit" in which we now consider the contingentist/inevitabilist issue—an argument that echoes with, and articulates much further, a thought only briefly suggested and superficially vindicated at the end of chapter 1, thus completing the loop of this volume. In the present configuration, regimented by the "philosophical-psychological ideal of monism," there is a "bias" in favor of inevitabilism. Inevitabilism is, without discussion, taken as the "implicit starting point" (or as the default position, see above, 21, point [1]). Accordingly, the claim that contingency could be the case, the suggestion that alternative systems of knowledge could have flourished and proved productive as well, is felt as a problem and a threat to scientific authority—at this level, Chang's message is very much in phase with Pickering's. What appears as normal, rational, or at least as the only conceivable regulative ideal, is that all specialists agree on one unique choice, that there is perfect unanimity about *the* best theory. As a corollary, philosophers of science write innumerable texts on the rational conditions of theory choice without even conceiving that theory choice is perhaps not the most valuable aim to pursue and might even be a bad thing. Now, had active pluralism been instituted, the intuitions and feelings would be, Chang insists, largely reversed. What would be astonishing and worrying would be complete consensus. Uniformity of opinions would raise suspicion: we would be anxious about "excessive herd instinct" or the imposition of factors that "suppress dissent." The problem of theory choice would be dissolved or at least deeply transformed in its formulation and stakes.

In this reflection, Chang's global attitude and strategy with respect to the contingentist/inevitabilist issue appears close to Pickering's: both use the inquiry on this issue as a means to specify, explicitly welcome, and actively try to foster a profound change of spirit regarding science and knowledge.

In the Guise of Conclusion

Taken globally, the contributions to this book overall suggest, each in their own style and in reference to the specific field under scrutiny, that the historical, social, and cultural matrix in which a given science develops has an effect at multiple levels: on the languages and vocabularies used and developed in scientific practices; on what is taken as a genuinely scientific question, problem, and explanation; on the collective attitudes toward the available alternative scientific viewpoints; on the identity and boundaries of scientific disciplines and their relations with other human activities; and on the whole, on the demarcation between science and nonscience. Together, they reinforce and specify the now common idea that science and society are integrated and entangled in constitutive ways, so that variations in one sphere are unlikely to leave the other sphere unaltered. This idea obviously militates for some form of contingentism, but a number of chapters argue that room remains for some varieties of inevitabilism.

My hope is that the methodologically and substantially diversified contributions of the present book will stimulate further work on what I take to be philosophically important albeit insufficiently attended themes. Among the corresponding multiple themes mentioned above, I am eager, in the last words of this introduction, to specifically insist on the need to be aware of and to reflect on the following important fact: our science, as it is conceived and practiced today, is dominated by two largely unquestioned ideals and commitments, namely, the inevitabilist and the monist ones, that come in relations of mutual reinforcement and mutual support. To insist on this need is to wish for further critical examination of the potential benefits of a more pluralist policy for our science, which would shake the credo that "there is," and that scientists should recognize and select, *one unique* best and thus inevitable available scientific option in a given stage of research. In brief, analysts of science would benefit by taking contingency, minimally understood as the *possibility* that our science could have been otherwise, more seriously, at least as a means to be better equipped to appraise pivotal features of our actual science and our scientifically based form of life.

PART I

Global Survey of the Problem Situation

CHAPTER 1

Why Contingentists Should Not Care about the Inevitabilist Demand to "Put-Up-or-Shut-Up"

A Dialogic Reconstruction of the Argumentative Network

LÉNA SOLER

Are the reliable achievements of science—for instance the value of the speed of light, the second principle of thermodynamics, Maxwell's equations, electrons, quarks, and so on—contingent or inevitable? To make the question more specific, I will start from a statement by Ian Hacking framed in terms of scientific *results*: "How inevitable are the results of successful science? Take any result R, which at present we take to be correct, of any successful science. We ask: *If the results of a scientific investigation are correct, would any investigation of roughly the same subject matter, if successful, at least implicitly contain or imply the same results?* If so, there is a significant sense in which the results are inevitable" (Hacking 2000a, 61). Hacking labels as "inevitabilists" those who would provide an affirmative answer to this question, and as "contingentists" those who would provide a negative one.

In this chapter, I first briefly clarify the initial question and the opposition between contingentism and inevitabilism. This helps anticipate the nature of some profound intrinsic difficulties that must be faced when dealing with this kind of question. I then devote the bulk of the chapter to discussing what is commonly viewed as the main and strongest inevitabilist argument, namely, what can be called, borrowing a telling formulation from Hacking, the "put-up-or-shut-up" argument (1999, 79, 89; 2000a, 70, see also 67). This discussion exemplifies and better reveals the intrinsic difficulties first put forward in abstracto. I conclude that the put-up-or-shut-up alleged argument is flawed: it is not able to support the conclusion according to which contingentism is highly implausible, if not false, so that inevitabilism is the most plausible position, if not the right one.

Clarifying the Contingentist/Inevitabilist Issue

Clarifying the Target of the Contingentist/Inevitabilist Issue

At first sight, contingentist/inevitabilist claims can be directed toward an indefinitely wide variety of targets on different scales, insofar as these targets have the status of scientific achievements that are taken to be reliable.[1] In this vein, we start by noting that "scientific achievements" broadly understood can include not only scientific *knowledge* (such as statements about the speed of light or the existence of quarks) but *also* scientific *methods*.

Contingency or Inevitability of Scientific Method

Concerning scientific method, contingentist/inevitabilist claims can, for example, be directed toward the very fact that human beings have developed an experimental physics; toward the development of this or that particular experiment; toward the development of any supposedly reliable technique involved in scientific processes of validation; and so on. A striking instance of a contingentist thesis about the experimental method in general can be found in Steven Shapin and Simon Schaffer's famous book, *Leviathan and the Air-Pump* (1985). Shapin and Schaffer (see, e.g., 1985, 13) claim it as a contingent fact that, today, almost everybody is committed to the idea of experimentation as *the* scientific method par excellence and the most powerful form of empirical proof that human beings can hope to have at their disposal. Another early attempt to elaborate a general argument in favor of the contingency of scientific method, referring to contemporary science rather than to the birth of modern science, can be found in (Collins 1981). In substance, Collins endeavors to show that "sound scientific method" is not something that is able to precipitate and impose one unique, inevitable solution to scientific problems. Contingency is constitutive of sound scientific method because sound scientific method is a social process.

Contingentist/inevitabilist claims can thus be directed toward scientific methods as well as scientific results. However, from an epistemological point of view, the crucial issue is, first and foremost, scientific *knowledge*. If in doubt, imagine a situation in which the scientific *methods* are profoundly different, but in which the *same knowledge* is established on the basis of these different methods. In this (highly implausible) case, the contingency of scientific methods would be *epistemologically* inoffensive.[2]

Taking that into account, we can start from a formulation of the contingentist/inevitabilist issue in terms of scientific *knowledge*, or equivalently, following Hacking, in terms of scientific *results* (say R)—even though we can anticipate that, sooner or later, the inevitability and compelling power of the scientific method will enter into play (as exemplified below).[3]

Contingency or Inevitability of Scientific Results

What kinds of scientific result R can be the target of contingentist/inevitabilist claims? As already stressed (see Soler 2008a, 222), the answer is: any kind.

Inside of a given discipline (for example physics), contingentist/inevitabilist claims can be directed toward multiple kinds of results R, such as:

- A whole theory (for example: was quantum mechanics inevitable?)
- A theoretical law (for example: was the law of gravitation inevitable?)
- A theoretical entity (for example: were quarks inevitable?)
- The value of physical magnitudes (for example: was the value of the speed of light inevitable?)
- Experimental facts (for example: given the experiments performed in the 1970s, was the experimental conclusion that weak neutral currents exist inevitable?)[4]

Contingentist/inevitabilist claims can also be applied *on higher scales*.

- On the scale of *one single scientific discipline* taken as a whole (for example: was a radically different physics possible?)
- Or even on the scale of the *whole cartography of scientific disciplines* (was the disciplinary mapping of our science inevitable? Or would a nonsuperimposable mapping of disciplines have been possible?)

Depending on the target, intuitions in terms of contingency or inevitability prove to vary (readers can test the matter on themselves, using the list proposed above). It would be instructive to examine whether such variations can be related to different types of arguments depending on the kind of target, but this would require another article.[5] In this chapter, I restrict myself to the scale of a given discipline, namely, physics. Three different kinds of targets will be involved in my discussion of the put-up-or-shut-up argument: physical *theories*; *experimental facts* of physics; and *whole subfields of physics*.

Clarifying the Opposition between Contingentism and Inevitabilism

Epistemically Benign Forms of Contingentism

The contingentist/inevitabilist opposition needs clarification because in a certain sense, indulging in a paradox, we can say that contingentism is inevitable. In a certain sense, everybody is a contingentist.

For example, we can assume that nobody would contest the following possibilities (but see note 6 for important qualifications):

> **Scenario 1**: Human beings could have been driven by different interests and ideals and thus might not have *developed anything that resembles our physics*.
> **Scenario 2**: Humans could have been driven by such interests, might have attempted to develop an investigation of the physical world, but might have failed to develop a *successful* physics.
> **Scenario 3**: Human beings could have developed a genuine successful physics, but physicists might have focused on *different questions* from the ones we have actually asked.

If this is contingentism, everybody is a contingentist—or so I shall assume in this chapter for the sake of argument.[6] Moreover, this kind of contingentism is compatible with inevitabilist claims such as "given a human interest in physics and given human questions about the speed of light, the value of the speed of light constant inevitably had to be close to the value we currently admit." Thus, the first task is to identify what I call "benign forms of contingentism" (Soler 2008b, 231), and to demarcate "benign contingentism" from nonbenign forms that are truly incompatible with specified forms of inevitabilism. Only in this way can we be left with a controversial opposition of philosophical significance, enabling us to turn to the issue of what arguments might support one side or the other.

Nonbenign contingentism will be more precisely delineated below (see 50–55), but one important specification about *the respect to which* I am ready to say that some contingentist positions are "benign" is worth mentioning from the start, so as to avoid a possible ambiguity. I have become aware of the possibility of such ambiguity only recently, reflecting on a draft from Ian Kidd (forthcoming) partly inspired by a paper from Henry (2008). In the corresponding works Henry, and Kidd following him, term as "radical" forms of contingency that I would categorize as "benign." In particular, they describe positions that assume a modification of "the entire cultural background," and the correlative possibility that modern science may never have emerged at all, as *radical* forms of contingentism (see note 6 for quotations), whereas I view such positions as one variety of benign contingentism (the variety involved in my scenario 1 above).

Putting aside quarrels over how to draw the boundary between cases in which "the entire cultural background" is supposed to have changed, and cases in which only contextual factors are taken to have changed, clarifications are required about the sense of the alleged "radicality" involved here. I agree that a modification of "the entire cultural background" might be called a radical form of contingentism, *in the sense that* the alternative form of life involved here would radically depart from our current form of life (for example, Hacking suggests a Zen culture in which nothing resembling our science would exist). However, from an *epistemic* point of view, such counterfactual thinking is, as argued below (and already in Soler 2008b), *benign* rather than radical. The reason for this is that such a counterfactual society, precisely insofar as it is *too radically different* from our own, will not be recognized by inevitabilist-inclined minds as a *legitimate competitor* vis-à-vis our *scientific* form of life. It will be viewed as a radically different human "choice" of lifestyle but not as a threat to our currently valued scientific methods and scientific knowledge.

The latter thoughts reveal a very important delicate point. Paradoxical though it may seem at first glance, and as already developed in Soler (2006a, 222; 2008b, 236–37), a historical alternative that is *too radically different* from our own history is *not* a good candidate, from an *epistemic* perspective, to support contingency in science. A delicate compromise always has to be found between, on the one hand, a story that is *too different* and has thus no chance of being seen *as a competitor* to our actual science (and therefore as a genuine *scientific* alternative) and, on the other hand, a story that is *too similar* and could thus not be seen as a *truly different* science (and therefore as a genuine scientific *alternative*).[7]

In this chapter, I am specifically interested in the issue of contingency versus inevitability *from an epistemic point of view.* Accordingly, the adjective "benign" used to qualify some forms of contingentism is intended to mean *epistemically* benign.

Inevitability as a Conditional Necessity

The (epistemically) benign forms of contingentism such as those involved in scenarios 1, 2, and 3 show why it is more convenient to talk about *inevitability* than *necessity.* All parties to the debate are prepared to admit that our science is not "absolutely necessary," in the sense that scenarios of types 1, 2, or 3 could have come true. Hence, if our science has any necessity about it, it is only a "relative necessity." This is what the term "inevitable" is intended to express and to hold constantly before our eyes: if there is any necessity here, it is a *conditional* necessity. But stressing this crucial point leads to the awareness that the very definition of the inevitabilist position will be associated with profound difficulties.

Inevitabilism with regard to some scientific result R must be formulated as a conditional of the type: If . . . , then, inevitably, R. Only by specifying what must come after the "if" can we get a precise definition of inevitabilism. This reveals a major challenge, namely, the challenge of avoiding tautology. The point has been stressed by Hacking. *If* other scientists had asked the same questions we did; and *if* they had worked hard to answer them, using our equipment, relying on our assumptions and tacit know-how in the implementation of the relevant experimental and theoretical resources; and *if* they had obtained an answer; and *if* they had not been led into error; and so on; *then,* they should have arrived at R, or at something that sufficiently resembles R. "We are close to an empty platitude, a tautology," writes Hacking (2000a, 66). A tautology such as: if all historical conditions had been the same, then all historical conditions would have been the same.

Moreover, if we leave definitions aside and turn preemptively to arguments, it is clear from the start that, in any polemical discussion of a particular scientific configuration, inevitabilists will always be in a position to escape from contingentist conclusions by claiming that one of the just-mentioned preconditions was missing.

Some Conditions Involved in the Definition of Inevitabilism

Despite the anticipated difficulties, we must nevertheless determine what conditions must be plugged into the inevitabilist definition. Only by specifying what comes after the "if" can we rule out scenarios of types 1, 2, and 3 so as to frame a *clear-cut opposition* between some well-defined forms of inevitabilism and some well-defined (epistemically) nonbenign forms of contingentism. I have already specified several of these conditions elsewhere (see Soler 2008b). In this chapter, I shall be content to briefly list some of the main conditions, before turning to a discussion of the put-up-or-shut-up demand in which a number of these conditions will come up as problematic points.

(a) The genuine physics condition

The first condition is suggested by scenario 1. It is what I call the "condition of a genuine physics." Suppose that some human activities have ended in a result R' incompatible with a given result R of our science. If the activities that have led to R' are dismissed as not being truly scientific activities pertaining to the physical world, the fact that these activities have arrived at R', rather than at the R our science takes for granted, does not threaten the claim that the result R of our science was inevitable.

What the "genuine physics" requirement actually means is far from trivial, and defining it is not easy (for more on this, see Soler 2008b). Moreover, judgments such as "this activity is/is not a *genuine* physics" are complex judgments about which different individuals, including scientists, might exhibit irreducible disagreements. As a result, we can suspect from the start that if this condition comes to be involved in arguments between inevitabilists and contingentists, it will be highly contentious.

(b) The similar questions condition

The second condition is suggested by scenario 3. I call it the "condition of similar questions." It is only on the condition that human beings ask about, say, the speed of light, that a given result R for the value of the corresponding speed can be claimed to be inevitable.[8]

In Hacking's formulation of the contingentist/inevitabilist issue, which I introduced in the first paragraph of this chapter, my "similar questions"

requirement is mirrored in the idea of the "same subject matter."[9] According to inevitabilism, "any investigation of the same subject matter," if successful, would lead to the same result R. I prefer to frame the issue in terms
of *questions*, as a reminder that questions are asked by people, and to avoid
suggesting a reification of these so-called subject matters.

(c) The equal-value condition

The third condition is suggested by scenario 2. I call it the "equal-value
condition."

Consider the following variation on scenario 2: the human beings in
the scenario have developed a genuine and successful physics; they have
asked the same questions that we have; but the answers they have provided
to these questions (R′) are *incompatible* with ours (R). Inevitabilists can
accept this scenario, while maintaining that our R answers were inevitable.
The proposed variation on scenario 2 is reconcilable with inevitabilism
insofar as the alternative physics is claimed to be *less good than* our own
physics (at least as far as the questions under scrutiny are concerned).

The value judgment inherent in this claim can be specified in reference to various virtues of the two physics under discussion (such as predictive power, simplicity, etc.). But insofar as a *hierarchy* is introduced, and as
soon as the genuine alternative physics is considered *inferior* to ours, in one
sense or another, with respect to the questions under scrutiny, the inevitability of the results R of our physics can be preserved. The incompatible
R′ will be dismissed as not being *genuine* results. They will for example
be attributed to human error—which can of course occur in any scientific
inquiry, given that scientists are fallible beings. Thus, any genuine alternative physical investigation of a given physical question must be recognized
to be *as good as* our actual physical investigation of the same question, in
order to become a potential contingentist challenger.

What I call the "equal-value condition" is often expressed in terms of
"successful science" (cf. in particular Hacking's quotation mentioned in
the first paragraph of this chapter: "would any investigation of roughly the
same subject matter, *if successful*," lead to the same results? [emphasis add-
ed]). I prefer to be explicit about the fact that value judgments are involved
here, so as to avoid suggesting any underlying universal or invariant conception or measure of scientific "success."

We can anticipate that the *meaning* of the equal-value condition ("as good" in reference to what idea of "good science"?), as well as *assessments and justifications* of whether or not this condition is satisfied with respect to particular cases (are these two physical investigations of the same question equally good, or is one better than the other, and why?), will constitute very sensitive points and frequent sites of divergence in any argument related to the contingentist/inevitabilist debate. Illustrations are provided below (in relation to the existence versus inexistence of weak-neutral currents, see 58–63, and in relation to Bohm's versus Bohr's quantum theories, see 73–75).

Definitions of Inevitabilism and Nonbenign Contingentism

Taking into account the thoughts developed above, we introduce the following definitions of inevitabilism and (epistemically nonbenign) contingentism—as a starting point in need of further qualifications, but nevertheless precise enough to begin discussing the put-up-or-shut-up argument.[10]

> **Inevitabilism: If (genuine physics) + (similar questions) + (equal-value), then inevitably R (or R′ different but reconcilable with R).**

In other words, inevitabilists claim that any genuine physics moreover recognized to be as good as ours must arrive at the *same* results as our physics, or to different but reconcilable results.

> **Contingentism: (genuine physics) + (similar questions) + (equal-value) does not uniquely impose R; an alternative history of science could have arrived at R′ incompatible with R.**

In other words, contingentists claim the historical possibility of an alternative physics as good as ours but associated with *irreducibly different* results from ours.[11]

With these clarifications in mind, let us now turn to an analysis of the "put-up-or-shut-up" inevitabilist argument.

Introduction of the Put-Up-or-Shut-Up Argument

Let me start with a warning. In what follows, I talk of "contingentist claims" or "inevitabilist claims" as if these phrases refer to well-defined positions of well-identified authors. This is a convenient artifice for the sake of clarity, and an innocuous one insofar as my aim is to reveal the *general structure* of the argumentative network that *can* be deployed in relation to the put-up-or-shut-up challenge. But of course, actual positions are diversified. More importantly, the reference to *inevitabilist* claims, replies, arguments, and the like is specifically problematic. This is because, as already stressed, although "there are inevitabilist scientists, scientists who express their inevitabilist faith" (Soler 2008a, 226, C), there are "few people, among professionals in Science Studies who explicitly advocate an inevitabilist position" (225).[12] However, any philosopher of science interested in the contingentist/inevitabilist issue quickly experiences that the inevitabilist instinct is deeply entrenched and widely shared, including within philosophical, sociological, and historical analysts of science, and that this instinct expresses itself through a set of paradigmatic claims and reactions. My aim, in what follows, is to reconstruct and to discuss prototypical—or at least possible—manifestations of inevitabilism and contingentism in relation to the put-up-or-shut-up challenge. More precisely, I propose a sort of "dialogic reconstruction" of what I take to be core points and typical replies of the corresponding debate.[13] This is the status of the argumentative network exhibited and analyzed in the rest of this chapter: a reconstruction of the space of reasons shaped as a dialogue between an inevitabilist and a contingentist philosopher.

Contingentists claim that science could have been otherwise. The structure of the contingentist claim is as follows: *had* different historical conditions prevailed—typically, had in themselves *indisputably contingent* material, intellectual, social, or historical factors been the case—scientific results *could* have been very different. In attempting to provide some plausibility to their position, contingentists almost always end up appealing to imagined, fictitious alternative scientific possibilities. Their arguments involve counterfactuals, what-if scenarios, and might-have-been stories about "other sciences."

Counterfactuals, however, suffer from a "credibility handicap"—as Steve Shapin puts it.[14] Inevitabilists are quick to exploit this.[15] The typical reaction of inevitabilists goes as follows:

> —Your alternative stories are just speculations, science fiction, gratuitous fantasies, mere logical possibilities. . . . But the point is their *historical* plausibility. In order to make your position plausible,

science fiction cannot do the job. What you would have to do is show an *actual* successful scientific alternative. This would be the only convincing evidence in support of your position. Can you put up an *actual* successful scientific alternative? Until further notice, you cannot. Thus, shut up.

The inevitabilist put-up-or-shut-up *argument* against contingentism is that contingentists are unable to provide the only kind of evidence that would be able to truly support their position, and that consequently, contingentism has no plausibility. Thus very schematically, the put-up-or-shut-up inevitabilist argument can be reconstructed as follows.

The put-up-or-shut-up inevitabilist argument against contingentism:

Premises:

> **(P1)** The only convincing way to make contingentism plausible would be to exhibit an *actual* (i.e., not just fictitious, but really existing) alternative science verifying the three conditions of genuine science, similar questions, and equal-value.
> **(P2)** Until now, contingentists have been unable to provide any such alternative.

Conclusion:

> **(C1)** Until further notice, contingentism has no plausibility.

Then, using (C1) and (often implicit) additional premises, inevitabilists frequently conclude that inevitabilism is secured, according to an auxiliary argument of the following type.

Auxiliary argument for inevitabilism:

Premises:

> **(P3)** Inevitabilism is the "default position."[16]
> **(P4)** Contingentism, if it was plausible, could threaten inevitabilism as the default position, but (C1).

Conclusion:

(C2) Inevitabilism is secured until further notice.

Some brief comments about (P3) may be helpful to begin with. *As a de-scription of widespread commitments,* (P3) can be taken to be correct: for many people, inevitabilism is, *as a matter of fact,* the default position or, in other words, a position that is intuitively (and often tacitly) assumed, that seems intuitively plausible, and for which no need is felt of any quest for explicit justification insofar as no plausible alternative comes to threaten it. Accordingly, it is assumed that the burden of the proof lies with contingentists.[17] *As a normative claim,* however, (P3) would call for discussion. Though it seems frequently assumed that inevitabilism *should be* the default position, under examination this assumption is a highly questionable one. I have discussed this point elsewhere (Soler 2006a) and will briefly come back to it at the end of this chapter (94–95).

In what follows, however, my focus is, first of all, on the put-up *demand* associated with premise (P1), namely, the demand to put up an *actual* successful scientific alternative—or more precisely, in my terms and relying on the analyses on pages 51–53, to put up an actual science *that satisfies the three conditions: genuine science, similar questions, and equal-value.*

At first sight, the inevitabilist demand might seem perfectly reasonable. Intuitively, the put-up demand looks like an *empirical test* of the contingentist claim. Given the naturalistic mood of contemporary thinking, the requirement of an empirical test appears both straightforward and legitimate. On examination, however, the situation is not so simple.

In what follows, I start with two *particular instances* of what can be reconstructed as contingentist attempts to meet the put-up-or-shut-up challenge: first, Pickering's attempt to put up, as an actual physics alternative to ours, the particle physics developed in the 1960s, which, contrary to the particle physics of the 1970s and contrary to current physics, assumed that weak-neutral currents did *not* exist; second, Cushing's attempt to put up Bohm's quantum theory as an actual alternative to the currently accepted "standard quantum theory." I start with particular instances, but the aim is to use them as exemplifications of two kinds of *general* epistemic configurations that present interesting distinctive features with respect to the discussion of the put-up demand. I characterize the main structural features of each general epistemic configuration (see 65–66 and 69–70), and I discuss their influence on judgments about whether or not the contingentist attempt to meet the put-up challenge has succeeded. I do this by exploring the space of the paradigmatic replies and strategies developed by inevitabilists in order to cope with each kind of contingentist attempt. In that way, I reveal what lurks behind the inevitabilist judgment that both kinds of contingentist attempts have failed. In the process, the inevitabilist and contingentist positions are further specified (see 79–82). Then I explain why the put-up demand cannot be satisfied and cannot be considered an *empirical* test of contingentism. Finally, I analyze why the put-up-or-shut-up so-called *argument* is null and void, and specify what I take to be the most appropriate contingentist reaction to an inevitabilist challenge of the put-up-or-shut-up type (93–95). I conclude by indicating delicate points that need further investigation (96–98).

A First Attempt to Satisfy the Put-Up-or-Shut-Up Demand: Pickering on Weak Neutral Currents

The first particular attempt I consider relies on Pickering's work on particle physics in the second half of the twentieth century (Pickering 1984a, 1995a).[18] More specifically, it relates to the historical episode of the so-called discovery of weak neutral currents (hereinafter NCs) in the mid-1970s. My presentation will inescapably remain very schematic.

Pickering's Contingentist Account of the Acceptance of NCs

To the question "do NCs exist?" physicists provided a positive answer (which I will call the "yes-NC-answer") in the mid-1970s, on the basis of several neutrino experiments. Today, the yes-NC-answer is still assumed to be the right one. In substance, Pickering (1) argues that the yes-NC-answer is contingent and (2) thinks that, in support of this contingentist claim, he has indeed put up an actual, incompatible, successful scientific alternative to the yes-NC-answer.

What exactly does Pickering put up that he takes to be a good candidate for an actual, incompatible, successful scientific alternative to the yes-NC-answer? The situation put forward by Pickering to support his claim is the following. *The same experimental data* from neutrino experiments (for example the same visible tracks on films from bubble chambers) have been, in the *actual* history of science, *actually* interpreted in two *contradictory* ways, which correspond, respectively, to a no-NC-answer and a yes-NC-answer. These two contradictory answers have been provided in relation to the adoption of what Pickering calls two different pragmatic "interpretative practices," say, IP$_1$ and IP$_2$.

Just to give an idea of what is at stake, Pickering's pragmatic interpretative practices refer to options such as the value of the "energy cut" experimenters choose to impose on their data. This value in turn determines which patterns on the films are excluded from subsequent data analysis. Such choices are never imposed by the data themselves; they always involve delicate compromises. Now, depending on the adoption of one or the other interpretative procedure, the *same tracks* are read—and have historically *actually* been read—*either* as *genuine* NCs, *or* as *pseudo*-NCs.

Pickering redescribes the situation in terms of what he calls "scientific symbioses." A "scientific symbiosis" refers to a robust fit between multiple ingredients of scientific practice—intellectual, material, practical, or other: all ingredients are mutually reinforced, and the resulting whole itself looks solid or, as Hacking writes, is "self-vindicating" (Hacking 1992, 30).[19]

In the episode under scrutiny, two scientific symbioses have *actually* been instantiated, schematically as follows:

The "no-symbiosis": IP$_1$ + no-NC-answer (+ other elements);
The "yes-symbiosis": IP$_2$ + yes-NC-answer (+ other elements).

Each of the two scientific symbioses has been found credible by physicists and has proven fruitful when used as a basis for physical inquiries.

Why Inevitabilists Dismiss Pickering's Attempt to Meet the Put-Up-or-Shut-Up Challenge

Did the contingentist succeed in putting up an *actual* incompatible scientific alternative? Can the actual incompatible scientific alternative put forward by Pickering satisfy inevitabilists? The answer is "no," and it is interesting to analyze why.

Inevitabilists readily concede that two contradictory interpretations have been associated with the same experimental outputs (for example, with the same tracks on films from bubble chambers). But several features of Pickering's case lead inevitabilists to dismiss it as a satisfactory answer to the put-up-or-shut-up demand.

1. The no-NC-answer and the scientific symbiosis that goes with it *historically precede* the yes-NC-answer and its accompanying symbiosis. The no-symbiosis dominated in the 1960s, whereas the yes-symbiosis was adopted around 1975. True, the no- and yes-NC-answers coexisted for a while during the transition in the mid-1970s, but one eventually won out over the other.

Moreover, the yes-NC-answer, which came after the no-NC-answer *is still assumed to hold today*. For this reason, on behalf of what could be called a quasi-irresistible "present-centered" instinct, inevitabilists are convinced from the very beginning that the symbiosis of the 1970s is *better* than that of the 1960s and *should* have been preferred *already in the mid-1970s*. In other words, using the distinctions introduced above (52–53), inevitabilists are tempted from the start to dismiss the proposed scientific alternative *by denying that the equal-value condition holds in this case*, that is, by denying that the no-symbiosis of the 1960s is *as good as* the yes-symbiosis of the 1970s.[20]

2. If we try hard to leave aside the quasi-irresistible feeling that currently accepted scientific results were better than the ones they won out over *already at the time they won out over them*, and if we attempt an "as neutral as possible" comparison of the merits of the no- and yes-NC-answers *during the mid-1970s transition*, we have to recognize that the task is, to say the least, not straightforward. The conclusions can easily be controversial. This is because when we compare two successive historical stages of scientific development, we are never, strictly speaking, in a position to circumscribe a small local difference while keeping "all other things equal." Inevitably, many things changed in the period from the 1960s to the 1970s. Accordingly, we cannot reason as if the *only difference* between the two scientific practices associated with neutrino experiments in the 1960s and 1970s were *just the two circumscribed packs*: the "IP_1 + no-NC-answer" and the "IP_2 + yes-NC-answer"—all other things being equal. Or in any case, if we try to reason this way and to argue that this was the only *relevant* difference, our argument will always be contestable (as is any historical reconstruction and its philosophical significance), and inevitabilists will easily find reasons to contest it.

3. With respect to the historical case under scrutiny, the difference between the IP_1 of the 1960s and the IP_2 of the 1970s is *not*, accord-

ing to Pickering himself, the only *relevant* difference. In the period
from the 1960s to the 1970s, it is possible to point out *additional*
changes that can be assumed to be *jointly responsible*, alongside the
differences in the IPs, for the reversal of the no-NC-answer into a
yes-NC-answer.

I cannot go into the details. But in that vein, relying on Pickering's own account of the episode, we can stress that in the 1970s the
theoretical configuration was very different from that of the 1960s. As
Pickering stresses, framing the point in terms of scientific symbioses:
"The communal decision to accept one set of interpretative procedures in the 1960s and another in the 1970s can best be understood in
terms of the symbiosis of *theoretical* and experimental practice" (Pickering 1984a, 188; emphasis added). In particular, in the 1970s, one
new *highly valued* theory—the renormalizable theory of Weinberg–
Salam and the unifying potential that went with it—predicted the
existence of weak neutral currents. Presumably, inevitabilists will categorize such changes as changes relating to *conditions of possibility*
—conditions that should be added after the "if" in the definition of
inevitabilism (see 50 above). They will narrate the story as follows:

—A variation in the theoretical context historically worked as
a strong incentive to investigate more deeply the issue of the
existence of weak neutral currents. This is indeed a *contingent*
initial historical condition with respect to the reversal of the "no"
into the "yes" in the 1970s. Of course, the Weinberg–Salam theory
could have been unavailable in the 1970s. Of course, the theory
could have been available and physicists could nevertheless have
had other priorities than the NC issue. However, none of this
calls into question the inevitability of what experiments "say"
once the question has been raised and seriously explored.

Another relevant difference we can stress between the 1960s and
the 1970s arises at the level of the *experimental means*. True, the experiments of the 1970s were, like those of the 1960s, neutrino experiments analyzed using bubble chambers. But the Gargamelle bubble
chamber of the 1970s was *much bigger* than its predecessors, and the
neutrino beams of the 1970s were *much more energetic*. Accordingly,
the new films contained more information; the statistics were better;

the genealogy of several reactions successively engendered one from the other could be followed; and so on. Presumably, inevitabilists will see these differences between the 1960s and the 1970s as something like the *cause*—or one *determinant condition* that has been causal in association with others—of the refutation of the no-NC-answer and the adoption of the yes-NC-answer in the 1970s.

Taking all that into account, at the end of the day, inevitabilists will tend to redescribe the historical episode as follows:

—In the 1960s, practitioners assumed that weak neutral currents did not exist. They were wrong to believe that. But of course, we cannot blame them for that. First of all, scientists are fallible. Second, in the 1960s, the available evidence relevant to the issue of weak neutral currents was insufficient and uncertain—or in any case less abundant and much more uncertain than in the 1970s. And third, before the 1970s, the existence of weak neutral currents, because it was not perceived as a crucial issue, had been the subject of only a few relatively superficial investigations. The no-NC-answer of the 1960s was not the result of a systematic and sufficiently deep inquiry. The yes-NC-answer of the 1970s was. In the 1970s, practitioners addressed the issue much more seriously and they obtained more evidence than in the 1960s. Given the additional evidence and the state of knowledge in the 1970s, the yes-NC-answer *inevitably had to be preferred*. Or, to reframe the point using Pickering's vocabulary, the scientific symbiosis of the 1970s was objectively superior to that of the 1960s.

Put differently in terms of the distinctions previously introduced, Pickering's attempt to meet the put-up-or-shut-up challenge is dismissed by inevitabilists via the equal-value requirement. Inevitabilists deny that the no option is *as good as* the yes option.

Are contingentists left with any argumentative resource at this point?

More about the Difficulties Involved in the Comparative Value of Two Scientific Symbioses

Contingentists can of course attempt to convince inevitabilists that they are wrong to deny the equal-value condition. They can attempt to articulate further arguments to support the claim that the equal-value condition held during the transition in the mid-1970s, and that the no-symbiosis of the 1960s, was as good as the one that emerged in the mid-1970s. Pickering has provided such arguments in relation to the particular case under scrutiny.

But attempts of this type are, in most cases, bound to meet deep and fundamental difficulties. As soon as we assume something like a symbiotic (or to use a perhaps more familiar term, "robust-fit") conception of scientific development—as many analysts of science, myself included, think it is most appropriate to assume—the evaluation of the comparative merits of two competing scientific symbiotic alternatives is, in most cases, a difficult task that can give rise to irreducible disagreements.[21] Since the corresponding difficulties constitute an integral part of the difficulties of the contingentist/inevitabilist issue in general, it is worth analyzing the very nature of what is at stake.

Even if, at first sight, the comparison is framed in a way centered on two *circumscribed* scientific pieces (such as the yes- and no-NC-answers), what actually has to be compared, at the end of the day, are two *more extended networks*—each including one of the circumscribed pieces as a local ingredient. The problem is that each such network, or symbiosis, borrows its force so to speak "from itself" or "from inside," in a sort of autarchic way. Each is a complex package that is, as Pickering says, "self-consistent," "self-contained," and "self-referential" (Pickering 1984a, 411), or in Hacking's already mentioned terms, a "self-vindicating structure" (Hacking 1992, 30). Both are "self-vindicating in the sense that any test of theory is against apparatus that has evolved in conjunction with it—and in conjunction with modes of data analysis. Conversely, the criteria for the working of the apparatus and for the correctness of analyses is precisely the fit with theory." As a result, each appears as "'a closed system' that is essentially irrefutable" (Hacking 1992, 30). Thus, each looks good in its own right if examined in isolation. But when confronted with another "self-vindicating structure," the ground is missing for obvious and straightforward comparative evaluations.

To get some idea of why this is so, consider the traditional problem of theory choice, and focus restrictively on the virtue that is traditionally taken as the decisive criterion with respect to theory choice, namely, empirical adequacy. It is traditionally claimed that we should choose the theory that shows the highest degree of empirical adequacy—the one that is able to explain and predict more phenomena or the same phenomena with better accuracy. But if we take for granted the symbiotic model of scientific development, we lack the common "phenomenal measure" that is required to perform this comparison. For in such a model, phenomena (i.e., what is taken as experimentally established fact, but also the "raw data"), like any other elements of science, are an emergent product of global co-stabilizations. Thus, except in very particular cases—that is, the case of *empirically equivalent* theories (see 67–79), or of theories that share a *large set* of relevant phenomena—the phenomenal strata of the two networks under comparison will be very different. The two phenomenal strata will show little overlap (they will correspond to largely disjoint sets of phenomena),[22] and they will possibly involve some local incompatibilities (as in the NC case, in which one phenomenon taken as existent on the one side is assumed to be inexistent on the other side). But then it becomes very difficult to apply the criterion of "superior empirical adequacy" to adjudicate between the two different scientific theories involved in each symbiosis. Each theory is empirically adequate with respect to its own phenomenal strata, but the two strata are largely disjoint and partly incompatible. And of course, empirical adequacy far from exhausts the virtues that are involved in comparative judgments about competing scientific options, so that the situation is actually much more complicated than this.

To generalize, the point is that it becomes very difficult to make sense of the "equal-value condition"—difficult to explain what the condition means and how to decide whether or not it is satisfied in particular cases.[23] Comparing the quality of competing scientific symbioses is not the kind of question for which universal compelling criteria can be provided. Accordingly, in relation to such questions, we can anticipate the occurrence of irreducible disagreements between individuals—and in relation to the NC case, we witness such disagreements (e.g., between Pickering and Peter Galison, see in particular Pickering [1984b, 1989a, 1995b] and Galison [1987, 1995]).

The Inevitabilist Appeal to "The Long Run"

As we have just concluded, in many configurations, and in particular in the NC case put forward by Pickering, inevitabilists and contingentists could go on indefinitely discussing the equal-value requirement without either side being able to convince the other. However, this impasse does not worry inevitabilists at all because they think they have another, stronger counterargument at hand. This counterargument consists in leaving aside the transition period during which the two contradictory NC-answers coexisted for a while, and in moving forward so as to bring the subsequent history of science into the picture. Inevitabilists are inclined to reply to contingentists something like the following:

—You claim that, in 1975, the no-symbiosis was as good as the yes-symbiosis. I do not agree, but anyway, even if I concede this point, the true issue is: was the no-NC-answer *viable in the longer run*? Could the no-NC-answer have been the basis of a *lastingly* successful physics? Had a no-NC-answer been accepted in 1975 instead of the yes-NC-answer, would the subsequent alternative physics have been as successful as ours? Would the subsequent alternative physics not sooner or later have come to the right answer about the existence of NC?

Faced with such a request to demonstrate viability *in the long run*, contingentists might attempt to argue, as Pickering attempted to do, that under the hypothesis of a no-NC-answer in the mid-1970s, a lastingly successful no-NC particle physics was indeed historically possible and plausible. They might, for example, develop arguments such as:

—The criterion of a "global robust fit" that drives the whole process of scientific development involves constraints that are not strong enough to impose one *unique* solution, either in the short run or in the long run. Moreover, the corresponding constraints are *global* or *holistic* constraints. Such constraints do not have the power to impose any *local* scientific item in particular. Thus, when the discussion is focused on the contingency of a *local* proposition such as the existence of NC, contingentist claims are still more plausible. As it so happens, physicists opted for the yes-symbiosis, and on this basis they developed a successful physics. But nothing was absolutely compelling in the adoption of the yes-symbiosis. Had physicists opted for the no-symbiosis, they would still have been able to develop a *different* physics, successful *in the same sense* as our physics is, that is, characterized by a satisfactory fit between multiple elements.

Obviously, this argument will not convince inevitabilists. For with such a reply, contingentists have been forced to leave behind *actual* scientific alternatives and to call for a *fictitious* history of science. Since contingentists have retreated back to the territory of counterfactuals, it is easy for inevitabilists to dismiss the whole attempt and reiterate the put-up-or-shut-up demand:

—This is just gratuitous speculation: show the *actual* alternative no-NC *lastingly successful* symbiotic development, or shut-up.

Revealing the General Epistemic Configuration Exemplified by the NC Case

The upshot of the previous discussion is that attempts of the NC type fail. This conclusion transcends the particular instance considered above. Any particular historical example characterized by the same epistemic features as the NC case will fail as well, for similar reasons and in relation to similar arguments. It is thus important to specify these features explicitly. Table 1.1 provides a synoptic characterization of the epistemic configuration exemplified by Pickering's case. The prototypical features of the epistemic configuration under scrutiny are indicated in italics.

Feature (2) can be involved, and is actually often involved in one form or another, as an *explicit* point of the debate (brandished by inevitabilists against their adversary). Feature (1) has a different status. It is rarely, if ever, explicitly brandished by inevitablists as part of an argument. But given the present-centered instinct, it nevertheless surreptitiously works as a condition that *more or less implicitly* feeds inevitabilist intuitions and the credo that the historically accepted option was superior *from the start* to the scientific alternative put forward by contingentists.

The two preceding features are the most determinant ones. We could however add a third feature, namely (3), less crucial but also underlying inevitabilist intuitions and possibly involved in the debate at certain points. This feature can provide some plausibility to the claim that at a particular point in the history of science, the adoption of a determined (local) option was *contingent* (in the NC example, the adoption of a no/yes answer; cf. above, the contingentist last reply, 64). But it also feeds inevitabilist intuitions according to which the difference between the two alternatives is epistemologically harmless because, *given its locality, it will be easily erased "in the long run."* It is indeed a common inevitabilist intuition that small differences with respect to the actual history of science would have been possible and even plausible, but that these differences are epistemically inoffensive because their effect on what had to be recognized as "the right answers" is thought to be null "in the long run."[24]

TABLE 1.1. A synoptic characterization of the epistemic configuration exemplified by the NC case

Taken-as-inevitable R: yes-NC-answer (weak neutral currents do exist) R = a *circumscribed (local) scientific proposition p* referring to a state of the matter that acquired the status of an *experimental fact* around 1975.
Actual incompatible scientific alternative candidate R': no-NC-answer (weak neutral currents do not exist) R' = *a circumscribed (local) scientific proposition not-p* that refers to a state of the matter taken as an *experimental fact* in the 1960s.
Sense in which the two options R and R' are irreducibly different: *logical contradiction: p versus not-p* The two scientific symbioses of the 1960s and 1970s go with two logically contradictory propositions: the existence/nonexistence of weak neutral currents.
Genuine physics condition: *satisfied* Particle physics and neutrino experiments in the 1960s and 1970s are recognized (including by inevitabilists) as genuine physics.
Similar questions condition: *satisfied* The common questions are: do NCs exist? Can NCs be detected by neutrino experiments?
Equal-value condition: *not satisfied according to inevitabilists*
Features that weigh against the proposed scientific alternative and contribute to leaving inevitabilists unconvinced (1) In the actual history: *R came after R'; R won out over R' at some point; R is still believed to hold today.* (2) *Problem with the equal-value condition:* inevitabilists deny that the scientific inquiry that led to R' was as good as the scientific inquiry that led to R, and hence that R and R' enjoy equal epistemic value. (3) The alternative physics differs from ours only with respect to a *small-scale* difference (in the NC example, with respect to a local proposition)

Identifying the Most Promising Epistemic Configuration with Respect to the Put-Up-or-Shut-Up Demand

Attempts of the NC type are clearly a failure. There is no chance of inhibiting, or even moderating, the inevitabilist instinct by calling for such an epistemic configuration. However, we are now better equipped to identify the kinds of epistemic configurations that are less likely to be rejected by inevitabilists as an inadequate response to the put-up-or-shut-up challenge, precisely because they are *not* subject to the characteristics and ambiguities associated with the previous epistemic configuration.

Contingentists should put up (1) currently available scientific alternatives rather than historically abandoned ones; (2) alternatives for which the equal-value condition is less subject to debate; and ideally, although not absolutely required, (3) alternatives that show large- rather than small-scale differences.

Can contingentists find some candidate of this kind? As hard as the imposed conditions are, the answer is: yes, they can! Quantum mechanics today can be taken to offer such an epistemic configuration. Quantum mechanics (hereinafter QM) will constitute the second example, and will exemplify the second epistemic configuration, considered in this chapter. The case of QM will be discussed relying on Cushing's book, *Quantum Mechanics: Historical Contingency and the Copenhagen Hegemony* (1994).

A Second Attempt to Satisfy the Put-Up-or-Shut-Up Demand: Cushing on QM

Cushing's Starting Point

The point of departure of Cushing's book is the fact that today, two quantum theories are available: "standard quantum mechanics" (SQM), also widely known as the "Copenhagen interpretation" and strongly associated with Niels Bohr's name (Cushing 1994 also talks of "the standard Copenhagen theory"); and David Bohm's theory (BQM, for Bohm's quantum mechanics), which Cushing also terms the "causal" "theory," "program," or "interpretation." SQM and BQM are two empirically equivalent theories: they make exactly the same predictions. But SQM and BQM are ontologically incompatible: in particular—using a rough widespread formulation—SQM is indeterminist, and BQM determinist. In Cushing's terms, we have two "observationally equivalent, alternative, and, indeed, incompatible descriptions or theories of our actual world" (Cushing 1994, xii). Both are "viable" and "logically consistent," but they are "conceptually incompatible" (xiii). SQM assumes the "currently accepted picture of an inherently and irreducibly indeterministic nature" and is "accepted almost universally by practicing scientists." BQM espouses the "apparently diametrically opposed view of absolute determinism" (xiii). On the whole, we "have two *actual* (not just fancifully concocted for argument's sake) empirically indistinguishable scientific theories that have diametrically opposed ontologies (indeterministic/deterministic laws and nonexistence/existence of particle positions and trajectories" [203]).

Starting from this fact, Cushing develops a detailed argumentation for the thesis that the adoption of SQM and its indeterminist ontology was contingent. I shall briefly come back to Cushing's argument below but, first of all, let me focus on Cushing's point of departure.

Why the Epistemic Configuration Exemplified by the QM Case Is Promising with Respect to the Put-Up-or-Shut-Up Demand

The epistemic configuration exemplified by Cushing's starting point is very interesting from an epistemological point of view, and specifically promising as an argumentative resource in support of contingentism. As in the NC case, there is a clear sense in which the two theories are incompatible: they are ontologically contradictory. But two other features of the QM case make it significantly different from the NC case and more promising in a contingentist perspective.

1. Unlike with the NC case, we are not in a configuration in which one of the two compared theories is a *past* scientific option once accepted by past scientists but *subsequently abandoned* in favor of the other option. The papers in which Bohm first introduced his theory were published in 1952 (Bohm 1952a, 1952b), at a time when SQM had already been in place as the orthodox theory for more than twenty years.[25] Since BQM was not first accepted and then superseded by SQM, there is *less room* for the "present-centered" instinct that pushes to dismiss from the outset superseded scientific options as *less good* than currently accepted scientific options.

Moreover, there is a sense in which we can say that SQM and BQM are two *living* coexistent contemporary theories. There is a sense in which we can say that the development of *both* theories is socially supported, even if the two theories are not socially supported on equal footing. SQM and BQM are *not* socially supported *on equal footing*, in the sense that, as it so happens, SQM largely dominates in all settings, in particular in educational curricula.[26] But despite this asymmetry, SQM and BQM can nevertheless *both* be considered socially supported, in the sense that there are (1) living physicists uncontested as good physicists who favor BQM over SQM, who prefer to work on BQM rather than on SQM, and who are paid with public funds to work on BQM; and (2) scientific publications on BQM, lively debates about BQM, and so forth. Consequently, there is less room for inevitabilists to dismiss BQM as a scientific alternative that is *not credible and not viable*.

2. The second feature that makes the QM case especially promising from a contingentist perspective is equally crucial: in the QM case, unlike in the NC case, there is a *clear* and *highly convincing* sense in which the two physical wholes (or robust fits) under discussion are *equally* good. BQM is as good as SQM, in the *clear* sense *that* the two theories make *exactly the same predictions*. And this sense of "good" is *highly convincing* because it corresponds precisely to what is usually put forward as *the* necessary condition for considering a scientific theory as successful and for accepting it, namely, *empirical adequacy* and *predictive efficiency*. Accordingly, the inevitabilists' easiest usual reply, namely, a denial that the equal-value condition applies, is not so easy in this case.

The General Epistemic Configuration Exemplified by the QM Case

Table 1.2 provides a synoptic characterization of the epistemic configuration exemplified by the QM case. The main features that are prototypical of the epistemic configuration under scrutiny are indicated in italics.

Discussions might arise in relation to the genuine physics condition, given that *some* physicists are tempted to assimilate BQM to a metaphysics. Wolfgang Pauli, among others, claimed in 1952 that, insofar as the hidden parameters produce no observable effects, hidden-variable theories à la Bohm are an "artificial metaphysics" (see Cushing 1994, 149). However, the reason usually put forward to view BQM as a metaphysics rather than a physics boils down to the claim that the ontology of BQM (in particular its determinist position) is not empirically testable. But if this is the reason that BQM is a metaphysics, then SQM is a metaphysics as well (its indeterminist claim is no more empirically testable). Accordingly, other reasons should be provided in addition to non-testability from an empirical point of view (for example, considerations of simplicity; see below 73–75). Here, I will not go further into the discussion of this point and will proceed as if the genuine physics condition is satisfied. This seems an acceptable global

TABLE 1.2. A synoptic characterization of the epistemic configuration exemplified by the QM case

Taken as inevitable R: SQM
R = a *whole physical theory*. SQM became a systematic, sufficiently well-developed theory in the late 1920s and then rapidly became (by 1927–1928) the new framework in which physicists worked instead of the "classical" one. Today, SQM has the status of *the accepted theory*.
Actual incompatible scientific alternative candidate R': BQM
R' = a *whole physical theory*. Embryonic versions of BQM-like frameworks were developed in 1926–1927 by Erwin Madelung and Louis de Broglie, but the well-developed and coherent BQM was first introduced by Bohm in 1952. Today, BQM is *considered by a minority of physicists to be preferable to SQM*.
Sense in which the two options R and R' are irreducibly different: *ontological contradiction*
The two scientific theories go with two contradictory ontologies: indeterminism of SQM versus determinism of BQM.
Genuine physics condition: *satisfied*
BQM is usually recognized (including by inevitabilists) as a genuine physics, or at least it is not widely condemned as a pseudo- or nonscientific theory.
Similar questions condition: *satisfied*
SQM and BQM aim to account for the same set of phenomena.
Equal-value condition: *satisfied as far as empirical adequacy is concerned*
SQM and BQM are *empirically equivalent*.

characterization of the situation, given that scientists who work on BQM are *not* socially dismissed by their colleagues and by scientific sponsors *as pseudo*-physicists who should be excluded from the profession.

By exhibiting BQM, have contingentists not exhibited an *actual, equally good, irreducibly different scientific alternative* to SQM? Have contingentists not indeed found a way to satisfy the inevitabilist put-up-or-shut-up demand? I think they have. So, are inevitabilists now prepared to recognize that an actual alternative physics, as good as our physics but incompatible with it, is possible, insofar as a real instance of it is now available? The answer is "no"—and at this point, we can begin to suspect that the put-up-or-shut-up inevitabilist demand is not a type of demand that can be satisfied.

What are the typical inevitabilist replies? They correspond to relatively familiar arguments in the philosophy of science for, at this point in our inquiry, the contingentist/inevitabilist debate intersects with more traditional debates about the underdetermination of theory by empirical facts, empirically equivalent theories, and Duhem–Quine-like theses. Accordingly, many of the traditional arguments developed by realist philosophers to cope with empirically equivalent theories can be borrowed by inevitabilists. With respect to contingentist claims specifically, I shall consider four typical inevitabilist ways of dealing with the BQM case so as to escape contingentist conclusions.

The first two consist of specifications and restrictions concerning the "about what," or the target, of the inevitabilist thesis.

Restrictions Concerning the Target of the Inevitabilist Thesis

A first version corresponds to what I will call the "restriction-to-phenomena" inevitabilist reply.

First Inevitabilist Reply: The Restriction-to-Phenomena Reply

According to this solution, inevitabilist claims are circumscribed to phenomena, so that contingentist claims can be accepted about high-level theories of these phenomena. A "constructive empiricist" à la Bas van Fraassen could endorse such position. Faced with the BQM case, this inevitabilist retreat consists of replying something like:

> —OK, you put up an alternative incompatible physical theory as successful as our currently accepted theory. No problem for me, because I do not deny that physical *theories* are contingent. I only claim that physical *phenomena* are inevitable. What I ask you to put up are alternative incompatible *phenomena* that are *taken as existing*.

A second, structurally similar version of this retreat is what I will call the "restriction to physical interpretations" inevitabilist reply.

Second Inevitabilist Reply: The Restriction to Physical Interpretations Reply

According to this reply, inevitabilists identify the common part of SQM and BQM as "*the* quantum *theory*" and claim that the QM theory *in this sense* is inevitable; as a corollary, they confer on the divergent parts of SQM and

BQM the status of two different physical *interpretations* of the QM theory in the aforementioned sense and concede that physical *interpretations* in this sense are contingent. The possibility of this inevitabilist retreat is, for example, indirectly suggested by the following remark, made by Joseph Martin as a preliminary to a discussion of Cushing's contingentist thesis about SQM: "Cushing uses 'theory' equivocally, as his prime example is the choice between Bohr's and Bohm's *interpretations* of quantum mechanics, which can be construed as *competing window dressings* of the theory of quantum mechanics *rather than as theories themselves*" (Martin 2013, 923; emphasis added).[27]

In this option, the inevitabilist reply to the contingentist attempt is, in substance:

> —OK, you put up an alternative incompatible physical *interpretation* of the currently accepted interpretation of QM theory, and this interpretation is as good as the standard interpretation. No problem for me, because I do not deny that physical *interpretations* are indeed contingent. I only claim that physical *theories* are contingent. What I ask you to put up is an alternative physical *theory* that is both incompatible and equally good as ours.

With such an inevitabilist reply, it might seem that only a question of words is involved—about what we are prepared to call a "theory." However, a deeper problem of *theory individuation*, associated with a judgment about the kinds of units that fundamentally matter in physics, is actually involved. In relation to the BQM versus SQM case, the problem of theory individuation has been stressed by Steven French (2008), in a paper concerned with the grounds on which we might assess the plausibility of a counterfactual history of science. French first notes that the formalism of "regular quantum mechanics" and empirically equivalent Bohmian versions of QM "differ only by a mathematical transformation" but concedes that "the ontological interpretation, of course, is very different." Then he asks: "is this difference enough to give us a different history?" He goes on: "Here we come to a crucial issue hitherto lurking in the wings: *theory individuation*. To put it crudely, the force of the claim that we have the genuine possibility of a different history is going to depend on whether we can delineate a *genuinely different* theory (or model, or paradigm, or whatever). Issues of theory individuation, of the identity conditions of theories, remain comparatively unexplored within the philosophy of science. The kind of formal account hinted at above that can accommodate the supposed openness of science can also capture the looseness with which we may want to individuate theories" (French 2008, 575; emphasis added).

This leads us to consider the criteria according to which "we may want to individuate theories." What do these criteria depend on? They depend on judgments about what essentially matters in physics. If what primarily matters is the formalism and its connection to experimental results, then SQM and BQM are essentially identical: they must count as *one* scientific unit (or theory, model, paradigm, or "whatever"); in this case, BQM is dissolved as an actual *genuinely different* physical alternative. Only if ontological interpretations are taken to be an essential part of what matters in physics will SQM and BQM be counted as *two* (i.e., genuinely different) scientific units. Thus, judgments about the kinds of units that essentially matter in physics, which in turn determine theory individuation—or more generally, whether some epistemic entities are categorized as *one and the same* unit, or as a *multiplicity* of different units—lurk behind the whole discussion and can be viewed as one possible source of disagreement between inevitabilists and contingentists.[28] Since the kind of judgment involved, namely, assessments of importance, is not a kind for which uniquely compelling criteria can be imposed, the resulting disagreement is likely to remain unsolved.

In the two inevitabilist replies above, concessions are made to the contingentist stance because some ingredients of science—high-level physical theories and physical interpretations of theories, respectively—are recognized as contingent. Notwithstanding, an inevitabilist position is essentially maintained, insofar as some fundamental stratum of science is preserved from contingency—physical *phenomena* and physical *theories* in a restricted sense, respectively. However, to circumscribe the target of inevitabilist claims is not the only possible inevitabilist strategy. A third option is to deny that the equally good condition is indeed satisfied—as in the NC case, and *despite the empirical equivalence* of BQM and SQM.

Third Inevitabilist Reply: Denying the Satisfaction of the Equal-Value Condition

In this vein, inevitabilists reply something like:

—OK, you put up a theory, BQM, which is equally good as SQM *as far as empirical adequacy is concerned*. But empirical adequacy is not the only virtue of a scientific theory. Other values must be taken into account, such as simplicity, coherence, beauty, naturalness, understandability, and so on.[29] Now, taking into account these other criteria, one of the two competing theories is better than the other. Consequently, this theory inevitably had to win out over its rival.

These types of argument are, as it so happens, at the heart of discussions among contemporary physicists about the comparative merits of SQM and BQM. The problem with these types of arguments based on "noneviden- tial grounds" (Cushing 1994, xiv) is that they are far from consensual—as has been well-known at least since Thomas Kuhn (see in particular 1977). Different physicists are often led to divergent, irreconcilable appreciations.

Concerning the relative merits *of SQM and BQM*, multiple examples of currently instantiated divergent evaluations could be provided.[30] As a single example, compare the following assessments of two living physi- cists. According to Hervé Zwirn, "the technical complexity [of BQM is] far greater than that of [standard] QM," and "all we get in return is the possibility of restoring an ontology that cannot be precisely known" (Zwirn 2000, 225; my translation). But according to Jean Bricmont, BQM "renders precise Bohr's intuition about the role of the measurement apparatus, gives clear physical meaning to the wave function, and eliminates all mystery surrounding the origin of probabilities in quantum mechanics. Moreover, it does all this just by adding one line to the usual formalism, thus making the theory perfectly deterministic" (Bricmont 2007, 40; my translation).

Taking into account such divergences between practitioners on the comparative merits of BQM and SQM, the inevitabilist attempt to call for nonevidential criteria to show that SQM is not *as good as* but *better than* BQM might well, at the end of the day, be turned against inevitabilism. For if nonevidential virtues determine actual theoretical choices in the history of science, and if nonevidential criteria are not uniform or not uni- formly applied among physicists, this works in favor of the contingency of actual theory choices rather than in favor of their inevitability. Cushing surfs on such a line of thought, among others, to build his contingentist argument (to which I will come back more systematically in the next sec- tion). In that vein, he first articulates the specific virtues of BQM (BQM "is arguably *more coherent* and *understandable* than the commonly accepted dogma" [Cushing 1994, 174; emphasis added]; "We are able to construct a *less incomprehensible, more nearly picturable*, representation of the physical universe with Bohm than with Copenhagen" [21–22; emphasis added];

etc.), before stressing: *"We do have the option* of giving up locality while maintaining a visualizable causality. *The choice is ours to make on purely pragmatic grounds"* (21–22).

In other words, the contingentist's answer to the inevitabilist denial of the equally good condition is, in substance:

> —You are perfectly right to point out that criteria other than empirical adequacy are involved in evaluating the comparative merits of two competing scientific theories. In particular, this cannot be otherwise when the competing theories in question are empirically equivalent. But such a nonevidential basis is *not powerful enough* to force any rational being to recognize *one* of the two competitors—say SQM—as *the* best one. On nonevidential grounds, to prefer BQM is perfectly acceptable and rational as well.

De facto, the current situation of QM is the following. Some physicists feel that SQM is obviously better. Other physicists feel that BQM is obviously better. Admittedly, a *majority* of contemporary physicists value SQM more. But what is the meaning of this imbalance? What conclusion can we draw from this fact? Can this fact be considered as inevitably imposed by some intrinsic characteristics of SQM and BQM? Global value judgments about the relative quality of theories are, no doubt, in part influenced by scientific education. If BQM were taught in scientific curricula on an *equal footing* with SQM, or if BQM were taught *instead* of SQM, would practitioners not differently assess the comparative merits of SQM and BQM?

Cushing's Argument for the Contingency of SQM as One of the Most Convincing Counterfactual Configurations

Cushing exploits the fact that global assessments of the comparative merits of two empirically equivalent scientific theories like SQM and BQM vary from one scientist or one scientific subgroup to another, as well as the idea that such assessments are influenced by scientific education, in order to argue that BQM could perfectly well have been preferred to SQM in the 1930s, such that today BQM would be judged superior to SQM by most scientists.

With this line of argument we are, of course, back to counterfactuals. Hence it is clear from the start that inevitabilists can ignore the whole argument just by reiterating the put-up-or-shut-up chorus.[31] However, Cushing's argument is worth considering because it instantiates one of the most—perhaps even *the* most—potentially convincing counterfactual configuration a contingentist can dream of. If this counterfactual configuration fails to convince inevitabilists, we can suspect that no one will ever be able to do the job. Space constraints prevent me from going into the details of the argument, but it remains possible to elucidate the reasons why Cushing's strategy carries a potentially high convincing power. I will put forward five reasons.

The first two reasons have already been stressed. Cushing's argument applies to a case that possesses two important advantages: it deals with two theories that are (1) currently alive and socially supported and (2) empirically equivalent. But that is not all. Cushing's argument also corresponds to the less contestable and at first glance apparently innocuous way of generating a counterfactual history, for the following additional reasons. (3) The counterfactual initial conditions and a large part of the subsequent counterfactual scenario do not invent fictitious historical conditions and events "out of the blue." They involve events that *really took place* in the actual history of science. (4) The counterfactual initial conditions are generated simply by *shifting through time* certain events that actually occurred at another *relatively close* moment in time, or by canceling some such events. (5) Nothing seems able to prohibit the temporal displacements or deletions of events involved in Cushing's scenario. This is uncontentious in many cases. Take the shifting or deleting of John von Neumann's so-called impossibility proof concerning hidden variables: what could have prohibited this "proof" from becoming available some years later, or not at all? Nobody is prepared to support the inevitability of such events—or so I assume. In cases for which some doubts might be raised, Cushing systematically provides arguments. For example, the claim that Bohm, or someone else, could have developed a BQM-like theory in the 1930s might be subject to further discussion. But Cushing shows that some physicists *actually* developed versions of BQM as early as the 1930s, and that all the theoretical resources Bohm used in his 1952 papers were already available in the 1930s.[32]

If any counterfactual scenario has a chance of being convincing, it should be Cushing's. However, as a matter of fact, many inevitabilist-inclined and realist philosophers, as well as many scientists, remain unconvinced by Cushing's argument. In particular, many physicists stick to the deep-seated intuition that SQM is intrinsically better than BQM and hence had to win out over BQM and to become the basis of our ontological beliefs. In any case, to repeat, inevitabilists can always dismiss the whole story because of its *counterfactual* character. As such, Cushing's argument, although worthy of consideration, cannot be viewed as an answer to the put-up-or-shut-up demand.

The fourth inevitabilist strategy intended to counter the contingentist claim that, with BQM, a satisfying answer has been provided to the put-up-or-shut-up demand—and the last inevitabilist strategy that will be discussed here—is what I call the "only-transiently-as-good" inevitabilist reply.

Fourth Inevitabilist Strategy: The Only-Transiently-as-Good Inevitabilist Reply

Principle of the Only-Transiently-as-Good Reply

In this strategy, inevitabilists call for the current coexistence of the currently equally-good SQM and BQM to be resolved *in the future*. They claim, in substance, the following:

> —I concede that as a matter of fact, it rarely but sometimes happens that incompatible, equally good scientific options coexist. I concede that at the *current stage* of scientific inquiry, BQM is as good as SQM, in the sense that BQM is empirically equivalent to SQM. But this is just a *transient* situation. This is just because QM is a very difficult matter, such that it inevitably takes time to fully explore the issue and to arrive at decisive conclusions. Physicists are still in the process of exploring the issue. But with *further time, further efforts,* and *further evidence,* one will prove superior to the other, as always. Ultimately, in the long run, the best theory will inevitably win out over its competitor.

The epistemic force of the only-transiently-as-good inevitabilist reply will be discussed below (see 81–84).

A Fifth Inevitabilist Reply?

For the sake of completeness, with respect to the particular case of BQM versus SQM, we also find what might perhaps be identified as a fifth inevitabilist reply. This fifth reply is actually a sort of combination of the third and fourth replies presented above. It consists in suggesting that BQM is not *quite* equivalent to SQM, to the extent that BQM is *anticipated* to be *more difficult to generalize* than SQM. Typically, the claim is that a relativist extension of BQM, or a BQM version of quantum field theory, will have to face profound if not insurmountable difficulties, such that the empirical equivalence holds only in a restricted domain (for an example of this type of criticism of BQM, see the position of Hervé Zwirn, and my own discussion of it, as developed in Soler 2006b, 147–70).

Such a claim can be characterized as a combination of the third and fourth inevitabilist replies listed above, for the following two reasons. As in the third reply introduced above, this claim denies the equal value of the two competing physical alternatives (SQM is supposed to be better as far as generalizations are involved). And as in the fourth reply introduced above, the conviction is expressed that future developments will decide in favor of one of the two theories (SQM in this case).

However, the claim that relativistic and quantum field theoretical generalizations face difficulties that are especially profound and probably too much so, although frequently put forward without justification, is contestable and is actually contested by several physicists who have examined the matter in detail, including physicists who are *not* supporters of BQM (e.g., Frank Laloë). According to them, BQM has already been successfully generalized to some relativistic and field theory cases; moreover, the principles of further extensions have been determined; no insurmountable difficulty is thus at work here (see, e.g., Passon 2005; and Laloë 2012).

The Only-Transiently-as-Good Inevitabilist Reply as an In-All-Circumstances-Usable Strategy and a Deeply Entrenched Conviction

Before discussing the epistemic force of the only-transiently-as-good inevitabilist reply, let me take it as just a common kind of reaction and develop two important insights in relation to such a reaction.

1. First, it must be stressed that the only-transiently-as-good inevitabilist reply is an *in-all-circumstances-usable strategy.* Whatever candidate contingentists put up as an actual, equally good, incompatible scientific alternative, inevitabilists will always have the possibility of dismissing this candidate as an alternative that is not *genuine,* by appealing to the temporary character of the situation.

In retrospect, we can see more clearly that a reply of this type was already involved, or at least kept in reserve, in the inevitabilist reaction to the NC case (see section The Inevitabilist Appeal to "The Long Run"). In the NC case, the involvement of an only-transiently-as-good reaction was concealed or delayed by the fact that inevitabilists had another at first sight more powerful argument at hand, namely, that the no-symbiosis was less good than the yes-symbiosis. In the BQM case, denying the equal-value requirement is less easy. Accordingly, the only-transiently-as-good reply comes more immediately and frequently to the foreground in debates on the relative merits of BQM versus SQM.

2. Second, as far as I can judge from my readings and multiple discussions with colleagues about the contingentist/inevitabilist issue, the conviction that, when competing scientific options are at stake, one option will sooner or later, "in the long run," manifest a clear superiority over its competitor is a deeply entrenched and widespread commitment. Accordingly, the inevitabilist reply of "only-transiently-as-good" appears natural and legitimate.

More about the Definitions of Inevitabilism versus Contingentism and about the Implicit Logic behind the Put-Up-or-Shut-Up Demand

Complements to the Definitions of Inevitabilism and Contingentism

Taking for granted the deeply and widely entrenched character of the credo according to which competing scientific theories will sooner or later be decided between, and treating this as recorded fact, I am led to complete the framework that I have articulated so far in order both to define a clear-cut opposition between inevitabilism and contingentism and to identify the problematic points that will inescapably be at the heart of the debate. Three conditions have already been revealed as problematic points constitutive of the contingentist/inevitabilist debate. Now a fourth problematic point should be added. I will call it the problem of "time-effort-evidence calibration." Implicit in the inevitabilist position, and in the put-up-or-shut-up demand, is a reference to the "long run" or to a clause of the type with "enough time and effort," with "further evidence," after a "sufficiently deep investigation," after "sufficiently hard and sustained effort," or the like. As vague and problematic as such references to time-effort-evidence might be, a condition of this kind must nevertheless be added to the definition of the inevitabilist position. Otherwise, the inevitabilist put-up demand could be claimed to already have been satisfied, and inevitabilism could be claimed to already have been refuted, by the very fact of the transient coexistence of empirically equivalent theories. Accordingly, we must complete the definitions provided above (53) with the clauses in italics below:

Inevitabilism: If genuine physics + similar questions + equal-value, then, inevitably, *at least in the long run, after enough time-effort-evidence*, R (or R′, different from but reconcilable with R).

Contingentism: (genuine physics) + (similar questions) + (equal-value) does not uniquely impose R, *even in the long run*; an alternative history of science could have arrived at R′, incompatible with R, *and stuck with it*.

The condition of "enough time-effort-evidence" is inescapably vague. There is obviously no way to provide a uniquely determined measure of the temporal interval supposed to be involved. It is difficult to claim that there is any absolute unique way of calibrating the duration for which the inevitabilist/contingentist positions *should* apply with respect to the discussion of a given case. As a consequence, when discussing any particular configuration, each camp will always be able, faced with undesirable conclusions from the other camp, to escape these conclusions by appealing to "further time," "further investigations," "further evidence," or similar intimidating phrases; by claiming that the relevant temporal interval has not been correctly taken into account in the analysis of the philosophical adversary; by replying "you did not wait long enough to be able to draw solid conclusions."

Regarding inevitabilist attempts to deny any plausibility to contingentism until the put-up demand has been satisfied, such a reply is an always available and easy way to dismiss any candidate option put up by contingentists. In other contexts (which I will not consider here), contingentists can also, symmetrically, call on the time-effort-evidence calibration condition to argue that inevitabilist conclusions are not convincing (cf. Soler 2008b, 238). All in all, a problem of time-effort-evidence calibration is almost always involved in discussions of the contingentist/inevitabilist issue, especially when the discussion relies on *particular* scientific results and episodes.

Let me now examine the epistemic force of the only-transiently-as-good inevitabilist reply more closely.

First Remarks about the Epistemic Force of the Only-Transiently-as-Good Inevitabilist Reply

Do inevitabilists have any argument when claiming that one of the two currently equally good incompatible scientific options will prove superior to the other with "further time-effort-evidence"?

The first thing to stress, in this respect, is that the claim is based on an *imagined* future science, and not on the exhibition of any *actual* instance, as inevitabilists themselves ask contingentists to produce. Inevitabilists simply express their confidence that future scientific development will decide between the currently as-good-but-incompatible scientific alternatives. Thus, inevitabilists seem to be less demanding with respect to their own arguments than with respect to the arguments of contingentists.

However, upon examination, the situation is not completely symmetric. Inevitabilists feel justified in their confidence that future science will impose an improved version of one of the two competing theories as clearly better, *because this has always been the case in past scientific history.* In other words, the grounds for inevitabilists' confidence is the *inductive support* provided by our past science. By contrast, the contingentist claim is, according to inevitabilists, lacking in such inductive support from the actual past history of science. When we look back to past physics, even if we concede that the coexistence of equally good scientific alternatives has sometimes happened, we must recognize that it is very rare, and that it is always a transitory situation that more or less rapidly disappears. This is why the situation is not symmetric for inevitabilists and contingentists, and why inevitabilists feel legitimized in asking contingentists to provide an argument stronger than their own. This is why contingentists should, according to inevitabilists, put up an *actual* equally good, incompatible scientific development in order to make their position credible, whereas for inevitabilists, it is enough to appeal to the still virtual *future* developments of our science.

Revealing the Implicit Logic of the Put-Up-or-Shut-Up Inevitabilist Demand

The preceding analysis of the only-transiently-as-good inevitabilist reply helps to reveal more explicitly the details of the implicit logic that lurks behind the put-up-or-shut-up inevitabilist demand. Now we better understand why, according to inevitabilists, the burden of proof lies with contingentists. This is because *according to inevitabilists*, the actual history of physics does not provide any grounds for contingency. Why? Because if *lastingly viable* equally good scientific alternatives were *truly plausible*—that is, not just plausible according to contingentist philosophers but credible for *scientific practitioners*—we should find some lastingly viable equally good scientific alternatives in the past history of science. In other words, if an equally good alternative physics were indeed possible, not just in principle but *in practice*, we would already have met some such instances in the past history of our physics. But it so happens that we do not. On the contrary, in the past history of our physics, physicists have always globally agreed on one *unique* physical theory, or one *unique, well-determined* answer to most physical questions they have asked. Discussions and controversies have sometimes occurred, of course, but eventually and almost always "quite rapidly," one unique option has imposed itself. When looking to the actual history of our physics, the striking fact is this *uniqueness*, and not the proliferation of alternatives that have looked equally good to practitioners.

Such a line of thought provides, to the inevitabilist eye, strong *empirical* evidence against the plausibility of the contingency thesis. Taking that for granted, the contingentist claim that physics could have been otherwise and profoundly different is perceived as *completely gratuitous speculation* by inevitabilists. Accordingly, inevitabilists feel that only an *empirical* counterevidence could make the contingentist claim credible. This is precisely what the put-up-or-shut-up demand is supposed to ask for.

Why the Put-Up-or-Shut-Up Demand Cannot Be Satisfied

If this is indeed the logic that more or less explicitly underlies the inevitabilist stance, and I think it is, then we can understand why the put-up-or-shut-up demand cannot be satisfied, and why, as a consequence, the put-up-or-shut-up "argument" collapses.

Two reasons that the put-up-or-shut-up demand cannot be satisfied will now be put forward.

First Reason That the Put-Up-or-Shut-Up Demand Cannot Be Satisfied: The Inevitabilist Reading of the History of Science and the Associated Way of Dismissing Any Contingentist Candidate

First, the put-up-or-shut-up demand cannot be satisfied, *in the sense that* inevitabilists will *never* be convinced, *whatever proposal* the contingentist philosophers may put up. Let me clarify why.

When attempting to answer the put-up-or-shut-up demand, the only possibility available to contingentist philosophers is to rely, as Pickering and Cushing do, on scientific alternatives *that were developed* in the *actual* history of our physics.[33] Otherwise, the alternative physics would not be "actual" and would be dismissed from the start by inevitabilists. In other words, the only possible strategy is to stress that, *upon examination*, the history of physics is not *so consensual*, not so strikingly characterized by uniqueness. The strategy is to point out that incompatible, equally good physical alternatives *have existed*, and for QM *still do exist*, such that multiplicity is *actually instantiated*.

Such a strategy, however, will never convince inevitabilists, since, as just explained, inevitabilists have *already decided* on a *certain reading* of the history of physics, according to which uniqueness is the striking dominant feature. As a corollary to this reading of the history of physics, inevitabilists keep in reserve a *ready-made* reply, which can be used *at any time* as a last resort if other arguments have failed, and as an infallible cheap weapon against any proposal put up by contingentists. I am referring to the time-effort-evidence calibration reply. The weapon works as follows:

INEVITABILISTS: Put up an actual, equally good scientific alternative, or shut-up!
CONTINGENTISTS: Here's one! (for example, BQM)
INEVITABILISTS: You are confusing a transient exploratory phase with a stable situation arising from a *sufficiently* long, profound, and evidence-based inquiry. With further time, effort, and evidence, one unique option always imposes itself as the best one. Conclusion:

you have not succeeded in putting up a *sustainable,* equally good, incompatible scientific alternative; thus, shut up!

At this point, we understand that the disagreement between inevitabilists and contingentists is related to a certain reading of the history of physics. Inevitabilists are more sensitive to uniqueness, uniformity, and consensus between physicists. As a corollary, they read all historical situations in which scientific alternatives are actually instantiated and actually supported by living physicists as *transient unstable situations* in which the matter has not been "sufficiently" explored. Contingentists are more sensitive to multiplicity, heterogeneity, and dissent among practitioners. As a corollary, they are prepared to interpret at least some historical situations in which scientific alternatives are actually instantiated and supported by living physicists as genuine examples of equally good, incompatible scientific alternatives.[34]

Is one of these two readings of the history of physics right and the other wrong? Do we have to decide the matter in order to opt for a well-determined conclusion as to whether or not the put-up-or-shut-up inevitabilist challenge has been met, and as to whether or not inevitabilists can use contingentists' inability to put up the required instance as an argument against contingentism (and as a possible corollary, for inevitabilism)?

No doubt, a certain reading of the history of science feeds both inevitabilist and contingentist intuitions, and no doubt, certain ways of writing the history of science lend support to one or the other of these two stances. However, I will contend that even taking for granted the inevitabilist reading of the history of physics, and even conceding the inevitabilist claim that contingentists have failed to comply with the put-up-or-shut-up demand, no argument is thereby provided against contingentism and a fortiori for inevitabilism.

Second Reason That the Put-Up-or-Shut-Up Demand Cannot Be Satisfied: The Monist Regime of Our Science

Let me concede to inevitabilists that, in the history of our physics, uniqueness indeed globally dominates the picture. Let me take this as a fact. Then, the next question is: what is the meaning of this fact? And is this fact itself *inevitable?* What I take to be the appropriate answer to such a question goes as follows.

As it so happens, our *actual way of conceiving and practicing* science is *monist*. It is monist in the sense that the development of a multiplicity of alternatives is *not valued* and *not socially encouraged and supported*—in any of the senses of "supported," in particular financially and materially. Our physics, and more generally our epistemic activities, are governed by a *monist ideal* and a *uniqueness commitment* that seem deeply entrenched. According to this ideal, all physicists look for *the* right option, or at least for *the best* option at the current stage of scientific development. Even if they disagree about what *the* right or best option is, *all* are convinced that *there is one*, and that they have to recognize and select it as *the* framework for future scientific inquiry.[35] Accordingly, as soon as some subgroup of physicists succeeds in imposing one option as the best, the development of others is no longer socially supported and thus *inevitably stops, whatever the will of individual scientists or minority groups of scientists*.

The monist regime of our science and the uniqueness commitment that goes with it are strongly related to the association of science with truth in the vague intuitive sense of correspondence truth. Providing that physical theories and experimental facts "(at least approximately) correspond to the physical world," and assuming that the world is unique and is what it is once and for all, there can only be *one unique* right physical theory. The uniqueness commitment thus works as a regulative ideal and is, *as it so happens*, translated into a uniqueness requirement throughout the scientific inquiry.

So it happens, but is this *inevitable*? In other words, could a more *pluralist* regime of science be instituted? Why not? In any case, it is not empirically impossible. It could be socially valued, encouraged, and supported through the attribution of resources to alternative scientific research. Could an efficient science be developed within such a regime? This point would require further discussion and the introduction of distinctions between several kinds of pluralist regimes.[36] But in any case this is an arguable position. A number of philosophers of science have even argued that we should prefer a pluralist regime for science, and that pluralism would be methodologically more beneficial to scientific progress than monism. Paul Feyerabend and, more recently, Hasok Chang are examples among many others.[37]

The monist regime, taken as a fact about our way of practicing science, is the most deep-seated reason that the put-up-or-shut-up demand cannot be satisfied and why the failure of contingentists to put up a satisfactory alternative should not be viewed as an argument against contingentism and even less for inevitabilism. Because of the monist regime, if we scrutinize the history of physics with the aim of finding configurations that might support the plausibility of an incompatible alternative to our current physics, there is no chance of finding anything straightforwardly convincing. Our science is indeed designed to eliminate competing scientific alternatives as soon as one acceptable option has been found.

Why Any Type of Alternative Physics that Contingentists Can Take from the Actual History of Science Is Unable to Meet the Put-Up-or-Shut-Up Challenge

Given that our science is designed to eliminate competing scientific alternatives, rather than cultivate them with respect to certain aims, any candidate for a scientific alternative that contingentists can find in the actual history of science is bound to appear "less good than" or "only transiently as good as" its competitor. There are, schematically, three possibilities:

1. First possibility: the scientific alternative is a *local* option, once designed and possibly accepted during a certain period in the history of science but subsequently disqualified as not the right or best one. Typically, the local alternative is a circumscribed descriptive claim about the subject under study (e.g., the physical world), incompatible with what is currently accepted as knowledge. Pickering's no-NC-answer and Weber's experimental claim about the existence of high fluxes of gravitational waves (extensively studied by Harry Collins [2004]) are examples of this type.

In this epistemic configuration, the alternative is, *when viewed from the standpoint of today's science,* indeed *objectively less good* than the incompatible hypothesis that has been retained by the history of science. It is objectively less good *on the only basis we have to assess the quality of a circumscribed piece of knowledge*: it is objectively less good in the sense that it is not well connected with multiple other pieces of our current knowledge.

Moreover, *given the monist regime,* the alternative is *inevitably* less good than the incompatible hypothesis that has been retained by the subsequent history of science. It is inevitably less good under the monist regime, because as soon as a local scientific option is superseded by another that is *officially* recognized as *the* right one, it becomes very hard and often impossible for individuals or minority groups of scientists who would disagree about the official "unique" viewpoint to cultivate or further investigate *alternative* scientific options. In particular, it becomes very hard to get any means of conducting alternative *experimental* investigations. Given the highly costly and essentially collective nature of contemporary physics, this implies, in most cases, the practical impossibility of developing experimental alternatives to scientific claims that are widely taken as already established (we are no longer in a situation where individual outsiders can tinker with alternative experiments in their basements).[38] Consequently, alternatives are bound to remain embryonic, poorly articulated, and poorly connected potentialities. As such, they are inevitably *objectively less good* than the options that won out at some point in the history of science, since the latter, contrary to the former, have subsequently been taken for granted until further notice and used as such in subsequent scientific developments, *thus becoming more and more entrenched.*

The official victory of one scientific hypothesis over its competitors at a given point in the history of science can occur against the backdrop of a strong or not-so-strong consensus. In some cases, no disagreement occurs between specialists, but in others, we find what I have elsewhere called "dissensual stabilizations" (Soler 2011). Dissensual stabilizations are cases in which scientific controversies are officially settled and some scientific conclusions are stabilized, although *some* of the scientists involved in the debate *never subscribe* to the officially accepted account of the situation (the gravitational waves controversy studied by Collins is a case in point). In such situations, because of the monist regime of our science, individual scientists or minority groups of scientists who favor options other than the officially dominant one, however strong their personal desire, *do not have the means* to pursue any investigation of the alternative option and to try to build a competing scientific symbiosis that would include the alternative option as a multitudinously and harmoniously connected piece of an extended whole. The situation would be different in a more pluralist regime that would encourage the culture of alternative options even when one option at the current stage of research is recognized as the right or best one.

2. Second possibility: the incompatible alternative to our current science is *an extended and well-articulated scientific paradigm*, once accepted and developed during a long period *in the past*, but then disqualified as not *the right* or *the best* one at the current stage of research. Two subcases can be distinguished, which convey different intuitions regarding the contingentist/inevitabilist debate, depending on whether the defeated paradigm can be viewed as a particular case of today's accepted paradigm (for example, as is often assumed of the Newtonian paradigm compared to the Einsteinian one) or not (for example, as is often assumed of the phlogiston paradigm versus the oxygen one).

In such an epistemic configuration as well, the alternative science exhibited by contingentists is unable to meet the put-up-or-shut-up demand. Inevitabilists will never see this alternative as equally good as our science, *and they will be right*. Although, intuitively, inevitabilists will be inclined to provide different types of reasons for the two subcases differentiated above, the common reason why they will be right to stress that, *from a present-centered vantage point*, relativistic physics is better than Newtonian physics, and that the oxygen theory in its present stage of development is better than the phlogiston theory in the most developed available version we currently have at our disposal, is that relativistic physics and oxygen theory are *more extended* scientific symbioses. Once again this is inevitable under the monist regime, for the same kinds of reasons as those put forward in point (1).

3. Third possibility: the incompatible alternative to the currently dominant physical theory is an *empirically equivalent* theory, currently available and preferred to the dominant theory by *some* (though a *minority* of) living scientists.

The BQM versus SQM case provides an example that is valid today. With this example, we have indeed found an actual scientific alternative, as good as the accepted theory in a well-defined and important sense. The very fact that we have found one such actual alternative, *despite the monist regime of our science*, could lend some plausibility to contingentism. However, precisely because of the deeply entrenched commitment to the uniqueness requirement that goes with the monist regime, almost all physicists and many philosophers of science irresistibly feel that a choice has to be made. Many of them think that one of the two competing options is *already obviously better* than the other.

And in any case, all are convinced that the coexistence of the two empirically equivalent theories can only be a *transient* situation; all are confident that with "further time, effort, and evidence," the balance will inevitably tip in favor of one to the detriment of the other.

Why the Put-Up-or-Shut-Up Argument Is Null and Void

Upshot of the Discussion of the Put-Up-or-Shut-Up Demand

One upshot of the above discussion is that, pressed by the put-up-or-shut-up demand, contingentists should refrain from wasting their energy in following inevitabilists in this direction. This attempt is bound to fail. The demand is actually impossible to meet under the monist regime of science as we know it and the exclusiveness commitment that goes with it. Whatever response contingentists may put up, it will never be enough to inhibit the inevitabilist instinct and convince the inevitabilists.

Another upshot is that, once the monist framework and its implications have been spelled out, contingentists are in a stronger position than before. Faced with the put-up-or-shut-up demand, contingentists are now in a position to reply:

—If what you want is a sustainable, nontransient, equally good, incompatible scientific alternative, you must recognize that the monist regime of our science prevents what you ask from being instantiated. Scientists simply do not have the practical possibility to develop incompatible scientific alternatives "in the long run." Given that, your put-up demand, though it might at first sight look like a legitimate demand for empirical evidence and an empirical test of the plausibility of contingentism, can actually *not* be considered an *empirical* test of anything. Give scientists the means to develop scientific alternatives; relax the monist regime of science as we know it, that is, value, organize, and support a more pluralist regime; create material, social, and psychological conditions that make concretely possible the actual development of scientific alternatives. . . . And then, we will reconsider the matter.

Before concluding with what I take to be the most suitable answer to the put-up-or-shut-up demand as well as the most appropriate reaction to the argument based on such a demand, let me briefly consider other critical discussions of the demand that exist in the philosophical literature and situate my own attempt with respect to these.

Brief Discussion of Other Existing Critiques of the Put-Up-or-Shut-Up Inevitabilist Demand

As far as I know, no extensive systematic discussion of the put-up-or-shut-up demand has been developed in the philosophical literature. However, some remarks can be found here and there. To my knowledge, the most extended critical discussion appears in one section of a recent paper by Ian Kidd (forthcoming).[39] Kidd's conclusion converges with my own, but his arguments are somewhat different from those articulated above.

Kidd's conclusion is that the put-up-or-shut-up demand "cannot be fulfilled. It is therefore not a real challenge but a *fait accompli* that takes the ostensible form of a choice—to *put up* or *shut up*—but which, in fact, only allows the contingentist to shut up, and any putative challenge which structurally excludes the possibility of successful defense ought of course to be rejected." As one can see, Kidd's conclusion and my own are in perfect agreement. However, the reasons put forward by Kidd for rejecting the put-up-or-shut-up inevitabilist challenge overlap only partially with those developed in this chapter. In particular, the monist regime of science as we know it is not considered.

Following Kidd, the put-up-or-shut-up inevitabilist "objection" should "be rejected on two counts."

1. The first reason is explicitly borrowed from a paper by Emiliano Trizio that appeared as part of a symposium on contingency, published under my direction in 2008. At the end of the paper, Trizio provides some brief remarks on the put-up-or-shut-up argument. He writes:

a science like physics is a highly collective enterprise. As a matter of fact, a single individual can develop his or her own interpretation of the French Revolution, but not an alternative high-energy physics. This is a basic fact about natural sciences, whose deep reasons haven't probably been entirely understood. Even if the historical development of those sciences is in part a consequence of *contingent decisions*, those decisions are always collective in character and therefore, in some sense, consensual. The very concept of stability of a laboratory science implies the reference to a complex system of interlocked communities of scientists and technicians, that could hardly exist, if members of these communities were to develop their own activities with the autonomy of historians or sociologists (let alone with that of philosophers). These considerations are able, in my view, to partly defuse the put-up or shut-up argument. (Trizio 2008, 258)

Reexpressed in Kidd's words: the put-up-or-shut-up request "implicitly incorporates a set of epistemic demands which ignore certain facts about the nature of science: namely, that it requires long and sustained processes of practical and intellectual activity which cannot be replicated or reproduced from the armchair." As one can see, Trizio, and Kidd following him, understand the put-up-or-shut-up demand as directed at a *single individual*. This single individual is a contingentist *scientist* in Trizio (2008), and a contingentist *philosopher of science* in Kidd (forthcoming). The contingentist individual in question would have to build an alternative physics so to speak "from scratch" (i.e., relying only on his or her own individual knowledge and skills, independently of any colleagues), and moreover possibly—in Kidd's version—uniquely "from the armchair." Thus understood, the task is obviously impossible.[40]

Two motives can be extracted from Trizio's and Kidd's texts. First, as stressed by Kidd, a philosopher cannot devise an alternative physics "simply through a quiet afternoon of armchair imagination." Here, the impossibility is primarily related to the material-practical nature of science: science develops by intervening in concrete, material situations, by manipulating and transforming (typically through experimentation), and not *just by thinking*. But that is not all. Second, as the quote from Trizio helps us to recognize, the task would be impossible even if the contingentist philosopher trying to meet the put-up-or-shut-up challenge *was also a practicing physicist*—no matter how talented—prepared to spend his/her entire life on the development of an alternative physics. For as Trizio insists, one *single* scientist *alone* cannot build a well-developed alternative physics. The *essentially collective* nature of contemporary physics is actually the *most fundamental* aspect of the first reason that the put-up-or-shut-up demand, interpreted as a demand directed at a *single* individual, "cannot be fulfilled."

2. The second reason that, according to Kidd, we should reject the put-up-or-shut-up "objection," is that "it relies upon a hubristic conception of human cognitive powers, and so cannot be fulfilled."

The exact relation of the second reason to the first is not made explicit in Kidd's text, but as far as I understand, the two reasons are logically hierarchized: the put-up-or-shut-up demand is rejected as hubristic (second reason) *because* it asks for something that is impossible given the essentially collective and material-practical nature of science (first reason). If this is the right interpretation,[41] the second reason is not independent of the first. Rather, the former is a more general characterization of the latter. Kidd's argument would then have the following structure: all hubristic demands should be rejected; the put-up-or-shut-up demand is hubristic (second reason) because an individual armchair philosopher cannot fulfill it (first reason); therefore, the put-up-or-shut-up demand should be rejected.

Let me now situate the analyses presented in this chapter with respect to Kidd's and Trizio's proposals. I agree that the put-up-or-shut-up demand, understood as a request directed at a *single* individual, and a fortiori at a single individual who would have to provide an answer *only* "from the armchair," should be rejected for reasons of the kind put forward by Trizio and Kidd. However, this is not the only possible understanding of the demand. We can also attempt to interpret it as a *collective* demand: as a request directed not just toward one *single* contingentist philosopher and/or scientist but toward a *subset of the scientific community*. Considered in the light of Kidd's and Trizio's perspectives, the analyses that I have proposed in this chapter can be redescribed as a discussion of the force of the put-up-or-shut-up argument when we open the door to a collective interpretation of the inevitabilist demand. In this respect, I have shown that, given our current monist way of conceiving and practicing science, the demand cannot be fulfilled and cannot be identified with a genuinely empirical test. I have revealed some conditions that would have to be satisfied in order to get *a minimum of relevant empirical data* with respect to the issue of a well-developed, sustainable alternative physics. The monist regime would have to be relaxed so as to examine what happens when scientific communities are actively encouraged and given concrete support to develop alternative physics. In terms of Kidd's categories, the question of whether the put-up-or-shut-up demand understood as a collective demand is hubristic or not (i.e., can or cannot be humanly fulfilled) equates to the question of whether a truly pluralistic collective scientific enterprise can be durably viable and productive.

The Most Appropriate Reaction to the Put-Up-or-Shut-Up Argument

Reaction to the Put-Up Demand

At the end of the day, the appropriate contingentist reply to the put-up-or-shut-up demand is, in my opinion, the following:

> —What you ask is impossible given the monist regime of science as we know it. This monist regime prevents multiple alternatives from being sustainably instantiated in our history of science. But the monist regime is not itself inevitable. A much more pluralistic regime could perfectly well be socially instituted. Given that, your confident induction—according to which equally good incompatible theories that might sometimes coexist for a while during some "short" historical periods can only *temporarily* coexist, and one will always win out over the other with time, effort, and evidence *because this has always been the case so far*—boils down to your faith that human beings will indefinitely continue to practice science within the monist regime. But since there is nothing inevitable about such human conservatism, your optimistic induction is not warranted. In any case, if scientists simply do not have the means to fully and lastingly develop alternative incompatible theories, the conditions are not met for your put-up demand to be an *empirical* test. Within the monist regime, the "protocol" is biased from the beginning.

Reaction to the Put-Up-or-Shut-Up Argument against Contingentism

From there, the appropriate contingentist reply to the put-up-or-shut-up *argument* as formulated above (55) is, I think, the following:

> —Since your put-up-or-shut-up demand cannot, given the current monist regime, be a significant empirical test with respect to the

contingentist/inevitabilist issue, the fact that I, as a contingentist, cannot provide the requested sustainably incompatible scientific alternative says nothing about the issue. Therefore, the fact that I cannot put up the requested alternative science is no evidence at all against the contingentist position. Your so-called argument against contingentism is flawed and ineffective. It collapses, because the demand associated with premise (P1) is impossible to satisfy and cannot be equated with a genuine empirical test. Consequently, (C1) cannot be considered established.

Contingentists might moreover go on as follows:

—To be complete, moreover, note incidentally that premise (P1) can also be attacked from a second angle. (P1) involves a strong presupposition, namely, that answering the put-up demand is the *only* way in which contingentism acquires some plausibility. But why would this be the case? No justification is provided for this strong claim. For (C1) to hold, you would have to show that there is *no other way* in which contingentists can attempt to lend plausibility to their position. To discuss this claim would be another story, of course. But a full discussion of this second problematic feature of (P1) should not be avoided.

Reaction to the Auxiliary Argument for Inevitabilism

Concerning the auxiliary argument in support of inevitabilism as stated above (55–56), contingentists would be advised to go on as follows:

—Let me concede for a moment that no sufficiently powerful evidence is available in support of contingentism. This does not imply that inevitabilism has a point. It provides indirect support to inevitabilism, *only under the assumption* that inevitabilism is the legitimate default position (i.e., P3), so that no specific argument is needed in favor of inevitabilism insofar as contingentists have not been able to provide convincing evidence in favor of the opposite stance. But why should we accept such an assumption? Suppose that under examination, *no more* powerful evidence is available in favor of inevitabilism *than in favor of contingentism*. In that case, the philosophical debate would be undecidable. Or suppose that, under examination, the meager evidence for contingentism, though weak, is nevertheless *comparatively stronger* than the evidence for inevitabilism. In that case, should we not conclude that contingentism is more credible or at least less implausible? In any case, the very existence of these possibilities shows that you *have the epistemic duty to face* the issue of the available evidence for your preferred position, inevitabilism. You should not evade the question just by comfortably resting on the perhaps intuitive but unquestioned commitment that inevitabilism is the default and a self-vindicated position. As philosophers of science, we should proscribe such cheap strategies. In other words, (P3) should not be taken for granted, and as a consequence, (C2) cannot be considered established. The burden of proof does not lie unilaterally with me, the contingentist. You, the inevitabilist, should provide arguments as well for your credo. Your put-up-or-shut-up "argument" against contingentism and its auxiliary argument in favor of inevitabilism are flawed and hence unconvincing: (C1) and (C2) have not been established. Do you have any other argument to put up in favor of the claim that certain components of our science were inevitable? If not, then, shut up!

It is often presupposed that the answer to the latter question is affirmative, or at least that inevitabilism ought to be the "default position amongst philosophers of science" (Kidd forthcoming, sec. 1). Accordingly, contingentists are asked much more frequently than inevitabilists to provide evidence for their stance: it is felt that contingentist-inclined philosophers have to exhibit strong evidence to make contingentism a plausible position—or even one worthy of philosophical consideration—whereas inevitabilists have nothing to do insofar as contingentists have not succeeded. But this feeling has no solid grounding. Inevitabilism and contingentism should be treated symmetrically, which implies fully developing the whole set of arguments that each side can produce and weighing one set against the other. This is all the more important because—although the point can only be claimed here—when one seriously scrutinizes the arguments inevitabilists have or could develop to support their position, it appears that beyond deeply entrenched intuitions, these arguments are, to say the least, very weak (see Soler 2006a; and Kidd forthcoming).[42]

Directions for Further Investigation

In this chapter, I have argued that the so-called put-up-or-shut-up challenge should be rejected as a genuine empirical test of contingentism and
that the contingentist failure to meet the challenge cannot be considered
as a support for inevitabilism. I take this conclusion to be an important step
in the contingentist/inevitabilist debate, but it is worth mentioning that
this is not the end of the story. To conclude this chapter, I would like to
mention two sensitive points in need of further investigation.

The first point concerns the claim that science as we know it is monist.
Determined critics could attempt to weaken my argument by questioning the very fact that science as we know it is indeed *sufficiently monist*
to ground the argument.[43] This could be questioned because science, of
course, is not "absolutely" monist as are, for example, totalitarian political
regimes. In scientific practices, dissent can be expressed and discussions
about multiple points actually occur. Accordingly, binary claims expressed
in terms of the monism versus pluralism of science actually hide *degree
judgments* about the *extent to which* alternative scientific options are, as a
matter of fact, encouraged and supported. But as soon as this is recognized,
delicate issues arise, such as: *How monist* is our science? *How hard* is it,
for scientific outsiders and minority groups, to have the concrete means to
pursue alternative research? How are we going to "measure" such things?
How should we understand the condition of "having the concrete means"
to pursue alternative researches? Many dimensions are potentially involved
and intertwined in such a condition: financial, material, psychological,
cultural, ideological, and so forth. Although it is out of the question to
discuss these issues here, I wanted at least to mention them and to suggest
that we should not minimize the associated difficulties.

The second point is about whether a relaxation of the monist regime
would have the power to transform the put-up-or-shut-up demand into a
genuine empirical test. Would the institution of a more pluralist regime
have the power to provide decisive empirical evidence for or against contingentism? Suppose that the famous motto usually associated with Feyerabend's philosophy of science, "let a thousand flowers bloom," has been
applied to science and more specifically to scientific theories. A significant
budget is dedicated to the development of a "thousand" types of "very different" scientific programs. A "thousand" scientific theories are systematically taught to children at school and to students in scientific curricula.

The culture and coexistence of multiple literally incompatible theories become part of scientific forms of life and are widely taken as an important scientific value, alongside other more classical virtues such as empirical adequacy, simplicity, scope, and the like.

In such a situation, how would the put-up-or-shut-up inevitabilist challenge be transformed? A great multitude of *actual* scientific alternatives would be available. Both the condition of "genuine physics" and the condition of (at least some) "similar questions" would apply to the multiple theories involved. Moreover, there is a sense in which each of the different options would be viewed, in some respect, as a *good* scientific theory— otherwise plurality would not be considered a scientific value. As a result, a much more extensive empirical basis would be available and could be used by contingentists, in response to the inevitabilist put-up demand. Much more could be said about the features of genuinely scientific valuable alternative frameworks able to cope with the same questions, because a multiplicity of such alternative frameworks would be universally considered actually available.

However, the empirical instantiation of such a configuration would still not provide any decisive empirical evidence in favor of contingentism. Why? First of all because the equal-value condition could still be denied. Despite the fact that a thousand flowers have bloomed, one of them could always be distinguished as the most beautiful—or more generally as superior to all the others in some important respect. In less metaphoric terms, even provided that a multiplicity of scientific frameworks is indeed available and that each framework is valued in some respects, one of them could nevertheless be viewed as *the* best one *regarding features that would make it a candidate for inevitabilist claims.* In other words, the third inevitabilist reply analyzed above (i.e., denying the satisfaction of the equal-value condition) could still be brandished. Moreover, in the event of hesitations or controversies about which of the multiple scientific frameworks is superior to the others in respects that designate it as *the* inevitable one, the fourth inevitabilist reply—the only-transiently-as-good strategy and the associated appeal to the "long run"—would remain an option as well. Yet no less in the pluralist than in the monist regime can we impose universal criteria that would prevent the kinds of judgments involved in these replies from varying from one individual or group to another. Thus, clearly, the institution of a more pluralist regime than ours, even if sustainably successful, would not provide decisive empirical evidence with respect to the put-up demand and more generally to the contingentist/inevitabilist issue (we may of course wonder if anything like this is ever available in philosophy, including when philosophy is conceived of and valued as empirically based as is often the case today, but I leave this question aside for the sake of the discussion).

Notwithstanding, it is likely that in a pluralist regime, the balance between inevitabilism and contingentism would be shifted and would become more favorable to contingentism than it is today. This is to be expected because there is an essential solidarity, and a relation of mutual reinforcement, although not a logical implication: (1) between the ideal and the fact of a multiplicity of copresent disparate scientific alternatives on the one hand, and philosophical positions like contingentism, constructionism, and relativism on the other hand, and symmetrically, (2) between the ideal and fact of a unique scientific framework, and philosophical positions like inevitabilism, realism, and "absolutism." When multiple discrepant voices can be heard and are not devalued, people become more sensitive to the possibility that none of them is inescapably imposed on any human being; they become more prepared to recognize variations from one individual or one group to another, and to accept these variations as legitimate; they become more responsive to conventions, human freedom, historical openness; so that overall, the idea that things could have been otherwise becomes more natural, intuitively more plausible. Such solidarity between multiplicity and contingency holds, whatever the domain—political, religious, scientific, or other. A democratic regime is much more amenable to relativist tendencies than a totalitarian one—we can see that in our world. Similarly, the development of several new geometries apparently incompatible with the previously lone Euclidean geometry has led people to question the truth and inevitability of the axioms of Euclidean geometry and has encouraged conventionalist interpretations of mathematics. It is thus likely that the institution of a pluralist regime in physics would have a similar effect, that is, would affect as well, and would weaken, the currently very strong inevitabilist instinct about physical results. To go further in the discussion, different kinds of pluralist regimes should be distinguished (see note 36). But overall, under a pluralist regime, inevitabilism would certainly appear less "natural," and contingentism more plausible. Taking all that into account, it is not impossible that in the background of a pluralist scientific culture, some philosophers would consider that we have "strong empirical evidence" in favor of contingentism.

CHAPTER 2

Some Remarks about the Definitions of Contingentism and Inevitabilism

CATHERINE ALLAMEL-RAFFIN AND JEAN-LUC GANGLOFF

O ur main purpose in this chapter is to propose some remarks about the definitions of contingentism and inevitabilism that have been presented over the past decade in the field of science studies. Our goal is basically to introduce the reader to some essential features of the topic that is under scrutiny in this book. Our main wish is to point out some problematic aspects of the definitions, without any pretention to being exhaustive. In the books and articles that have shaped the current debate, we find several definitions of contingentism and inevitabilism, with some slight differences (Hacking 1999, 72; 2000a, 61; Soler 2006a, 2008a, 2008b; Trizio 2008, 254).[1] The definition by Léna Soler is our starting point: "Contingentism claims that it is possible for there to be a science that is, at the same time, as successful and progressive as ours but radically different in content, especially at the ontological level. By contrast, according to inevitabilism, any science which is as successful and progressive as ours and which has addressed the same questions as ours, would inevitably yield answers essentially similar to those that have been actually offered by our own science" (Soler 2008b, 230).[2] As a newcomer in the field of science studies, the contingentism–inevitabilism debate takes place in an old tradition of philosophical debates. J. R. Searle has brought to light the specific logical structure of such debates: "How can it be the case that p, given that it appears to be certainly the case that q, where q apparently makes it impossible that p?" (Searle 2001, 4). Searle reminds us that the most famous example of this kind of debate in the philosophical tradition is the one about free will and determinism. How can we act with free will in a universe in which every event is causally necessitated by antecedent events? Of course, to argue for one term of the alternative contained in such debates is not the only way to conceive the philosophical activity. There are many others. For example, some philosophers have conceived their activity in order to dissolve problems. In order to do so, they have suggested that only a careful and rigorous analysis of our use of words can be helpful. Perhaps some participants in the present debate will invite us to follow that path, instead of defending a position relative to the thesis p (a science as successful and progressive as ours but radically different in content is possible, especially at the ontological level) or q (a science as successful and progressive as ours can only be the one we have). In any case, the emergence of a new philosophical debate presupposes the existence of determined goals from its promoters.

What are the goals that we can identify in the contingentism–inevitabilism debate (hereafter the C-I debate)?

We can roughly distinguish two of these goals, which are not mutually exclusive:

- To refine our understanding of another debate, the realism-constructivism debate. More precisely, the C-I debate is related to "irresoluble differences between realists and constructivists." It reminds us of "profound and ancient philosophical disputes" (Hacking 1999, 68).
- To preserve a conception of science as a rational activity while leaving aside the scientific realism viewpoint—by assuming the relevance of the contingentist thesis (Soler 2006a, 365; 2008b, 232).

In Ian Hacking's case, the goal is to obtain a deepening of the reflection, without any hope of ultimate solution. This is completely in accordance with Searle's analysis of such debates.[3] In Léna Soler's case, the aim seems more specific: it is possible to conceive an alternative view to the realist one concerning science: "some contingentist approaches may offer a more modest position than realism, without giving up rationality and scientific

progress altogether" (Soler 2008b, 232). The underlying question is whether it is possible to elaborate such an alternative image of science.

In this chapter, we will focus on the following points, which we consider problematic. First, we will discuss the links between contingentism and
ontology. Second, we will try to circumscribe the scope of the C-I debate.
Third, we will ask the following question: Which science are we speaking
about? Fourth, we will underline that the whole argumentation developed
within the C-I debate goes back and forth between a logical level and a
historical (or sociological) level. In particular, it is not so easy for contingentists to find clear cases in favor of their thesis in the history of science.
Fifth and final point, we will emphasize the fact that the definition of the
word "success" is highly problematic.

Contingentism and Ontology

The definition of contingentism we adopted above includes the expression:
"at the ontological level." This means that the contingency thesis is not
reducible to an epistemological thesis. It also has ontological implications.
In his book *The Social Construction of What?* (1999), Ian Hacking has
proposed a different framework. The C-I opposition, defined as a sticking
point between realists and constructionists, is conceived in relation to two
other oppositions:

(1) contingentism versus inevitabilism
(2) inherent-structurism versus nominalism
(3) internal explanations versus external explanations of the stability
of the sciences

Oppositions (1) and (3) are situated mainly at an epistemological level.[4]

The contingency thesis tells us that our scientific knowledge could
have been radically different than it is. The inevitability thesis denies the
relevance of such a claim. The C-I alternative is clearly about scientific
knowledge. The opposition between external explanations and internal explanations of the stability of science refers to the genesis of scientific knowledge: What are the crucial factors leading to the fact that our knowledge
has a specific given structure? This is again an epistemological alternative.

In fact, there is only one opposition—between inherent-structurism and
nominalism—which implies explicitly an ontological questioning. The

main references in this case are Latour and Woolgar (1979) and Pickering (1984a).

Is it possible to dissociate the three sticking points proposed by Ian Hacking in *The Social Construction of What?* Such an approach seems well founded in some case studies. But the ontological questioning will probably reappear sooner or later. This is perhaps why Léna Soler has explicitly included the "ontological level" in her own definition of the contingency thesis.

How Can We Define the Scope of the Debate?

According to Soler, the challenge is to build an epistemically harmful distinction between contingentism and inevitabilism, that is, to identify versions of contingentism that really threaten inevitabilist claims about scientific knowledge. For example, if someone claims that radically different alternative scientific trajectories would have been possible but would nevertheless have led to the same scientific knowledge, this kind of "contingentism" is compatible with, and therefore epistemically perfectly benign with respect to, inevitabilist claims about scientific knowledge. In this sense, contingentism is not epistemically harmful. In this perspective, Soler is led to define what could be called radical forms of contingentism and inevitabilism. These radical forms could be viewed as two extremities of a whole spectrum of less radical and possibly compatible versions of inevitabilism and contingentism. In this respect, an analogy could be established with the debate about free will and determinism. Beside the development of incompatibilist theses that defend either the free-will thesis or the determinist thesis, we find at the same time compatibilist theses. It seems that we are facing a similar scenario within the framework of the C-I debate.

The incompatibilist positions can be formulated as follows:

- Radical contingentism (or hard contingentism): this corresponds to the initial definition of Léna Soler: it is possible for there to be a science that is, at the same time, as successful and progressive as ours while radically different in content.
- Radical inevitabilism (or hard inevitabilism): inevitabilism claims that any science that is as successful and progressive as ours and has addressed the same questions as ours would inevitably yield answers essentially similar to those that have been actually offered by our own science. This version of hard inevitabilism may be con-

trasted with another that we could call "very hard inevitabilism." The latter would be the refusal of any form of contingency. It can be seen ironically as an occurrence of the "fatum mahometanum" or of the inflexible necessity we can find in some masterpieces of Western literature or in some Hollywood movies.[5] What the inevitabilist tells us is that only one path leads to successful science. We can reach the next step on this path more or less quickly, but we will never find a crossroad on it. So, in a certain sense, everything is written. Such a view of scientific progress is nothing less than a form of fatalism. Obviously, nobody subscribes to that kind of fatalism in the field of science studies.

Let us also observe that, again carefully reading the definition of contingentism, we could identify an asymmetry between what we have called radical contingentism and radical inevitabilism. Indeed, the definition of contingentism indicates only a set of alternative possibilities that should be taken into account to renew our understanding of the history of science. By way of contrast, very hard inevitabilism leaves no room for discussion because it claims that there is only one right way to understand the development of scientific activities and results. Only hard inevitabilism is compatible with contingentism.

Despite this asymmetry, we can maintain the assumption that it is possible to be more or less contingentist, more or less inevitabilist—in other words to be compatibilists. Hence the points of view of the contributors to the debate seem to be distributed within a complex argumentative space. They cannot be reduced to the dichotomy between radical contingentism and radical inevitabilism. To specify these various positions within the area of our debate, we can distinguish four axes:

- **Axis 1**: the "contribution of the world," the more or less direct causal relations with the material world;
- **Axis 2**: the specific structure of a given science, the constraints exercised by the theoretical, practical, and instrumental forms of knowledge (Galison's constraints, Hacking's styles of reasoning, etc.) defining the standards of objectivity;
- **Axis 3**: the specific structure of the human cognitive apparatus; and
- **Axis 4**: the social dimension of any human endeavor, the weight of the social factors.

These four axes (relating to reality, knowledge, the knowing subject, and society, respectively) determine the space of the arguments that all may bring to the debate in favor of their position. Let us note that only axis 1 indicates a clear orientation in favor of inevitabilism: the more weight we put on the contribution of the world, the more inevitabilist we will be.

The other three axes are more ambiguous:

> Axis 2 could allow an emphasis on the contingency of a style of rea- soning at the time of its appearance, but it can also incline toward inevitabilism once this style is well stabilized.
>
> Axis 3 could allow underlining the limitations of our cognitive device, the character inevitably aspectual and perspectival of our cognitive relation to the world. In this way, it is an argument for contingentism. It can also serve as an insistence on the existence of essential limitations of our human capacities and thus on a neces- sary determination of the configuration of our knowledge.
>
> Axis 4, finally, can also be run in opposite directions: we can find microsocial and macrosocial factors in the sciences that will consti- tute arguments for contingentism, or we can insist on the binding character of these factors within the scientific field, and thus it can lead us to inevitabilism.

Each contributor to the debate, philosopher, historian, sociologist, or even scientist, will adopt a position within this complex area. Some authors will focus on one of these axes rather than on the others.

Which Science Are We Talking About?

The C-I debate is generally framed in reference to physics, or more pre- cisely contemporary physics. In some papers (such as Soler 2008a, 2008b; Trizio 2008), the word "science" is quickly replaced by the word "physics."[6] Why such a shift? Several hypotheses are possible. They are not exclusive but, rather, complementary.

Hypothesis of the Key References

Is it because the central reference in Ian Hacking (1999) is constituted by the work of Andrew Pickering (1984a), *Constructing Quarks*? And is it because the work of James T. Cushing (1994), *Quantum Mechanics: Historical Contingency and the Copenhagen Hegemony*, constitutes another often quoted reference?

Hypothesis of the Epistemological Primacy of Physics in the Philosophy of Science

Could it be because physics is considered, in the philosophical tradition, the most representative science? Can we extend the epistemological conclusions drawn from physics or from a particular branch of contemporary physics to all kinds of science? For some philosophers, this extension is based on an explanatory reductionism—in other words, the idea that all genuine explanations must be couched in the terms of physics and that other explanations, although pragmatically useful, can or should be discarded as knowledge develops. But this reductionism itself has yet to be proven and this will not be our issue in this chapter.

Hypothesis of Physics as the Most Favorable Case for the Contingentist Thesis

Could it not rather be because contemporary physics constitutes the best candidate in favor of the contingentist thesis, for more essential reasons that remain to be clarified? But maybe this privileged status of physics will be disputed or even rejected. In any case, we cannot avoid asking whether, among all the sciences, some of them are perhaps more relevant for the contingentist thesis.

If this is the case, we have to assume the heterogeneity of sciences in regard to the contingentist issue. In other words, the degree of contingentism we can attribute to what we call "science" and the "scientific results" of a science can vary, depending on the science under consideration. The degree of contingentism could be "science-dependent." Instead of defending a universal thesis (either the results of the sciences are contingent or the results of the sciences are inevitable), would it not be more fruitful to turn toward the idea of a continuum? This continuum would go from a degree of maximal contingentism to a degree of minimal contingentism, depending on the science under scrutiny. In this case, what are the criteria to estimate the degree of contingentism? How do the various sciences distribute themselves on the continuum and depending on what—on their

object, their purposes, their methods, the universality of their conclusions? Besides, in this continuum, do clear cases of contingentism and
inevitabilism exist on which everybody could agree? Can we find clear
and nontrivial cases of inevitabilism? Can we find clear and nontrivial
cases of contingentism?

In our C-I corpus, the heterogeneity of the sciences does not appear,
apart from some exceptions, and remains widely to be explored.[7]

Another point we will not develop is the question of the scale of "science"
under scrutiny. To be more precise: for example, should we consider the
whole of biology or would it be more relevant to take a more specific example
inside the field of biology (e.g., immunology or immunology in the 1930s)?

A third point concerns the scale of the temporal sequence taken into account. The temporal scale can be a determining factor for the conclusions
of the analysis.

Let us take an example. The study we want to talk about briefly is a
study by Ian Hacking (forthcoming) concerning syphilis. His main reference is the famous book by Ludwig Fleck, *Genesis and Development
of a Scientific Fact* (1935). According to Hacking summarizing Fleck, the
concept of syphilis in 1905 is relatively vague and encompasses different
kinds of symptoms. Fritz Schaudinn in 1905 identifies the pathogenic
agent, *Spirochaeta pallida*. During the same years, John Siegel identifies
structures that he considers responsible for smallpox, foot-and-mouth dis-
ease, scarlet fever, and syphilis. Both theories could support therapeutic
developments allowing remission but not complete recovery. According to
Fleck, we would have a different concept of syphilis if Siegel's view had
been supported by a thought collective. But in fact Schaudinn's views were
developed, mainly for social—and hence contingent—reasons. Therefore,
at the time of Fleck, in 1935, it was possible to support a contingentist thesis
about these two alternative theories.

For Hacking, Fleck's study of syphilis is a perfect example of the contingentist thesis. We quote him:

> By taking this example of John Siegel and Fritz Schaudinn as two
> alternative ways in the possible evolutions of the knowledge—and
> the recognition of the agent of syphilis—Fleck gives us a perfect
> example of the thesis of the contingency applied to syphilis. . . .
> [T]he criterion of equivalence of Fleck requires only a
> harmonious system of knowledge. It does not require a success

comparable to ours, because in 1935, there is no real success against syphilis. (Hacking forthcoming)

But the affirmation of a contingentist thesis—the idea that both theories are "harmonious systems of knowledge" in the sense of Fleck—is not possible anymore today. Why? Because since Siegel and Schaudinn, we have penicillin and many other antibiotics. Penicillin, developed in an independent research program, enables us to cure syphilitics and leads us to reject Siegel's theory. For Siegel, some cases of syphilis are similar to smallpox. But penicillin does not enable us to cure patients affected by smallpox. According to Ian Hacking, Siegelian syphilis could only be an alternative to our medicine today if the physicians had abandoned their obligation to cure.

This case study shows that the assumption of a contingentist thesis for a given case is itself subject to variation over the years. It can vary in the long run for the same discipline or the same theory, according to the temporal sequence taken into account. In other words, what *seems* to be contingent in the short run *could be seen* as inevitable as a last resort if the temporal sequence taken into account is longer and vice versa. Reciprocally, what *seems* to be inevitable during two centuries, for example, *could be seen* as contingent (or at least, historically situated) if the temporal sequence considered is longer.[8] The question of the temporal scale can thus be crucial.

To conclude on this point, it seems that there will be no unique answer to the question: "Are the results of our science contingent or inevitable?," even for the radical contingentist or for the radical inevitabilist.[9] Maybe we will find a plurality of answers:

- depending on the science we want to study: botany, high-energy physics, biomedical sciences, mathematics, sociology, and so on;
- depending on the scale of science we take into account: a scientific domain in general; a theory within a domain; a definite result;
- depending on the temporal sequence under scrutiny.

Logic versus History

It is not always clear on which level the C-I debate is formulated. In order to draw epistemological conclusions, the argumentation goes back and forth between a logical level and a historical (or sociological) level. It is not so easy for contingentists to find clear cases in favor of their thesis in the history of science. Obviously, it is always possible to imagine that things could have gone another way. But when the historian shifts from the proposition "Things happened that way" to the proposition "Things could have happened in another way, *if . . .*," he/she is not doing history anymore. We are then on a logical level. Hence the inevitabilist could address the following question: is it not paradoxical for a contingentist to call upon pure logical possibility, considering the fact that the contingentist position relies mainly on case studies performed in constructivist sociology? These case studies have particularly stressed day-to-day scientific practice and rejected purely logical arguments.[10] This paradox leads to several interrogations:

- Which science are we talking about? Is it the pure and clean science of the philosopher or the effective science used in the laboratories? From a strictly logical point of view, it is quite easy to argue that we could have a different science. But if we consider the practice of science, much remains unclear. When a scientist tries to understand a new phenomenon, he makes use of the available set of current theories, instruments, experimental resources, and so on. He also follows the rules of communication and the standards of proof accepted in his domain. Seen from the inevitabilist's perspective, to fulfill these methodological requirements is more and more compelling. So, it is highly improbable that an alternative and irreconcilable science will emerge in the usual context of scientific activity. Seen from the contingentist's perspective, irreconcilable local hypotheses with a common theoretical and experimental background can be asserted simultaneously. Then, the contingentist might argue that step by step, these local differences may lead to a completely different science.
- How are we supposed to evaluate the rationality of two alternative sciences? Can we consider that rationality is unhistorical and universal or should we take into account its historical development? If we admit that the accepted standards of rationality are intimately intertwined with the history of science, we could argue that the radical contingentist thesis leads to incompatible conceptions of rationality. For this reason, there would be no common basis on which to evaluate the rationality of alternative sciences.

Definition of the Term "Successful"

The last point we want to stress is the problematic definition of the term "successful" as it is used within the C-I debate. The challenge is to find a definition of "success" that can be accepted by all contributors to the debate. Otherwise, the debate itself cannot even take place properly.

The notion of success implies the existence of a goal that is or is not reached. According to a traditional philosophical view, the most appropriate candidate as a goal, in this case, is truth or approximate truth. In other words, successful scientific work contributes to the corpus of known truth. The problem is that this view corresponds to the position of the scientific realists and is rejected by the antirealists. The antirealist view of scientific success is formulated as follows by Larry Laudan: "A theory is successful so long as it has worked well, that is, so long as it has functioned in a variety of explanatory contexts, has led to confirmed predictions, and has been of broad explanatory scope" (Laudan 1981, 226). As Miriam Solomon has emphasized it, "Realists agree with the content of this account of scientific success, as far as it goes. They regard it as *incomplete*: typically, realists argue that truth explains this success and is the ultimate goal of science. Truth is an *interpretation* of scientific success that realists and antirealists do not accept" (Solomon 2001, 16). Thus, truth is not mentioned by contributors to the C-I debate as a relevant candidate for scientific success.

In the C-I debate, Emiliano Trizio (2008) has proposed another solution: he identifies "success" with "robust fit." The idea is to consider the consistency between the various components of a given science as a criterion for its success. This minimalist definition presents a certain number of obvious advantages. First, it fulfills a fundamental requirement to which all contributors to the debate can subscribe: the internal consistency between the various features of the science under scrutiny. Second, it could also enable us to avoid the opposition realism/constructivism, because "robust fit" can be accepted by both camps. Everybody can admit its relevance.

But we can observe that some authors have conferred very different argumentative status on robust fit or robustness. It is used as an argument for realism by, among others, William C. Wimsatt (1981), Sylvia Culp (1994, 1995), Peter Kosso (1988, 1989), Ian Hacking (1983, 2000a), and Allan Franklin (1986, 1990, 1998).[11] On the other hand, for Andrew Pickering or Léna Soler, it gives support to a constructivist position.

Hence exactly what we understand by "robust fit" remains to be clarified. We will rely mainly on the definition proposed by Emiliano Trizio, inspired by Hacking's paper "The Self-Vindication of the Laboratory Sciences" (1992) and by Pickering (1984a, 1995a).

According to Hacking (1992), robust fit is the adjustment in the laboratory between three fundamental categories (ideas, things, and marks) that enable one to obtain reliability and repeatability of the results. This definition is explicitly correlated to the assumption that "the phenomena scientific theories account for are not independent from our theoretical assumptions and experimental activities. In other words, phenomena are not out there, ready for us to discover and describe, for what we call phenomena are actually complex patterns of results that emerge in a process of stabilization of a certain branch of laboratory science" (Hacking 1992, 257).

Phenomena are theory-laden and practice-laden. This dependence can be understood in a more or less strong way, especially by people who do not use the word "phenomena" in a rigorous Kantian or phenomenological sense but as a synonym for "objects."

Proposition (1). "Scientific activity in laboratories produces phenomena."

Proposition (2). "Science constructs its objects."

Proposition (3). "The natural world is a product of scientific knowledge."

Proposition (1) results from the work of the ethnographers of laboratory since the end of the 1970s. The ethnographer insists on the drastic preparation and "reconfiguration" that are needed to make "natural" materials suitable for laboratory investigations. Besides this activity of reconfiguration, scientists create phenomena in the laboratory that do not exist in the natural world. This means that the concrete material of the experiment has nothing to do with the macroscopic things of everyday life. This remark had already been made by Gaston Bachelard with his concept of "phenomenotechnics" ("phenomenotechnique," Bachelard 1934, 1951).[12]

Proposition (1), which is not opposed by any philosopher, sociologist, or scientist—whether realist or constructivist—can be modified into proposition (2), "Science constructs its objects."

This thesis remains epistemologically innocuous if we conceive the activity of construction as the determination of a domain of relevant objects, the elaboration of experimental protocols, the choice of theoretical tools, and so forth. But it is when we shift from this epistemological thesis to proposition (3), "The natural world is a product of scientific knowledge," that the traditional oppositions reappear because this proposition corresponds to an ontological thesis. Then the borderline is not between realists and constructivists but between the radical constructivists and the others.

Among the authors who subscribe to proposition (3), we find Latour and Woolgar who developed a "splitting-and-inversion model" in their book *Laboratory Life* (1979). According to them, the solid existence of a fact "out there" results from the settlement of a scientific controversy. As long as controversies are still raging, there is no stable reality to which scientific statements refer. Once agreement sets in, something strange happens: "The statement becomes a split entity. On the one hand, it is a set of words which represents a statement about an object. On the other hand, it corresponds to an object in itself which takes on a life of its own" (Latour and Woolgar 1979, 176–77). Subsequently, history is rewritten, as it was. Now, it is held that the object has been there all along, waiting to be discovered. Thus, an inversion takes place: "the object becomes the reason why the statement was formulated in the first place" (176–77). According to Latour and Woolgar, an object enters into existence precisely at the time of its discovery. The hormone TRF (thyrotropin releasing factor) began to exist at the time of its official discovery, in 1969, but did not exist before that year.

The splitting-and-inversion model contains a thesis that is highly controversial: the idea that objects are created out of negotiation and eventual consensus or, in other words, that representations, once generally shared, give rise to their objects. Latour and Woolgar propose here the strongest version of the theoretical and practical dependence of phenomena. Phenomena do not exist outside the laboratory. They are actually created in the laboratory.

But this version cannot be endorsed by a realist, or a perspectival realist, or even a moderate constructivist.

To summarize, as long as we reduce, without any further precision, the meaning of "robust fit" to a mutual adjustment between the various items of a science, this definition could be accepted by all contributors to the debate. We are then on an epistemological level, corresponding to our aforementioned axis 2. But the notion of "robust fit" that generates this collective agreement is a giant with feet of clay. It all depends on the nature of the items one wishes to include in the definition of "robust fit." If those items are sociological or historical factors, strong oppositions reappear, as we have just seen. These oppositions are related to our axis 1: the contribution of the world.

Some incompatible positions can be summed up as follows:

1. Science provides knowledge about the world, because it is obtained through interactions with some aspects of the world.
2. Science does not provide knowledge about the world in itself.
3. Science creates the world.

These three positions are compatible with the requirement of a "robust fit." So the consensus holds only as long as the expression is not more precisely defined. And we can really wonder if "robust fit" is the best candidate for defining "scientific success."

Finally, it is not sufficient to define "successful science" if we limit the "robust fit" to items like ideas, things, and marks within the laboratory. For example, could we evaluate the success of bacteriology in terms of "robust fit" only? Is it not necessary to consider the effects that a science produces outside the laboratory when we want to evaluate its success? In this context we can refer to the proposal of Ian Hacking who suggests evaluating success not only in terms of Lakatosian research programs but also in terms of "sheer material success," such as, for example, "the smallpox vaccine, the plutonium, the transformation of agriculture," and so on. According to Hacking (2000a, 63), we have here a "more mundane but equally important idea of success" than the one we find in Lakatos's methodology of scientific research programs. Following this proposal, we could try to address the problem of "success in science" within an explicitly pragmatist framework. In the pragmatist tradition, scientific knowledge is conceived as something producing tools that are used to serve various human goals. In such a framework, the success of a science has to be evaluated on different levels: empirical successes (including experimental and instrumental successes), theoretical successes (including simplicity, consistency, explanatory and predictive power), and material successes (including technological developments and day-to-day applications). More generally, the pragmatist framework can enrich the C-I debate because it encourages us to understand human knowledge as a process and not as a static relation of "correspondence." According to William James, for example, "no theory is absolutely a transcript of reality, but any of them may from some point of view be useful. . . . A scientific theory is to be understood as 'an instrument: it is designed to achieve a purpose—to facilitate action or increase understanding'" (James 1907, 33).

The main point we wanted to highlight in this section of our article concerns the very possibility of a common definition of the word "success" in the C-I debate. It seems to be very difficult to establish such a definition considering the incompatible presuppositions to which the participants to the debate subscribe.

The aim of this article was to raise the following questions that underlie any reflection on contingentism and inevitabilism:

- What are the possible positions within the framework of the C-I debate? Should we reduce them to a dichotomy between the two "radical" positions defined above or should we accept compatibilist positions?
- Which kind of science are we talking about? What are the relevant temporal scales?
- Can we clarify the relationships between the logical and the historical (or sociological) levels of argumentation?
- Can we agree on a common definition of what makes a science successful? Would a pragmatist perspective be suitable for this purpose?

These questions are highly problematic and will not lead to any obvious answers. But these questions are unavoidable for anyone interested in the C-I debate. What is certain is that the notions of contingentism/inevitabilism seem to be the most promising heuristic device of the past decade for a better understanding of what science actually is.

PART II

Contingency, Ontology, and Realism

CHAPTER 3

Science, Contingency, and Ontology

ANDREW PICKERING

Léna Soler (2008a) makes a nice contrast between contingentist and inevitabilist conceptions of science. The latter is much easier to grasp and much more intuitive. It is the idea that any community that embarks on a project akin to science will end up believing what we do. Ian Hacking (1999) made this idea a bit more specific by saying that anything like physics must eventually arrive at Maxwell's laws. None of this is very clear. What are we thinking of when we imagine "project akin to science" or "anything like physics"? How can we imagine the space of possibilities? What's in and what's out? I'm not going to agonize over such points, because I want to develop a contingentist vision, drawing on the history of science instead of imagining other sciences. But I think it helps to start with inevitabilism, and I want to begin by talking about where our intuitions about inevitability come from. While most discussions of inevitability, contingency, and associated issues are usually couched in an epistemological idiom, my suggestion is that our intuitions have an importantly ontological aspect—they are intuitions about the sort of place the world is. The rest of this chapter then circles around questions of how and why I think our ontological understanding of the world needs to be modified to help us come to grips with contingency.

Where do our inevitabilist intuitions come from? By and large, they are implanted in us from a very early age by people like parents and teachers, certainly before we have developed much of a faculty for critical thought. We teach our children that, in general, the world is a fixed and lawlike place and that, in some respects at least, we already know its structure. When they were young, I used to expound the principles of heliocentric astronomy to my defenseless offspring. The earth goes around the sun, I told them, in an elliptical orbit while spinning on its axis: hence day and night and the seasons—things like that. This sort of thing goes on endlessly in homes and classrooms. Everyday features of the made environment conspire with its vision: door handles and clocks have a fixed regular structure that we can know, too. And, of course, this ontological vision just intensifies if one moves further into the arcana of scientific education. I was about eleven when I was first introduced to the idea that the world is built out of atoms and molecules, and I was taught how to associate that idea with some aesthetically seductive reactions in the chemistry lab, and how to do calculational tricks with it.

Growing up, at least in the mainstream of the modern West, is thus an intensive indoctrination in the ontology I just mentioned, that the world has a fixed reliable structure and we can know it: atoms and molecules, quarks and DNA, and so on. And that ontology then serves to underpin our inevitabilism—it makes inevitabilism inevitable. If the world is built out of definite fixed entities and structures then inevitably anyone who explores the world will run into and know them. From wherever one starts, from whatever angle, sooner or later one will run into atoms and molecules—they are what is there in the world; there is nothing else to run into. Likewise Maxwell's equations: that is how the electromagnetic field is. This, I believe, is where we are all coming from when we try to think what the world is like.

All this is obvious enough, though phrased the other way around I find it more striking. Thinking about my own education, and my childrens', and what I know about school and university curricula in general, I am struck by an almost total absence of ontological visions that do not conjure up a regular and knowable world, that conjure up something different. We have nothing to set against what I could call the modern ontology of fixity and knowability. At some level that is hard even to recognize, we are defenseless against it. I think this is a sad fact about child rearing and education, but that is not the topic of this essay. Instead, as stated, I want to build up here another ontological vision, more adequate to what I know about the history of science.[1]

How might one begin to dislodge inevitabilist intuitions? For me, it began with a fascination with notions of incommensurability that grew
out of reading Thomas Kuhn (1970; see also Pickering 2001) and Carlos
Castaneda (1968) in the early 1970s. I was attracted to Kuhn's idea that
scientists working within different paradigms somehow inhabit different
worlds, and Castaneda's account of his initiation into the magical world of
Don Juan just made the idea more appealing. What puzzled me about it,
as an erstwhile physicist indoctrinated along the lines just indicated, was
the question of how incommensurability could be possible. What sort of a
place is it that we inhabit, that we can conjure it up in such different ways
as modern physics and Yaqui sorcery? I wanted an ontological picture that
would make sense of that, and my first inspiration came from crystallog-
raphy. I thought of the world on the analogy of a crystal that could be split
along different axes, and whose different faces displayed different patterns.
Each axis, on this model, would stand for a specific paradigm that, so to
speak, produces its own specific world.

I like this crystal ontology. To a degree, at least, it makes sense of some
striking features of the history of science, and it also constitutes an ontological antidote to inevitabilism. It makes it possible to think that the angle
from which we approach the world matters. Depending on the angle, the
world shows itself in this way or that, and these ways do not lead into one
another—there is no inevitability that they will arrive at some common
point; in fact the inevitability is that they will remain forever different. Or,
to put it the other way around, since the angle we come at the world is not
given by the world itself, it is, in that sense, contingent. The crystal ontology is an ontology of contingency. Contingency, on this account, is not so
much to do with us as human knowers as it is an attribute of the material
world—the world just is a crystal that can split along several different axes.
But more needs to be said here.

If we want to stay in this crystalline world a bit longer, we need to think
more about the splitting. Just how does the world-crystal get cleaved along
this or that axis? At this point mainstream philosophy of science might start
talking about the Duhem–Quine thesis, the idea that many different accounts can be given of any set of data, or, more imaginatively, about Norwood Russell Hanson's arguments that scientists perceive gestalts rather
than particulars and that many gestalts are possible (Duhem 1962; Hanson
1958). But although these arguments are interesting, they remain in the
ontological space that I am trying to get away from. They point to a line of

thought that remains faithful to the idea that the world has a single fixed and knowable structure, but add to that the idea that on some higher level we are capable of picking out all sorts of different patterns in it. This leaves the crystal world behind and shifts all of the burden of contingency back onto us. I think we need something more radical to do justice to the history of science.

In science, we latch on to the world not via our unaided senses but through the use of machines and instruments and all sorts of contrived setups, and my suggestion is that we think of specific fields of machines as cleaving the world along specific axes. Instruments latch on to the world and elicit it in a certain way and, thus, so to speak, translate it into the world of science. *Constructing Quarks*, for example, includes a chapter I called "Producing a World," in which I documented the ways in which different fields of instrumentation in high-energy physics elicited different fields of phenomena, which in turn sustained different understandings of elementary particles themselves (Pickering 1984a). If we follow this line of thought, the ontological picture becomes more interesting and complicated. The original crystal metaphor locates contingency in the world itself, in nature. Duhem–Quine arguments move it back into the human realm. But if we take this instrumental aspect of science seriously, the picture starts to decenter itself. The splitting at the heart of contingency has to do with both nature and instrumentation and, lurking on the far side of instrumentation from nature, ourselves as the designers and builders of machines. We move, that is, toward a nondualist and posthumanist ontology. I want to move further in this direction now, though this entails saying good-bye to the crystal world.

What is wrong with the crystal ontology? No doubt many things but, most seriously, it lacks any dynamics. You split the crystal, there is a pattern on the exposed face, and that is it, nothing further happens. But science is not like that. It evolves, continuously and discontinuously, in time. What sort of an ontology can accommodate that? How can we get dynamics into the picture?

I want to continue to focus on the way we latch onto the world through machines, and I want to emphasize that this is a nontrivial business, and especially that it is a continuing process, extended in time, not a one-off event. We can find a classic description of this process in Ludwik Fleck's (1979) history of the Wassermann reaction as a blood test for syphilis. At some early point in its history, the test had something like a 15 percent success rate; at some later point it went up to 70 percent or 80 percent. The test, we could say, latched on to syphilis more effectively over time. And how did this latching-on happen? It happened, according to Fleck, in a sort of trial-and-error tinkering. The scientists tried varying the prototypical recipe in all sorts of ways and eventually arrived at a recipe that was medically useful. Fleck describes this process as one of tuning—tuning the experimental procedure to hone in on a signal for syphilis. Following Fleck, and on the basis of my own studies of scientific practice, I am inclined to think of it as a *dance of agency*—a constitutive back and forth between human agents who contrive specific material setups and the agency of those setups themselves—what they do. I documented several of these dances in *The Mangle of Practice* (1995a): Donald Glaser assembling all sorts of setups en route to the bubble chamber, then standing back to film what they would do, then reconfiguring the apparatus in response to its performance, and then around the circle again; Giacomo Morpurgo doing just the same in the development of his quark-search apparatus.

These dances of agency are, I think, endemic to and constitutive of scientific practice in all sorts of ways, and I now want to think about them ontologically. The first point to grasp is that they conjure up an image of the material world not as fixed, static, and knowable but as endlessly lively. The world performs—does things—that continually surprise us. My reading of the history of science is that the world is a place of endlessly emergent performativity; I can see no reason to think we shall ever reach the bottom of it. And it is worth remarking that this is an ontological discovery. It seems, at least, that we do not, for example, live in a virtual reality simulation where novelty comes to an end when we get down to the level of the fundamental pixels—though perhaps we just have not gotten there yet.[2]

A corollary of these observations can get us back to our theme. The only way to get along in a world of endless emergence is to be light on our feet. Just as Glaser or Morpurgo never knew how their apparatus would perform next, so their responses to such material performances were also emergent and made up on the spot, rather than given in advance. My argument in *The Mangle* was that one can discern a temporal pattern in scientific practice, centered around a notion of modeling, but that contingency and chance are an integral part of the pattern. In the end, something is inexplicable, though not at all mysterious, about scientific practice. It just so happened that Glaser thought of building a detector like the cloud chamber but different; it just so happened that he responded to resistances in his practice the way he did, rather than some other way. What happened in this dance was not preordained; it could have gone differently.

This is the basic argument I made in *The Mangle*: science is built in dances of agency; chance is endemic to these dances; therefore one should see the state of scientific culture at any given time—meaning its fields of instruments and the facts, theories, and social relations that surround them—as genuinely historical, as the product of contingent rather than necessary developments. I think this is how it is. But the picture needs a bit of elaboration.

First, a simple point: if one stays with the old ontology, of the world as a place of fixed, knowable properties, then contingency seems immensely threatening. In that kind of world, you either get the story right or you get it wrong, and to speak of contingency is to conjure up the latter, as if the scientists must be just making their stories up if they are not latching on to the structure of nature. (This is the point at which philosophers start talking about method: we have to police the scientists to rein in the contingency of their practice. From another angle, it is the standard opening into social constructivism.) But on the ontological account I am trying to develop, things look different. The sort of contingencies in scientific practice that I am pointing to are necessary counterparts of the endless emergence of the performativity of matter. They are integral to our struggles with the otherness of nature, not something to be feared or regretted. They are the mark of the fact that scientists are not in control of their own endeavors, that they are not just inventing their culture from whole cloth. We should admire science for its dances of agency, including all of these dances' contingencies—these are where scientists genuinely grapple with their object. This is the position I called "pragmatic realism" (Pickering 1995a).

My second point is less straightforward. I have so far described dances of agency as if science is just an endless struggle through an unmappable and continually mutating jungle. I think this a good image to start from, but more needs to be said. In science, dances of agency have a specific structure that we need to think about. It seems to me that they—and science itself—are characterized by a certain telos, which is what I want to discuss now. What is this telos? It is, in the first instance, that dances of agency in science aim at their own self-extinction. Scientists do not enjoy them much; they want to get out of them. I can remember doing my PhD in physics, trying to write a big computer program to fit a lot of data. Every night I would leave the latest version to run, and every morning something

would have gone wrong, which I would then have to try to fix—which an-
noyed me immensely: why won't the damn thing just run? So what does it
mean to extinguish a dance of agency? In *The Mangle* I talked about special points of "interactive stabilization," which, when achieved, are places
where practice can rest for a while, and facts or whatever be reported, when
the program runs—moments when the dance is temporarily over.

But still, what does this interactive stabilization amount to? I described
it as various cultural elements fitting together in some way, in contrast to
the mismatches that are the usual state of affairs. I also insisted that it was
impossible to give any closed definition of what this "fitting together" in
general amounted to. But here I want to make things difficult for myself
by admitting that in this instance I missed a trick, namely, that in these
special moments of stabilization some things hang together in such a way
that some other things—namely, the human and the nonhuman—are split
apart. When my computer program would not run, our lives were bound
up together: every day I would tend it and work on it; every night it would
disappoint me. When it did eventually run, we could go our separate ways.
Our relationship had changed. The program had at last achieved its independence from me—it had become something I could use, what I would
call a freestanding machine—and I had achieved my independence from
it—I was once more, in this respect at least, my own man, a freestanding
human being, and not the other half of a stack of punched cards: likewise
Glaser and his bubble chamber, and Morpurgo and his electrometer.

So, if one follows this line of thought, one arrives at a more tightly specified account of the dance of agency in science: it is a dance structured
through and through by an invariant telos, that of splitting the human
from the nonhuman—a telos of dualization, of making the world dual.[3]

What should we make of this? First, it is a major ontological discovery
of modern science (and engineering, of course) that this dualization can
be done. I cannot see that the world has to be such that dualization would
work; it just turns out to be that way.[4] Second, we have here another reason
to admire science. Going back to the earlier discussion of the dance of
agency, finding these islands of dualist purity is hard and uncertain work,
entailing arduous searches through spaces of material agency (and much
else). Third, there is something objective and noncontingent about these
islands, which needs further discussion in the present context.

Let us try a different example, for the sake of variety (Pickering 2005).
In 1856, William Henry Perkin patented a recipe for the production of the

synthetic dye mauve, a recipe that began: "I take a cold solution of sulphate of aniline, or a cold solution of sulphate of toluidene, or [etc.] . . . and as much of a cold solution of a soluble bichromate as contains base enough to convert the sulphuric acid in any of the above-mentioned solutions into a neutral sulphate. I then mix the solutions and allow them to stand for 10 to 12 hours, when the mixture will consist of a black powder and a solution of neutral sulphate. I then throw this mixture upon a fine filter [etc., etc.]" (Pickering 2005, 365). Here Perkin is describing one of these islands of stability that science aims at, and I would be happy to read him as describing a property of the world that exists quite independently of us. My ontological intuition (though maybe I am still under the spell of my indoctrination as a physicist) is that were any being, at any time, anywhere in the universe, to mix the named chemicals and leave them to stew, they would end up with a black sludge from which one could extract a pretty colored dye. Similarly, given the right components, any being could build a bubble chamber or a quark detector.

So what? If I wanted to support our intuitions about the inevitabilty of science, I would not point to rarefied theoretical constructs like Maxwell's equations; I would start with visible material achievements like Perkin's mauve recipe. Look, I would say, the world is this way and there is nothing contingent about it all. I would even be prepared to elaborate the ontological picture and start talking about "attractors" here, as if the world pulls us into zones of stabilization. Give this hypothetical being some aniline and bichromate and tell her to find out how to make a purple dye, and if she messes around long enough I think there is a good chance she would find something like Perkin's recipe. (There is an argument here against Harry Collins's [1974] famous discussion of building a TEA laser.)

Let us, then, imagine, this more sophisticated ontology of a world of endless emergence that nonetheless is characterized by all sorts of basins of attraction that we can, as it happens, settle upon and exploit. This sounds like a good description of the world to me. The question then becomes: is this ontology consistent or inconsistent with contingency in science? It seems to point to inevitability rather than chance, but does it?

Of course, I would not ask that question if I did not think the answer was no. The key questions here, I suppose, concern finitude and uniqueness. How many of these islands of dualist stability are there in the world? Is there some finite list that all beings operating a dualist telos are destined necessarily to run into? Everything I know about the history of science suggests this is not the case, that there are, in fact, indefinitely many such islands and that which ones we settle upon are, again, matters of contingency.

One way to motivate that thought is simply to go back to instances of incommensurability large and small. At the micro end of the scale, I think of the Morpurgo–Fairbank controversy about the existence of free quarks. Within a shared cultural context, these two physicists found different material ways of latching on to the world that pointed to diametrically opposed conclusions. At the macro end, I think of the difference between the "old" and the "new" paradigms, as they were called, in particle physics. Both of these achieved the dualist telos of human–nonhuman separation, but using quite different fields of machines quite differently tuned to display quite different phenomena. This last example, of course, moves us beyond the question of individual islands of stability to overall patternings, without, as I said at the start, pointing to any necessary uniqueness.[5]

Another way to argue the point is to look more closely at the idea of an attractor. If we want to speak this way, we have to ask where attractors are. Are they just there in the material world? This is a tricky point. I stated earlier my conviction that the material world just does behave the way Perkin's mauve recipe suggests. But from my ontological perspective, the material world must have indefinitely many ways of behaving—however we configure it, it will reliably do something or other. But most of those somethings we do not care about at all. When I conjured up an imaginary being who would find the dye mauve, I had to supply her with both some named chemicals and an objective: produce a dye. Without these special material elements and a culturally situated goal I do not suppose she would ever arrive at mauve.[6] So despite the telos of dualist purification—of making a clean split between people and things—somehow these attractors fail to respect the split. They remain themselves decentered things, existing in a nondual space that is a joint product of the material and social worlds. Only for diseased aliens obsessed with ideas of bad blood could the Wasserman reaction count as an attractor. In effect, then, we have come around in a circle. What counts as an island of dualist stability in scientific practice still depends on all the contingencies of how we approach the world that I talked about before.

Though I want to address some subsidiary points, this completes my basic line of thought, so I should briefly sum up where we have arrived. I do not suppose I have settled the issue of contingency in science, but I have tried to shift the terrain on which we think about it, from epistemology to ontology. I find this shift useful and attractive because it (1) gets us closer to the practice of science itself, (2) helps us see the multiple contingencies of science more clearly, and (3) defangs the threat of contingency: seen from an ontological perspective, contingency is not the sort of thing we need to worry about; it is intrinsic to getting along with an indefinitely lively, emergent, and always surprising world.[7]

Now for the subsidiary points. There are four of them. First, I have to recognize that in my shift from epistemology to ontology I have moved the goalposts away from the usual obsession with knowledge. The traditional question about necessity has to do with things like Maxwell's equations rather than bubble chambers and mauve recipes. In this connection, I just want to note that in *The Mangle* I did try to map out how scientists move between the machinic and the epistemic, and nothing much changes in my conclusions if we follow this transit. If anything, the position just looks bleaker for the necessitarian. The islands of dualist stabilization I have discussed so far have been simple and obvious ones: a bubble chamber that produces particle tracks; a recipe that produces pretty clothes. My argument concerning the production of facts and theories is that they, too, are produced in dances of agency, though much more intricate and delicate and more complexly situated ones than those discussed here. If one goes into the details, more contingencies appear, not fewer.

My second thought remains at the epistemic level. In *The Mangle* I argued that articulated knowledge is built in the creation of alignments between machinic performances and conceptual structures. And what I can see more clearly now is that the conceptual structures of science have themselves a specific and peculiar structure, precisely in that they refer only to a fixed and regular material world having a knowable structure. Another aspect of the telos of science is, then, to find islands of dualist purity that can be aligned with dualist accounts of the world—accounts of how the world functions quite independently of us, as a giant freestanding machine. The old ontology thus reappears here as something imposed on the world by science, and this, indeed, is where our ontological intuitions about science come from, from mistaking a given telos with how the world is. Again, I should say that there is an ontological discovery here: that the world is the sort of place where these kinds of assemblages of machines and a very particular form of knowledge can be built. The Scientific Revolution was, I suppose, precisely this discovery. We can admire the construction of these assemblages as highly nontrivial achievements. But, as I have said already, we would be mistaken if we thought they somehow efface historical contingency or obviate the case for the alternative ontological vision I have been trying to conjure up.

My third thought returns to the material plane. I have been talking about "islands" of stability in our relations with nature, but I want simply to raise the question of whether this is the right metaphor. I have in my mind, for example, the idea that Perkin's discovery of the mauve recipe was not in itself sufficient to lead to a transformation of nineteenth-century organic chemistry (or the establishment of a new synthetic dye industry). It was crucially important that, as it happened, extensions of Perkin's work quickly turned up other islands of stability, other recipes, which could collectively be caught up in Kekulé's theories of molecular structure and the benzene ring. It seems appropriate here to think of the science of chemistry developing in a process of following a "seam" or "vein" of attractors, located, as before, in a hybrid human/nonhuman space (Deleuze and Guattari 1987). The idea I want to float is that perhaps all sciences are founded not on single islands of stability but on finding a vein of attractors. Robert Kohler's (1994) classic description of drosophila as a "breeder reactor" for genetics might point in this direction.

My fourth and last point goes in a different direction. I have, in effect, described science as a singular stance in the world—a stance that insists on imposing a specific structure on the dances of agency in which we are all enmeshed—an insistence on finding islands of dualist purification that can be connected up to accounts of a fixed and regular nonhuman world. This teleological structure is not given by the world (though the world certainly conspires in it) and the whole scientific enterprise might itself thus be described as contingent. Furthermore, this scientific stance has certain characteristics that are worth noting. As Martin Heidegger (1977) argued, the modern sciences conjure up the world as a calculable "standing reserve" awaiting human domination, and I think he was right in describing science and technology as integral to the mode of being that he called "enframing."

Ten years ago, that observation would have seemed largely meaningless to me, because I would have found it hard to imagine any other mode of being in the world. What I have since come to realize is that there are, in fact, alternatives to the scientific stance, and I have become especially interested in projects that, so to speak, dwell on dances of agency rather than trying to bring them to an end and expunge them from our imaginations (Pickering 2008). My recent work on the history of British cybernetics has been precisely an attempt to explore what this stance of "revealing" can look like in all sorts of fields and endeavors (Pickering 2010).

My closing remark is, then, that ontological reflection does not have to remain in the realm of ideas. Just as a dualist ontology of a fixed and knowable world comes down to earth in modern science and engineering, so an ontology of emergent performativity and dances of agency manifests itself in all sorts of cybernetic projects and artifacts. By their fruits ye shall know them. If I wanted to discuss the practical merits of the ontological visions I have discussed here—which is it better to believe?—I would start from Heidegger's idea that enframing is a "supreme danger" to humanity, but that is a topic for another essay (Pickering 2008, 2009, 2010).

CHAPTER 4

Scientific Realism and the Contingency of the History of Science

EMILIANO TRIZIO

In the Maze of Possible Histories of Science

The temptation to imagine alternative historical developments of political, social, and cultural phenomena has always been a strong one. It is thus not surprising that one might try to include the history of science in this exercise and wonder whether it could have ended up differently than it actually did. Yet this attempt becomes problematic as soon as we impose epistemological conditions on the alternatives whose possibility (and plausibility) we are trying to evaluate, because, in order for the imagined scenarios to be interesting from an epistemological point of view, (1) we need to refer to an at least roughly defined specific subject matter such as optics or high-energy physics, (2) the imagined alternative histories of science must arrive at a different and yet *equally successful* stabilized stage, and (3) they must imply some fundamental disagreement on the subject matter in question. We are, therefore, putting heavy constraints on the alternatives we wish to consider, for most of the possible histories of science differ from our own in ways that are epistemologically uninteresting for one or more of the following reasons: they are about the investigation of subjects other than the actual ones; they are histories of failure, not of achievement; they lead to results that are not incompatible with ours and therefore can be combined with them. The epistemological relevance of these three conditions should not be missed. Indeed, there is little epistemological interest in comparing what our science says about planets with what one might have ended up thinking about viruses, or with what a bunch of fools unable to conduct any scientific research could have dreamed about planets, or, finally, with a planetology differing from ours as to the nomenclature only. Keeping this in mind, we can now turn to the relevant definitions.

Contingentism, as it has been defined by Ian Hacking (2000a), is the claim that the history of a particular field of science could have turned out differently than it actually did, and that it could have resulted in a science as successful as the actual one but, in a nontrivial way, incompatible with it. Inevitabilism consists in the denial of this claim.

All crucial terms involved in this definition are affected by a certain degree of vagueness and can be defined in multiple ways,[1] but probably the trickiest of them is the term "successful." There are of course different definitions of scientific success, depending on the aim that one assigns to science.[2] Following Hacking, it is reasonable to include in the idea of scientific success a certain degree of progressiveness. However, the idea of progressiveness already implies a number of positive features that admit of improvement (such as predictive power, technical achievements, etc.). We shall see that the notion of success, along with the even thornier idea of evaluating "degrees of success," can be better analyzed while examining specific contingentist scenarios.

If we believe that the history of the scientific investigation of a subject matter could have led to a stabilized stage as successful as our own, but incompatible with it, we also believe that different mutually incompatible and equally successful scientific accounts of the subject matter in question are possible. I call the latter claim the *multiplicity thesis*. Contingentism, therefore, implies the multiplicity thesis. More precisely, contingentism can be equivalently reformulated as the conjunction of the multiplicity thesis with the claim of the possibility of an alternative history of science leading to one of the successful alternatives that are incompatible with our own science.

A logically weaker form of contingentism (and, correspondingly, a stronger form of inevitabilism) is sometimes evoked when one approaches the issue by asking whether *any* scientific investigation as successful as ours of a given subject matter would need to lead to roughly the same results as ours. For instance, one can ask whether, had modern science developed outside the Western world and had it reached a level of success comparable to ours, it would have necessarily achieved the same results, or, to push the example to the extreme, one can ask whether the results of an alien science as successful as ours would have to look pretty much like those that are familiar to us. The claim that, say, an alien science could be different from ours while enjoying the same degree of success is logically compatible with the idea that we could not possibly come out with that science either for want of material and intellectual resources or because of the intrinsic features of our historical starting point. The multiplicity thesis is thus compatible with the denial of contingentism as previously defined: some alternative successful sciences might be possible but simply de facto out of *our* reach, given our cognitive structure and the cultural and scientific stage at which our research into a certain subject matter developed or, in short, given the cognitive, cultural, and scientific background underlying the research. In this article, unless otherwise stated, contingentism will be intended in this stronger sense, that is, as implying that some successful alternatives remain open even once the background of the research is fixed.[3]

One further point requires discussion. It will be argued that the opposition between inevitabilism and contingentism, thus defined, somehow presupposes a more or less fixed notion of *science*. This is certainly the case. We are trying to understand what *degrees of freedom* are left to the historical evolution of that cognitive activity that we call *science*, no matter how difficult it is to specify its nature, in general. In other words, we are concerned with the extent to which successful science is *bound* to evolve the way it does; we are not concerned with the deeper issue of whether, as cognitive subjects, we are *bound by* the standards of scientific rationality, let alone with the even more fundamental problem of whether there are universally binding standards of rationality at all. If we drop these constraints and thereby also any shared notion of success, we also give up any epistemological criteria restricting the family of cognitive activities we are considering, and we end up comparing epistemic "forms of life" that may have little or nothing in common.

These introductory remarks suffice for the purpose of this article, which is twofold: first, to circumscribe and analyze the conflict between contingentism and scientific realism; and second, to characterize it from a methodological point of view. The term *scientific realism* will cover the family of theses according to which our successful scientific theories make claims that are true (or approximately true) about the aspects of the world that they describe. I will adhere to the canonical distinction between scientific realism and *metaphysical realism*, which is the thesis according to which the aspects or parts of the world investigated by science have a given, intrinsic nature, whether we succeed in acquiring knowledge about it or not. According to metaphysical realism, there exists a *true description* of the entities and processes inhabiting the world, a description that our theories try to approximate. I also assume that scientific realism implies metaphysical realism, but the contrary does not hold.[4]

More precisely, in the next section I will reconstruct the antirealist motivations of the classic contingentist scenarios developed by James Cushing and by Andrew Pickering; in the subsequent section, by taking into account some versions of scientific realism that are more sophisticated than those discussed by contingentists up to now, I will clarify the logical relations of compatibility and incompatibility existing between contingentism and inevitabilism on the one hand and scientific realism and antirealism on the other; furthermore, I will try to spell out the specific contribution of contingentist historical reconstructions to the critique of scientific realism; finally, in the last section, I will recapitulate the results of the article and argue that the conflict between contingentist antirealism and scientific realism can be seen as a clash of inferences based on interpretations of the history of science. This article will thus consist of a philosophical meta-analysis of a controversy existing between different meta-scientific investigations.

Contingentist Scenarios as Challenges to Scientific Realism

Questions about the contingency of the history of science can in principle be discussed without reference to the debate over scientific realism; nevertheless, most of the works that have raised the issue were written with the intention of challenging standard realistic standpoints (e.g., Pickering 1984a, Cushing 1994) or, at any rate, providing a framework for discussions over antirealist constructivism (Hacking 2000a).

More specifically, significant work has been aimed at describing two different scenarios that give a more precise content to the idea of alternative successful developments of the history of science. The two scenarios correspond to two ways in which the multiplicity thesis can be declined. This first is the good old *underdetermination scenario*, according to which a given subject matter could be described by different mutually incompatible theories that, nevertheless, equally succeed in accounting for some of all the relevant phenomena.

The multiplicity thesis in this case would boil down to the underdetermination thesis. In order to find examples of this type, we would need to look for a successful theory that was developed at a moment in which an alternative underdetermined theory could have been conceived, given the historical background existing at the time. Let us further notice that the strongest possible argument for contingentism would be based on the very existence of an alternative incompatible development,[5] therefore, ideally, we should also be able to produce the alternative theory or, at least, the embryo of it.

As a matter of fact, there is one detailed example of such a contingentist scenario based on underdetermination and on a historically plausible reconstruction of a counterfactual history of science, an example that also meets the strong demand about the possibility of producing the core of an alternative theory. This is described by James Cushing (1994) in his book *Quantum Mechanics: Historical Contingency and the Copenhagen Hegemony*.[6] Cushing clearly illustrates that during a scientific controversy over a new theory or experimental result, what matters is not only the very fact that somebody comes up with an idea (a contingent factor that should not be underplayed) but also at what point in the controversy that happens. Indeed, in any debate, the temporal order in which arguments and counterarguments are given can turn out to be decisive. Let us recall the essential traits of this story. In 1952 Bohm developed a version of quantum mechanics empirically equivalent to standard quantum mechanics but radically different at the ontological level. As Cushing puts it, the two theories share the formalism but not its ontological interpretation (Cushing 1994, 9). In particular, at the ontological level Bohmian mechanics is much more similar to classical physics because it ascribes to each particle at each instant a defined position evolving in a deterministic way. Furthermore, it explains the collapse of the wave function as a consequence of the equations of motion, thus abolishing the special status that standard quantum mechanics assigns to the observation process and the thereby related paradoxes of quantum measurement. Bohmian mechanics does imply paying the price of nonlocality, but so does standard quantum mechanics.[7] Cushing's thesis, which is supported by a careful reconstruction of the scientific debate between the 1920s and 1950s, is that it is only a matter of historical contingency if that theory was not put forward in the 1920s. For that, only a few other results would have been needed, such as the proof that the instantaneous collapse of the wave function does not contradict the no-signaling principle of special relativity. All of these results could have been obtained with the theoretical resources available at the time.

"The choice would then, early on, have been starkly clear: *either* a realistic, nearly classical worldview based on a theory like Bohm's, with the price of non-locality, *or* an indeterministic and nonlocal Copenhagen worldview with its truly bizarre ontology and a radical, revolutionary departure from any comprehensible 'picture' of physical process. The causal quantum-theory program could have been off and running" (Cushing 1994, 186).[8]

This counterfactual scenario is particularly interesting because it is based on the modification of historical occurrences that one could hardly consider inevitable such as the temporal order of events that actually took place in the minds of a handful of researchers. The choice Cushing refers to could not have been made on the basis of logical coherence and experience alone, and, in cases like this one, it is legitimate to suppose that social and cultural factors, not to mention subjective preferences, play an important role in the final decision. The case at hand is particularly interesting also because what now appears to us, accustomed as we are to the oddities of standard quantum mechanics, as a bizarre quasi-classical quantum theory would have looked much more palatable in the early twentieth century precisely on the grounds of its conservative character with respect to the paradigm of classical physics dominant at that time.[9]

Historical contingency is thus used by Cushing to destabilize the belief in literal truth of the "worldview" deriving from physical theories. There is, though, a second, more radical, way to draw antirealist arguments from contingency: it is what can be called the *robust-fit scenario*. In a nutshell, the idea is that successful science is not based on a predictive or explanatory match between theories on the one hand and fixed phenomena on the other. The so-called phenomena emerge from a complex interplay of several practical and theoretical items ranging from raw data, techniques of data analysis, and methods of approximation, to background theories, accepted experimental facts, and phenomenological laws, and including the very material aspects of the relevant equipment as well as its expert use. According to this account, there is no experimental bedrock invariant throughout history that all rival theories would have to predict and explain. Rather, experimental activities and theoretical beliefs must co-stabilize in such a way that they produce a robust fit. The key aspect of this process of co-stabilization consists in a sort of generalization of Duhemian holism to the ensemble of the aforementioned ingredients of experimental science,[10] ingredients that, let us stress once more, are not restricted to intellectual items but also include material ones. According to this extended holism, when the researchers' expectations are disappointed by the upshot of the experiments, all the items on the list can, in principle, admit of modifications, in view of restoring coherence among them. Crucially, the so-called experimental data, whether raw or interpreted, are no more fixed and given than any other items. In this sense, the evolution of experimental science implies always a coevolution of intellectual, material, and practical elements whose aim is the achievement of a robust fit, that is, a configuration in which each element works well in the system of all other elements.[11] It is precisely the need to achieve virtuous adjustments among the components of experimental science that puts constraints on the scientists' choices.

The multiplicity of possible robust fits even within the investigation of a single subject matter would now amount to a new version of the multiplicity thesis. This is, in short, the conception emerging from Pickering's sociological history of particle physics. Pickering's work constitutes the constructivist approach to science that is more explicitly tied to the notion of contingency.[12] His historical reconstruction is explicitly presented as a contingentist alternative to the way in which scientists tend to view the history of their own field, that is, on the basis of a belief in the truth of the theories that have ended up being accepted and in the existence of the entities postulated by them. This ontological bias retrospectively renders unproblematic, to the scientists' eyes, the choices made in the past, which were responsible for the emergence and acceptance of what came to be their worldview (Pickering 1984a, 7). Note once more that, as follows from the thesis of extended holism, those choices concerned both which experimental results had to be accepted as established "facts" or "phenomena" and what theory should be retained as more capable of explaining them. Pickering argues that the choices made in the history of particle physics were in no way determined either by experimental "facts" or by any available method: "Historically, particle physicists never seem to have been *obliged* to make the decisions they did; philosophically, it seems unlikely that literal obligation could ever arise. This is an important point because the choices which were made *produced the world of the new physics*, its phenomena and its theoretical entities. As we saw in most detail in the discussion of the neutral current discovery, the existence or nonexistence of pertinent natural phenomena was a product of irreducible scientific judgments" (Pickering 1984a, 404).

Pickering's final point is that there is no obligation to "take account" of the ontology of particle physics, on the grounds of its being a contingent cultural product (Pickering 1984a, 413–14). The realism defended by certain scientists is, in his view, a mistake that fosters an inevitabilist view of the history of science, reinforcing the mistake itself.

One should not miss the sharp difference separating Cushing's and Pickering's brands of contingentism, a difference that clearly surfaces in the two passages quoted above. Cushing talks of the contingency of a *world view*, whereas Pickering refers to the contingent production of the *world* of the new physics.[13] Those terminological choices mark different if not opposing attitudes toward *metaphysical realism*. Cushing does not question metaphysical realism; rather, he seems to presuppose it, for he defends a form of skepticism about the power of successful physical theories to yield a reliable ontological picture *of* reality. According to Cushing, physical world *views* can prove to be untrustworthy representations of the *world itself*.[14] His conclusions, as we have seen, rest on a fully representational analysis aimed at showing that physical ontology is underdetermined by empirical evidence. Pickering, on the contrary, was heavily influenced by Kuhn's notions of world change and incommensurability and by the metaphysical antirealism that he sometimes associated with them.[15] Pickering not only claims that the old and the new particle physics predicted and explained different sets of data and different phenomena but also tries to build on this interpretation a nonrepresentational, agency-based account of the very notions of world change and incommensurability.[16] He believes that a careful analysis of historical case studies undermines "the intuition of uniqueness" motivating metaphysical realism and, in turn, inevitabilism.[17]

Note further that the difference between the two scenarios is not without consequence for the clause "equally successful" contained in the definition of contingentism. Indeed the predictive success of two theories can be compared to a certain extent as long as they both try to account for the same phenomena. This comparison becomes more and more difficult, however, if, as one envisages under the robust-fit scenario, discrepancies appear between the data or the phenomena themselves. In this case, I suggest that the clause "equally successful" be construed as "both very successful, without any way to decide which is more successful than the other."[18]

These examples illustrate what is at stake in many discussions concerning the contingency of the history of science. They both explicitly imply a criticism of realistic standpoints: Pickering focuses on scientists' spontaneous realism, whereas Cushing challenges an unqualified belief in the ontological reliability of physical theory. However, what is still missing is an analysis that takes into account a more elaborated version of scientific realism and investigates the logical relations between the latter and contingentism. To what extent is scientific realism *incompatible* with the contingency of science? Or, more generally, where would a scientific realist stand on the contingency issue?

To What Extent Is Contingency Compatible with Scientific Realism?

Scientists' scientific realism is not philosophers' scientific realism. The former is voiced at times by some members of the scientific community not rarely in the form of a blunt faith in the unshakable truth of scientific achievements. It surfaces mainly in debates that take place outside laboratories and academic institutions, and it is not even clear that it is very widespread among scientists themselves.[19] Normally for such realists, the experimental evidence available for a scientific theory is enough for them to believe in its literal truth. In contrast, professional philosophers of science who advocate one or another version of scientific realism do not argue for their position by simply pointing out the evidence scientists purport to have for their theories, nor do they defend a theory against a rival one: their analyses take place at the meta-level, where one wonders which epistemic attitude it is rational to adopt toward something that science, in general, or, more frequently, a specific branch of it, teaches us about the observable and the unobservable aspects of the world. To say that the properly philosophical debate about scientific realism takes place at a meta-level with respect to the level of working scientists in no way means to claim that this debate is entirely and necessarily based on purely philosophical, a priori arguments. Quite the contrary: today the vast majority of those who are occupied with the issue, whether in the realist or the antirealist camp, share one or another variety of epistemological naturalism or, even when that is not the case, tend to be skeptical toward the possibility and legitimacy of a foundational philosophy. Their philosophical contributions owe a lot to the traditional logical analysis of the relation between theory and evidence but are also nourished by the results of empirical disciplines such as history and sociology of science (or even cognitive science), results that are often produced by highly specific case studies or detailed reconstruction of historical episodes.

The latter point is particularly important for us. Most versions of scientific realism (and scientific antirealism) are developed from a philosophical standpoint according to which the epistemic import of a scientific discipline becomes understandable only if that discipline is considered as embedded in its history (see Psillos 1999), if not in the wider social context surrounding it, or even in the natural history of humans as cognitive agents endowed with certain mental capabilities. Philosophy of science and science studies in general develop, in this way, a critique of a more or less broadly conceived *scientific* worldview, which in most cases is nourished by the empirical results of one or another kind of (more or less broadly conceived) *scientific* investigations. Today's trend thus contrasts with more formal and a priori approaches such as logical empiricism, neo-Kantianism, and phenomenology, which were predominant before the Second World War. As we shall see, this has important consequences for the very nature of the answers presently given to traditional philosophical questions such as those concerning scientific realism.[20] It is outside the scope of this article to develop a complete account of the different brands of scientific realism and antirealism. I will only single out the theses and arguments that are more significant for understanding the relation between contingentism and scientific realism.

It is by no means a coincidence that the whole contingency debate is framed in terms of successfulness because, at least with respect to the field of natural science, the common starting point of scientific realists is the argument based on success. It is the impressive predictive, explanatory, and technical success of natural sciences that promotes epistemic optimism about the truths of their claims. In a word, how could our view of the natural world be entirely wrong, given the outstanding theoretical and practical accomplishments that derive from it? Furthermore, the realists' acceptance of a fallibilist epistemology allows them to revise at least part of their beliefs in the light of new evidence or new theoretical developments, without having to drop their epistemic optimism altogether, for, in a fallibilist perspective, knowledge does not equate to certainty.

As is well known, the usual responses to this argument rest on the underdetermination of theory by empirical evidence and, most of all, on the so-called pessimistic meta-induction (Laudan 1981). Indeed, in past years, in the absence of a consensus on the actual import of arguments based on underdetermination, the debate has focused on the threat that the pessimistic meta-induction poses to the arguments from success.[21] Nevertheless, a refined version of realism will take up the challenge deriving from a pessimistic reading of historical records. A realist knows or at any rate expects that our current scientific theories will be modified by future scientific research in ways that cannot simply be equated to emendation, completion, or improvement. The way out of the difficulty of mediating between the realist intuition that the success of science is a sign that its theories cannot be completely false on the one hand and the various arguments akin to the pessimistic meta-induction on the other is often given in terms of positions that can be defined as "selective or preservative realism." According to the latter, past theoretical changes must be taken seriously when evaluating the kind of epistemic warrant that our well-confirmed scientific theories can enjoy. The result is the attempt to specify what parts or aspects of scientific theories have been retained through theoretical change and are also likely to be retained by future successful science. These parts or aspects will in turn be considered to be true or, more often, approximately true. There are several different versions of partial realism, but they all share the features of being based on a discussion of actual historical case studies and of being compatible, to a certain extent, with the prospect of future major theoretical changes. Here is a short presentation of it based on the work of John Worrall (1989).

As we have seen, the historical records indicate that past predictive and explanatory successful theories, like Newton's mechanics or Fresnel's optics, have been superseded by successor theories that postulate a very different ontology: curved space-time instead of the gravitational forces, electromagnetic field instead of the ether, and so on. The history of science teaches us, therefore, that there is no continuity at the ontological level when a major theoretical change takes place. The specific kind of selective realists called *structural realists* accept this conclusion, but they do not accept that the success argument in favor of the partial truth of science must be given up altogether. A form of realism can survive even if we give up the idea that the central theoretical terms of our successful theories must refer to real entities. By looking at actual historical cases of theory change, the structuralist aims at highlighting the existence of continuity at the structural/syntactic level in spite of the discontinuity at the ontological level. As is well known, the most famous example of structural continuity was given by Poincaré: when the ether was replaced by the electromagnetic field, what was retained, according to Worrall, was not only predictive power, for the forms of the equations governing optical phenomena were preserved by the new theory. The interpretation of the symbols appearing in the formulas changed because the oscillations of the particles of the ether were replaced by the oscillations of the electromagnetic field, but, crucially, the mathematical laws governing these phenomena have the same forms. In conclusion: the predictive success of science does not guarantee knowledge of the entities that really inhabit the world, but only knowledge of the relations among them. There are of course several possible criticisms to this approach,[22] which remains, by and large, an incomplete research program that should be developed through a careful analysis of a huge number of different examples possibly issued from more recent scientific developments. Here I will not try to evaluate the plausibility of structural realism per se, for my aim is, rather, to address the relation a realism of *this kind* bears to the contingency issue. These brief indications can suffice for our purpose. We can now return to the problem of the relation between contingentism and realism.

As we have seen, the very notion of contingentism has emerged in the context of history-based critiques of scientific realism. However, it is important to understand that the inevitabilism/contingentism pair does not overlap with the realism/antirealism pair (see Soler 2008a and Sankey 2008). To perceive this it suffices to realize that inevitabilism can coexist with both scientific realism and antirealism. Let us recall that, according to the inevitabilist, it is impossible that the history of science could have yielded a scientific account of a given subject matter as successful as ours but incompatible with it. Now, this tenet is of course logically compatible with the view according to which it is rational to believe that some or all theoretical components of our current science are literally true, but it is also compatible even with the gloomiest version of the pessimistic meta-induction. One could endorse the view according to which all our scientific theories will be abandoned in due time and replaced by new, wildly different ones and still claim that there is a certain fixed pattern in the succession of successful theories that the history of successful science must inevitably follow. For instance, the shift from Newton's theory to Einstein's could be, according to this view, just as inevitable as the prophesized future shift from Einstein's to the "who-knows-what" theory that will supersede it and wash away its ontology of curved space-time manifolds, more or less in the way in which the latter ousted the classic ontology of gravitational forces acting in absolute space and time. To be sure, the anti-realist inevitabilism would prompt reactions different from the realist one. The latter sounds as if it is a controversial but fairly complete account of science, insofar as we can here generalize what Pickering has shown in the case of the scientist's realism, that is, that their realism fuels an inevi-tabilistic reading of the history of science and provides a sort of post facto intuitive (albeit unrigorous) explanation of the inevitability of successful science. In contrast, the antirealist inevitabilists could be at a loss about how to argue for the inevitability of the historical trajectory of successful science, given that the success of science, according to them, is not a sign of the truth of its theoretical claims about the world. As far as I can see, there is nevertheless a strategy that one could follow to render antirealistic inevitabilism as more than an ungrounded logical possibility: one could (1) endorse metaphysical realism, and (2) stress the role of the initial start-ing point of a research as a constraint on its future development.[23] As we have already indicated, given a certain subject matter, scientific research always develops on a cultural and scientific background. One could argue that, in particular, the scientific background mediates our access to the subject matter in question, or else that that objective domain appears to us in a certain way also because of the technical and theoretical resources allowing us to access it. This theoretical and technical mediation would thus constrain the way in which science will further develop, *given* the way reality is. Under this perspective, reality might well admit of different successful scientific accounts of it, and therefore there would be no guar-antee that our account is the true one, and yet, given a certain theoretical and instrumental way of access to it, only one such account is possible. In some sense, scientists will be doomed by their own scientific background to *read* reality in a certain way, even when that very same scientific back-ground or, more realistically, a part of it is abandoned in the course of the research: the past of science plus the way reality is would thus determine the future of science.[24] This position is admittedly very speculative, but it is worth mentioning it in order to correctly map the differences between the contingentism/inevitabilism and realism/antirealism pairs.[25]

Given the obvious fact that a contingentist *can* be an antirealist, we still need to settle the issue of whether a contingentist *must* be an antire-alist, or, equivalently, whether, notwithstanding the antirealist inspiration of contingentist accounts of the history of science, there is a form of scientific realism that can be reconciled with contingentism.[26] In order to defend the view that scientific realism is compatible with contingentism one might recall that most contemporary versions of scientific realism are not committed to an uncritical belief in whatever claim is derivable from successful theories. One might then argue that, after all, scientific realists can concede the possibility of alternative incompatible successful sciences that account for a certain domain of objects, but then they would add that all these alternative routes, when successfully pursued, would progressively converge toward a unified final account of that domain. However, in contrast to antirealist inevitabilism, which is a position that, though logically coherent, is hard to establish, the combination of contingentism with scientific realism does not seem to qualify even as a logical possibility. The reason is that scientific realism *does not amount* to what could be called *eschatological realism*, for it does imply more than a vague confidence that science will eventually yield a true account of the world. Again, I have to insist on the fact that the alternative routes must differ in what are called stabilized stages of science. Any realist would admit that a host of practical and theoretical aspects of today's scientific practices could have been different, and that the same theoretical or practical results could have been achieved in many different ways, even once the historical series of stabilized stages is fixed. However, realism implies an epistemic optimism about our present successful, stabilized science, and that optimism cannot live up to the idea that another, wildly different science could be or could have been just as successful as ours. Incidentally, only a very weak form of contingentism is compatible with eschatological realism, a form according to which, as science progresses, its development becomes less and less affected by contingency. Indeed, how is it to be possible that all successful histories of science have to converge toward a final unified account of the world, if at each temporary stabilization of a particular field of research several bifurcations are always possible, as the original definition of contingentism requires?

Thus, at a very general level, there is no easy way to reconcile scientific realism and contingentism. A more fine-grained analysis is needed, one that takes into account both a specific version of scientific realism and the differences between the two previously discussed contingentist scenarios.

Let us consider each contingentist scenario in turn, starting with that based on underdetermination. What could a structural realist say about it? As we have seen, structural realism does not consider the success of our present scientific theories as a reason to regard them as literally true. It is rather a meta-approach aimed at finding elements of theoretical knowledge that have shown to be more or less invariant under actual theoretical changes. We sense already that any kind of selective scientific realism, whether structural or not, any kind of realism, that is, that would be based on the comparative analysis of actual successful scientific theories would not be troubled by talk of possibilities. It is a central feature of this approach to come to a conclusion about the realism issue and, in general, about the evidential basis for inferring the correspondence to reality of a constituent of theoretical knowledge, only after a careful comparative examination of *actual* successful theories. In order to see this in detail, let us return to the multiplicity thesis. This thesis has to be made more precise if its implications for structural realism are to be worked out. In particular, we need to be more precise as to the nature of the supposed incompatibility between the rival underdetermined theories. A conflict at the level of the entities posited by the theories, for instance, would not trouble the structural realist at all. In this respect, the often-cited example of an imagined nonquarky high-energy physics, when reformulated in the framework of the underdetermination scenario, would not imply, in principle, a deep incompatibility at the structural level. In general, two underdetermined theories with a different ontology could share a deep structural similarity, and, if this were the case, their existence, far from constituting a threat for structural realism, would instead provide further evidence for it. In a sense, structural realism is designed to cope with the situation of empirically equivalent theories that postulate different kinds of entities, although it was based on the comparative historical analysis of different successor theories retaining their predecessors' empirical content, rather than on imagined globally empirically equivalent alternatives. In order to be harmful for this rather cautious form of realism, the multiplicity thesis must be sharpened in the following way.

Given a certain subject matter, different scientific accounts of it are possible that are (1) equally successful, and (2) incompatible at the structural level.

Now, the multiplicity thesis thus formulated is incompatible with structural realism, and, hence, the truth/plausibility of the former would imply the falsity/implausibility of the latter. A fortiori, therefore, structural realism is incompatible with a contingentism based on this version of the multiplicity thesis, but it is fully compatible with a contingentism restricted to the ontological implications of scientific theories.

The incompatibility between structural realism and a qualified version of contingentism has been discussed, so far, in the framework of the so-called underdetermination scenario. What can be said about the second contingentist scenario that we have considered, the one based on the notion of interactive stabilization and robust fit? Again, I will take structural realism as a representative of any kind of selective realism intended to draw consequences about reality from the comparative consideration of different successful theories. When turning to the robust-fit scenario we fully appreciate the philosophical consequences of the so-called practice turn, with its insistence on the importance of the generative process of the experimental activities. As we have already noticed, the antirealist arguments based on underdetermination need not challenge the solidity of the empirical evidence produced by experimentation. This has motivated philosophical analyses that focus almost solely on the representational aspects of scientific inquiry. Many types of scientific realism have been developed in this vein, and the various kinds of selective realism are no exception to the rule. Discussions about scientific realism have been based on fine-grained analyses of the parts of the representational content that are deemed to account for the predictive success of science and that, furthermore, appear to be retained through theory change. A multiplicity thesis based on the notion of robust fit could pose a very serious threat to this approach.

In the first place, it becomes more difficult to imagine a contingentist scenario of this kind that could be compatible with the kind of scientific realism I have considered. As we have seen, in this case, all the ingredients of scientific practices are allowed to vary—instrumentation, know-how, techniques of data analysis, theoretical hypotheses, and the data themselves. Now, it is certainly possible to imagine that two groups of researchers might get to the same theoretical result while using different instrumentations, know-how, and techniques of data analysis. On the other hand, it is harder to see how they might get to the same conclusion from a theoretical point of view if the data and the models of data are different. The whole idea of looking for historically invariant components of theoretical knowledge that are responsible for predictive success and its retention through theory change becomes problematic. Does it really make sense to look for structural similarities between theories that make different predictions verified with different experimental techniques? Here we come to a somewhat stronger opposition between contingentism and scientific realism.

In conclusion, it is impossible to reconcile "realism about X" with "contingentism about X," if to be realist about X means to hold the view that the success of the theory implying X gives us rational grounds to believe that X or something similar to X actually exists or is true.[27] A realist about structures, as we have seen, although allowing contingentism about entities, would certainly be against contingentism about structures. Hence, with the previous qualifications, contingentism cannot be reconciled with scientific realism. But the result is also that this cannot be taken as an unqualified thesis. An unqualified contingentist thesis, that does not make explicit reference to the level at which the scientific investigations are deemed to be mutually incompatible (empirical basis, entities, structures . . .) is harmful only for a wholesale realism that takes virtually all our scientific claims as literally true, it is harmful at bottom for the realism of some working scientists.

The Specific Contribution of Contingentist History of Science to the Critique of Scientific Realism

As we have seen, contingentism amounts to the conjunction of the multiplicity thesis with the claim asserting the possibility of a history of science leading to one of the supposed successful alternatives incompatible with our own science. The short discussion just presented should suffice to persuade us that the part of the contingentist thesis that is problematic for scientific realism is the multiplicity thesis, which can be seen as a sort of generalization of the doctrine of underdetermination. In sum, the scientific realists who recommend an optimistic epistemic attitude toward the ingredients of successful scientific theories, which they deem preserved through theoretical change, cannot at the same time be contingentist about that ingredient, that is, they cannot consistently endorse a multiplicity thesis involving it. This does not mean that scientific realists have the burden of the proof that the multiplicity thesis involving the components of scientific theory about which they are realists is *false*. It would be an unreasonable demand. The situation here is, once more, just a generalization of the one we are used to in the debates about realism and underdetermination. As long as scientific realists base their recommendation of optimism on the available historical records (or, at any rate, on the performances of actual scientific practices), they cannot be required to prove the impossibility of rival alternatives, unless they were claiming certainty for their realist tenets.[28]

This being the situation, one might formulate the following doubt: if scientific realism is, at bottom, threatened by the multiplicity thesis, and if contingentism, as we know, *implies* the multiplicity thesis, that is, if the multiplicity thesis is logically weaker than contingentism, it then becomes unclear what the specific contribution of *contingentism* as such to the realism/antirealism debate might be. In other words, one might argue, if scientific realism is jeopardized whenever a consistent case is made that there exists a plurality of equally successful accounts of a given subject matter, considerations concerning actual or potential *historical paths* are redundant, insofar as already the actual reality or established possibility of the *stabilized stages* to which they lead count, by themselves, as powerful threats to scientific realism. However, the structure of the debate cannot be portrayed in this way. True, from a logical point of view, contingentism says something *more* than the multiplicity thesis, something specifically *historical*; nevertheless, historical reconstructions do have their own peculiar function in the critique of scientific realism, for they can enhance the degree of *plausibility* of successful alternative developments, and, thereby, the degree of *plausibility* of the multiplicity thesis itself. In this way, contingentist historical reconstructions can at least weaken the position of scientific realists, even of the moderate kind epitomized by structural realists. Going back to Cushing's and Pickering's examples will help us to understand it.

Cushing's analysis, as we have seen, provides probably the most complete example of contingentist account of the history of science, an account that contains not only a plausible counterfactual history but also an alternative theoretical development in flesh and blood. However, even Cushing's analysis does not really provide a full-fledged alternative development. The reason is that Bohmian quantum mechanics, as discussed by Cushing, is a nonrelativistic theory, that is, an empirically equivalent competitor of nonrelativistic standard quantum mechanics only. Bohmian mechanics does not account for particle creation and annihilation; this is done, instead, by quantum field theory. Some attempts to develop a Bohmian quantum field theory are under way (e.g., Dürr et al. 2004, 2005), however, as of today, there is no consensus on a single quantum field generalization of Bohm's theory. Indeed, at the moment, one would be right in claiming that Bohmian mechanics is, strictly speaking, *less successful* than standard quantum mechanics: (1) it is less progressive, in the sense that the rate at which it produces consensus on new results is far slower, and (2) it only tries to keep up with the advances obtained by mainstream quantum physicists. Yet, instead of weakening Cushing's analysis, this fact foregrounds the real import of its *historical* dimension. Cushing shows us that Bohmian physics, which is at present a minority view among physicists, could have occupied center stage from the very beginning. In that case, as we have already stressed, Bohmian mechanics not only would not have *looked* so odd and far-fetched after all but also would have provided the shared theoretical background for the vast majority of the community of theoretical physicists, who would have produced a massive amount of theoretical work based on it. Standard quantum mechanics, consequently, could have been a minority view among researchers (or even an unsettling dead-end in the history of physics). Of course, there is no absolute guarantee that a given scientific research program *could have been* successful, and this general rule also applies to the causal program in quantum mechanics, for there is no guarantee that it would have proved *as fertile as* standard quantum mechanics in the extension to field theory. But does it now really look so difficult to imagine an alternate present in which the balance of success is reversed and Bohmian mechanics both has a wider empirical scope and enjoys a higher degree of progressiveness than standard quantum mechanics?

Let us also note that Cushing's example is also particularly dangerous for structural realism. Structural realists would have to show that Bohmian and standard quantum theories, both in nonrelativistic and relativistic form, are compatible at the structural level, and this does not seem to be very simple. For instance, does it really make sense to say that standard quantum mechanics and Bohm's mechanics make *similar* claims about the structures existing among the real entities inhabiting the universe (entities that cannot be equated with the Bohmian particles of course, for this would imply the acceptance of a full-blown kind of realism)?

If one turns to Pickering's brand of contingentism, as we have already noted at the end of the previous paragraph, we certainly find less clearly delineated alternative scientific developments. This is not surprising after all, for alternative developments involving a sharp difference at the level of material and practical resources are unlikely to cohabit for long periods of time, given the nonpluralistic ideology that has so far dominated the scientific community.[29] An actual example of alternative development of this kind, such as Bohm's theory in the case of underdetermination, is less likely to be available. The reason is that the scientific community can and sometimes does tolerate the existence of deviant theoreticians trying to subvert the dominant views of their research field; but it is very unlikely that it should tolerate the coexistence of two different and conflicting experimental traditions, both of which would require the support of several interconnected communities of technicians and manufacturers, a related process of standardization of tools and instruments, and a network of recognized institutions in which experimenters could be trained to use them. There certainly is a strong tendency to preserve the unity of the material infrastructure of scientific research. And this tendency is likely to hide the contingent factors at work in the history of science. Therefore, the historical examples of actual alternatives at the level of laboratory practices are bound to be very local, especially when the most recent episodes of the history of science are taken into account, for contemporary science involves a huge amount of financial, technical, and human resources (see Trizio 2008, 258).[30] Nevertheless, we do find in studies such as Pickering's a specific historical element that lends credibility to contingentism—the plasticity of the so-called empirical basis of science. Indeed analyses such as those of Pickering, insofar as they make it plausible that the so-called phenomena can stabilize in a number of different ways, threaten to undercut the very project of any preservative realism, which always presupposes the invariance of the phenomenal basis of science.

In sum, contingentist histories of science pose a challenge to scientific realism (whether global or selective), which, although akin to that of the more familiar arguments based on the doctrine of underdetermination and on the pessimistic metainduction, is logically distinct from them. In the case of underdetermination, the discussion is twofold: whether there is a general argument to the effect that each theory admits nontrivial underdetermined alternatives, and whether there are actual cases of rival, radically underdetermined theories. In the case of the pessimistic meta-induction the debates heavily depend on examples of past successful theories that were subsequently superseded. Thus, in both cases the philosophical discussion is fed either by *logical possibilities* or by specific actual historical *facts*. In contrast, contingentist reconstructions of the history of science, by striving to enhance the plausibility of alternative successful developments, occupy a space that is intermediate between sheer logical possibility and historical factuality. In that lies the specificity of their challenge to scientific realism.

Conclusion

At the beginning of this chapter, I argued that the critique of scientific realism is the driving motive of contingentist reconstructions of the history of science. This fact has called for an analysis of the relations of logical compatibility between the inevitabilism/contingentism and realism/antirealism pairs, an analysis that also takes into account structural realism as a representative of preservative (or selective) variants of scientific realism. It has appeared that contingentism and scientific realism, when referred to a specific component of scientific knowledge, are incompatible. On the other hand, inevitabilism could in principle coexist with both realism and antirealism, even though the latter theoretical configuration appears difficult to substantiate and defend. Furthermore, I have suggested that contingentist histories pose a sui generis challenge to scientific realism, consisting in enhancing the degree of plausibility of alternative scientific developments. The alternative scenarios presented in the examples of contingentist history of science that I have considered also threaten the "continuistic" strategy of structural realism.

One should not forget, however, that historical reconstructions are by definition local in character and can provide no general argument for a claim such as contingentism. They cannot rule out the possibility that only some scientific disciplines or only some aspects of some scientific disciplines are contingent, whereas others are inevitable.[31] Philosophers of science working in the realism/antirealism debate are familiar with this situation. If we focus on the way in which the confrontation between contingentist antirealism and preservative realism has developed so far, we can observe a clash of empirical inferences resting on evidence mainly derived from the history of science: on the one hand, scientific realists, from the standpoint of their meta-approach, posit structures, entities, properties (or whatever is the case depending on the specific type of scientific realism) on the grounds of their enduring role throughout the historical succession of successful theories, and on the other hand, contingentist antirealists, from their own meta-approach, posit possible alternative successful sciences.[32] On the one hand, we find hypotheses about the natural world and, on the other, hypotheses about possible sciences and possible social arrangements supporting them. Realists view the history of science as a smooth and uniform land, on which an external force has left readable and persistent signs that we can decipher and tell apart from our own prints; contingentists contemplate a varied landscape, rich in breaks and discontinuity, and disseminated by signs of unfulfilled possibilities of human intellectual and practical life. On both sides we find theoretical constructs whose legitimacy does not imply any mutual contradiction: even the demonstrated possibility of an entire maze of alternative successful incompatible sciences is not logically incompatible with the *existence* of the entities, structures, or properties realists believe in. The world may be *one-way* at the level of its deep constitution, as metaphysical realism claims, while still supporting many conflicting scientific accounts of it. The conflict between these two types of hypotheses is not at the ontological level but at the epistemic one.[33] The more one believes in the possibility of alternative successful developments the less one feels entitled to believe in the reality of a given component of our successful theories.

This methodological characterization of the debate is not intended to denounce its inconclusiveness, but it does indicate that, until a general argument is at hand, the controversy that I have reconstructed in this chapter will probably evolve on the basis of case studies supporting more or less *local* claims of antirealist contingentism or scientific realism.

CHAPTER 5

Contingency and Inevitability in Science

Instruments, Interfaces, and the Independent World

MIEKE BOON

A Viable Philosophical View about Science

In her introduction to the workshop "Science as It Could Have Been: Discussing the Contingent/Inevitable Aspects of Scientific Practices" (Les Treilles, August 31–September 5, 2009), Léna Soler (2008a) states that the issue of contingency versus inevitability in science is of great epistemological significance. She suggests that in order to enrich the space of viable philosophical views about science, the debate on this issue must be distinguished from the discussion on scientific realism. In line with this suggestion, my broader aim in addressing this issue is to ensure that the resulting philosophical view about science is prolific for scientific practices, especially those that focus on experimental research in application contexts.

My guiding question in developing a prolific philosophical view of science is this: If philosophers of science have the opportunity to teach science students, what should they tell them? My short answer is that the philosophy of science has a role to play in explaining "what science can do and what it cannot do." Science, in close collaboration with technology, has been incredibly successful. At the same time, it is an empirical fact as well as a philosophical insight that science does not provide certainty. Against this background, scientific researchers must learn to ride the waves of overly high expectations and overly low confidence in what science can do. A philosophical understanding of science should facilitate critical as well as creative reflection on methods of producing and testing scientific knowledge. Discussions in the ongoing realism/constructivism debate have not always been productive in this respect. But the refinement of this debate through accounting for the issue of contingency versus inevitability within science may contribute to a more viable philosophical view of science.

Inevitable versus Contingent Aspects: "Content" versus "Form"

In his seminal paper "How Inevitable Are the Results of Successful Science?" Hacking (2000a) has added this extra dimension of *contingent* versus *inevitable* aspects of science to the realism/constructionist debate in the philosophy of science. Soler (2008a, 222) proposes that one way to understand the contingency of science is well expressed by Steve Shapin, who wrote: "Reality seems capable of sustaining more than one account given of it" (Shapin 1982, 194). Hacking acknowledges the contingency in science to the effect that the scientific results of a science may have a different "form"—but he rejects the constructionist contingency claim that no scientific result is an inevitable part of successful science: "The '*forms*' of scientific knowledge could have been different, yet still, we would be recognizably exploring the same aspects of nature [i.e., its '*content*']" (Hacking 2000a, 71; emphasis added). Hacking argues that there is a significant sense in which the results of a successful science are inevitable (that is, noncontingent), namely, in the sense that any investigation of roughly the same subject matter, if successful, would at least implicitly contain or imply the same results. In this regard, his bête noire is the "boldest construction title in the natural science arena" of Andrew Pickering's (1984a) monograph *Constructing Quarks*. As the basis for his argumentation, Hacking uses the standard model of physics in which quarks are the building blocks of the universe; whereas Pickering's provocative title suggests that quarks are constructed. Hacking analyzes Pickering's ideas by distinguishing between "objects, ideas, and more abstract items, arrived at by semantic ascent, such as facts, truth, and reality" and concludes that "quarks, in that crude terminology, are *objects*. But Pickering does not claim that quarks, the objects, are constructed. So, the *idea* of quarks, rather than quarks, might be constructed" (Hacking 2000a, 61; emphasis added). Below, I will argue that Hacking's interpretation is inadequate.

Distinguishing Inevitability/Contingency from Realism/Constructivism

Although the contingency/inevitability issue is closely related to the realism/ constructivism debate (e.g., Hacking 2000a; Giere 2006), Soler (2008b), who aims at precise definitions of the inevitabilist and contingentist positions, argues that in order to address epistemologically significant aspects, the 'contingentism versus inevitabilism' debate must be disentangled from the 'realism versus antirealist constructivism' debate. In Contingency/Inevitability Disconnected from Realism and Constructivism, I will review ways in which inevitability and contingency—as semantic notions that concern the relation between knowledge and world—have been analyzed. I agree with Soler (2008b) that, in spite of their frequent association in the writings of philosophers, the two oppositions do not coincide. As I will argue, it is very well possible that a scientific realist accepts the role of contingency in science, whereas an antirealist constructivist admits inevitable aspects. But I will also propose that the meaning of inevitability versus contingency is framed by the position one takes in the realism/constructivism debate.

Accordingly, I will propose (see table 5.1) a matrix that frames the inevitability/contingency issue within two philosophical stances, which I call *metaphysical realism* and *epistemological constructivism*.

My construal of these stances (in Two Philosophical Stances) will be such that they facilitate the articulation of significant controversies important for understanding the mentioned scientific practices. Furthermore, it aims to steer away from some of the unbridgeable controversies between scientific realism and *social* constructivism (see also Hacking 1999). First, in my construal, both stances take account of the central role of an independent material world (including instruments and apparatus) and share the idea that this independent world sets limits to our knowledge. Second, as an alternative to strong *social* constructivism, the image of construction I endorse is that of constructing and using structures in mathematics for describing patterns and performing mathematical operations (e.g., math-

TABLE 5.1. A matrix for the inevitability/contingency issue

	1. Metaphysical realism	2. Epistemological constructivism
a. Inevitable aspects	1.a	2.a
b. Contingent aspects	1.b	2.b

ematical equations, axiomatic systems, mathematical transformations), together with the construction and use of concepts and metaphors (e.g., force, acceleration, reversibility) that enable the description and modeling of physical phenomena (see also Boon 2011, 2012b; Rouse 2011).

The views on inevitability/contingency in column 1 are framed by metaphysical presuppositions about the character of nature, such as, that nature contains or consists of fundamental building blocks and/or a fundamental (causal) structure (cf. Putnam 1981). Position 1.a entails the belief that some results of science (its "content" as Hacking [2000a] puts it) are inevitable. This content may be referred to as objects such as quarks, electrons, and proteins (e.g., Hacking 2000a). Other authors assume that structures such as the laws of nature are the inevitable results of science (e.g., as in Worrall's "Structural Realism" [1989], but also physicists such as Glashow [1992] and Weinberg [1996a]). Position 1.b accounts for the whimsicality revealed in the history of science (such as radical changes in scientific paradigms) through the admittance that some aspects are contingent (e.g., the "form" of scientific results, as suggested by Hacking). Hacking's (2000a) position is therefore covered by 1.a and 1.b in this framework.

Constructivist positions may be motivated by a *metaphysical* presupposition claiming that there is no independent order or structure in the world (e.g., Cartwright 1999), or by *epistemological* presuppositions according to which the question of whether there is a pregiven order in nature cannot be answered in principle because we do not have epistemic access to confirm this claim (e.g., van Fraassen 1980, 2008). Pickering's (1984a) view fits with 2.b, but it is hard to tell whether his constructivism is either metaphysically or epistemologically driven. Still, Hacking's (2000a) suggestion that *constructing quarks* means that the *idea* of quarks is constructed rather than the *object itself* disagrees with the presuppositions of both metaphysical and epistemologically driven constructivism. Indeed, as I will argue in Observing the World, the divide between knowledge about the real existence of an object versus *ideas* of the object is obvious and intelligible for a scientific realist, but not for a constructivist.

Contingency and Inevitability according to Epistemological Constructivism

The position I aim to defend in this debate can be classified as a combination of 2.a and 2.b. My constructivist position is motivated by the episte-

mological presupposition that we do not have knowledge of the world independent of the apparatus and instruments we use. Our motor system and
technological devices enable access to and interventions with the world,
whereas our perceptual apparatus and cognitive faculty, together with
the technological instruments needed to perform measurements, enable
epistemic access as well as the construction of epistemic results. The way
in which scientists construct epistemic results is dependent both on the
data they have gathered by means of a contingent assembly of instruments,
apparatus, and procedures and on contingent ways in which they structure
data and form concepts. Important to my view is the idea that apparatus
and instruments used in experimentally investigating the world form an
inherent part of our knowledge of the world (see also Floridi 2011). More
specifically, I will argue that we cannot get beyond them in such a way that
we acquire *noncontingent knowledge,* which is knowledge that somehow
reflects the inherent structure of the world independent of our instruments
and apparatus (including our motor system, perceptual apparatus, and cognitive faculty).

In explaining my view of the role of instruments and apparatus, I will
use Giere's (2006) *Scientific Perspectivism* as a productive starting point (in
Observing the World). Giere's goal is to develop an understanding of scientific claims that mediates between the "objectivist realism" maintained by
physicists such as Glashow and Weinberg and/or the hard realism of many
philosophers of science, versus the contingency of science as held by social
constructivists. According to Giere: "Full *objectivist realism* ('absolute ob-
jectivism') remains out of reach, even as an ideal. The inescapable, even if
banal, fact is that scientific instruments and theories are human creations.
We simply cannot transcend our human perspective, however much some
may aspire to a God's-eye view of the universe" (Giere 2006, 15; emphasis
added). However, Giere's scientific perspectivism draws on an epistemological picture that sometimes suggests the involvement of a metaphysical
realist stance. In order to bring the role of "perspectives" into accordance
with the epistemological constructivist stance, I will propose that, rather
than to be understood as having different scientific perspectives *on* the
world, the way in which "perspectives" (which encompass our perceptual
apparatus, cognitive faculty, technological instrument, and theories) make
the world epistemically accessible has the character of the workings of *interfaces,* which transform aspects of the world (the input) into perceptions
and knowledge of the world (the output).

The Inevitability/Contingency Debate

Contingency/Inevitability Disconnected from Realism and Constructivism

Soler (2008b) argues that, although a connection is often made between inevitabilism and realism, and between contingentism and antirealism or constructivism, it is worthwhile to define the contingentism/inevitabilism issue as separate from the realism/constructivism issue. Soler defines inevitabilism as follows: If more or less the same initial conditions exist as those in our own history of science, for which a successful and progressive physics has indeed been developed, then, inevitably, physics in this setting, at least in the long run, yields (a¹) more or less the same results or (a²) different but reconcilable results, and (b) the same ontology as our own. Conversely, contingentism involves the possibility, at least in the long run, of an alternative physics, as successful and progressive as ours, which yields (a') results irreducibly different from ours, and (b') involving an ontology incompatible with ours (Soler 2008b, 233). Next, Soler aims to elucidate the problematic notions "different but reconcilable" and "irreducibly different" results or ontologies, by examining these options in thought experiments with two different physics.

Soler's approach to articulating an "epistemologically significant controversy" makes sense, and the resulting definitions in which contingentism is contrasted with inevitabilism, as well as her attempt to elucidate how we could possibly decide between them on empirical grounds, are clarifying. Nevertheless, I doubt whether her suggestion about the contingentist position is correct, or at least, the only possible interpretation of this position. I agree that contingentists believe that: (1) their position would be empirically supported if it turns out that "after a very long time" the two physics are irreducibly different in the sense that they are incompatible. But I doubt that contingentists assume that: (2) this empirical finding thus supports the idea that two physics can be "essentially incompatible." Indeed, Soler concludes that "such a contingentist position about the results of physics requires, as a precondition, the adoption of an inevitabilist stance (inevitably, the two physics had to remain disjoint, unreconciled, because of their very nature)" (Soler 2008b, 240).

The apparent contradiction of the contingentist view expressed by Soler can be clarified within the previously proposed matrix. This schema says that metaphysical realism and epistemological constructivism are two stances within which contingent and inevitable aspects of science are interpreted differently. From an epistemological constructivist stance, the empirical finding that two successful physics are incompatible is explained epistemologically rather than metaphysically. An epistemological constructivist understands at a meta-level why the two physics are each sound but incompatible (e.g., in the sense of Kuhn's meta-level understanding of the incommensurability of two physics in terms of distinct disciplinary matrices). Conversely, drawing the conclusion that they are "essentially incompatible," as expressed in (2), typically agrees to a metaphysical realist stance. Hence, Soler's (2008b) suggestion that a contingentist accepts this latter conclusion is only correct for those who have adopted a metaphysical realist stance (e.g., position 1.b in the matrix). Accordingly, in my view, definitions of contingentism/inevitabilism are entangled with positions in the realism/constructivism debate, and epistemological constructivism is a way to escape the realism/constructivism dichotomy.

Empirically Supported Contingency

The purpose of my contribution to the contingentism/inevitabilism debate is to reconcile two seemingly contradictory intuitions of inevitability and contingency, respectively: (1) that an independent world determines (or sets limits to) scientific knowledge, and (2) that "Reality seems capable of sustaining more than one account given of it." From a scientific practice point of view, an argument in favor of the contingency thesis is the crucial role of constructive activities and conceptual work. Accepting this role involves acknowledging that scientific theories are not "discovered" but "constructed," yet without claiming that an independent reality does not have a role to play (also see Boon and Knuuttila 2009). Moreover, it involves the idea that inventing or developing or "radically changing" scientific *concepts* is essential to the construction of theories—these concepts are an ineliminable part of the final epistemic result, rather than being mere heuristic means that enabled the discovery of the theory but subsequently can be eliminated from its central core (see also Boon 2012b).

The contingentist thesis also finds empirical support in the history of science (e.g., Pickering 1984a; Cushing 1994), and I assume that in this very sense it is unproblematic for those inevitabilists who, like Hacking (2000a), admit the role of contingency. The central issue of the contingency/inevitability debate relevant for a viable view of science is, then, whether at least some parts of the content of scientific claims are inevitable. Hacking, who raised this issue, puts it this way: "If the results R of a scientific investigation are correct, would any investigation of roughly the same subject matter, if successful, at least implicitly contain or imply the same results?" (Hacking 2000a, 70–71). Hacking's inevitabilist position consists in an affirmative answer to this question, whereas an epistemological constructivist will object that this would very much depend on the conceptual framework adopted in the investigation.

Inevitability: Existence and Knowability

Hacking's (2000a) inevitabilism seems to be at odds with his contingentist position in "The Self-Vindication of the Laboratory Sciences" (Hacking 1992; also see Trizio 2008n19), which can be summarized by the following quotes:

> It is my thesis that as a laboratory science matures, it develops a body of types of theory and types of apparatus and types of analysis that are *mutually adjusted to each other.* They become . . . "a closed system" that is essentially irrefutable. They are self-vindicating in the sense that any test of theory is against apparatus that evolved in conjunction with it. . . . The present picture suggests that there are many different ways in which a laboratory science could have stabilized. *The resultant stable theories would not be parts of the one great truth,* not even if they were prompted by something like the same initial concerns, needs, curiosity. Such imaginary stable sciences would not even be comparable, because *they would be true to different and quite literally incommensurable classes of phenomena and instrumentation.* . . . *Our preserved theories and the world fit together so snugly less because we have found out how the world is than because we have tailored each to the other.* (Hacking 1992, 30–31; emphasis added)

Trizio (2008) concludes that Hacking (2000a) seems to have become less enthusiastic about his own contingentism. Indeed, the epistemological constructivist interpretation of contingency and inevitability I propose in this article for the most part agrees with and has been inspired by Hacking

(1992), which suggests a serious incoherency in Hacking's (2000a) ideas. It must be kept in mind, however, that Hacking calls himself a materialist. His inevitabilism concerns the materially existing entities and phenomena, which he thinks of as identifiable and/or recognizable "aspects of nature" that *exist* independent of us, and which he refers to as the "content" of scientific knowledge. The way in which I understand Hacking's (2000a) position regarding inevitabilism, is that once scientists have adopted certain questions as relevant and thus opened up a specific scientific field—for instance, questions about fundamental matter, or questions about the material functioning of human bodies—they will (at least, if they are successful and do not make mistakes) inevitably find the entities that are relevant to those questions, very similar to the inevitability of discovering America once discoverers leave port to chart the world, simply because these objects are out there in the world. This situation of 'how the world is' warrants that answers to questions about the natural world have nothing to do with us, as Hacking (2000a) puts it. In addition, if we can *intervene* with the theoretical entities represented in theories, we know that these entities *exist* (Hacking 1983).

Hacking's (2000a) inevitabilism/contingentism involves a metaphysical realist stance as it presupposes (1) the existence of an independent material structure in the world that is knowable to us, and (2) that the content of this knowledge is inevitably true, whereas the form in which it is represented is contingent and does not qualify as having truth-content. Although Hacking's position is attractive since it reconciles plausible aspects of both contingency and inevitability, I am critical of the two presuppositions.

My critique of the first presupposition agrees with Trizio (2008, 254), who criticizes the suggestion that discoveries of theoretical entities and phenomena are similar to geographical discoveries. Geographical discoveries are inevitable because there simply is no alternative history: "all conceivable alternatives lead either to the same discovery, or to no discovery at all." Regarding the 'discovery' of theoretical entities and phenomena, Trizio builds on the ideas of Hacking (1992): "Phenomena are not out there, ready for us to discover and describe, for what we call phenomena are actually complex patterns of results that emerge in a process of stabilization of a certain branch of laboratory science" (Trizio 2008, 257). Given the epistemological constructivist stance I endorse (and will explain in more detail), my difficulty with the second assumption concerns how we can conceive of the "noncontingent" part of knowledge. On a more rigorous take, the noncontingent part of knowledge is the knowledge that remains after liberating it from all conceptual content (i.e., its "form"). Yet, how could knowledge liberated from its conceptual content (i.e., its contingent part) tell us anything about the world?

Two Philosophical Stances

A viable picture of science must counter contraintuitive consequences of strong forms of social constructivism as well as the philosophical puzzles that arise from naive forms of scientific realism. This is why the distinction between inevitability and contingency of scientific knowledge may be productive. As was discussed above, Soler (2008a, 2008b) proposed to disentangle the definition of the inevitability/contingency issue from the realism/constructivism debate. Whereas the latter is primarily philosophical, her approach to the former is also empirical (including historical analyses and thought experiments). It turns out, however, that drawing conclusions from possible empirical outcomes involves a philosophical stance. Against this background, I propose to analyze the issue within two distinct stances: *metaphysical realism* and *epistemological constructivism* (see the matrix in Distinguishing Inevitability/Contingency from Realism/Constructivism, table 5.1). The two stances are constructed such that the first agrees to the kind of metaphysical realist position I attributed to Hacking (in the section above), whereas the second presents my own epistemological constructivism.

According to my construal, the two stances share the conviction that some aspects of science are inevitable/contingent but disagree on what those aspects are. Furthermore, the philosophical presuppositions of each stance seem to play a crucial role in explaining *why* those aspects are inevitable/contingent. They also share the idea that the contingency of scientific knowledge—in the sense that "Reality seems capable of sustaining more than one account given of it"—can be explained in part by the underdetermination of theories by empirical data and contingent metaphors and concepts in terms of which empirical findings are interpreted. In addition, both agree on the existence of an independent world that puts constraints on our knowledge. Yet, they disagree as to whether we can know that the independent world has a well-ordered "inherent" (material or causal or abstract) structure (also referred to as "intrinsic nature"). What is more, as has been discussed in Contingency/Inevitability Disconnected from Realism and Constructivism, they disagree on why different accounts of reality are incompatible.

In relation to the latter, metaphysical realists deny incompatibility regarding the inevitable, true part of scientific knowledge, which they believe refers to the independently existing objects or structure of the world. Our *representations* thereof may be incompatible, but this does not change the object or structure. What they may have in mind, metaphorically, is that we can point at them (and/or intervene with them, as Hacking [1983] puts it) in a similar way to how we point at observable things and give them names such as "apple," "America," "protein," and "electron," as well as gather knowledge about their properties and behavior. This latter epistemic activity does not change what "the thing itself" is. Hence, according to this picture, the inevitable, noncontingent part of our knowledge ties up with the independently existing thing we point at, whereas our representations of it are contingent. The contingency of our representations of things is due to the contingency of what we pick out when describing them, and to the contingent metaphors and concepts scientists employ.

Although, as I will show in Observing the World, Giere's (2006) scientific perspectivism agrees in many respects with the epistemological constructivist stance, his metaphor of mapmaking can be used as an illustration of the picture of the metaphysical realist stance regarding the epistemological distinction between the object under study, which we can point at, and our knowledge of it. By presenting mapmaking as a metaphor of modeling the world, Giere aims to explain the incompatibility of knowledge in terms of the incompatibility of distinct representations of a thing (e.g., the Earth) due to the distinct perspectives we have of it. In mapping the three-dimensional surface of the Earth onto a flat surface: "Every projection gives a different perspective on the Earth's surface. But these projections are all incompatible. They cannot, for example, simultaneously preserve shapes and areas everywhere" (Giere 2006, 80). Clearly, these incompatible, contingent maps do not coincide with what the Earth inevitably is, let alone that the mapmaker constructs the Earth. In line with Hacking's (2000a, 61) phrasing, the mapmaker constructs *ideas* of the Earth, rather than the Earth. In the eyes of a metaphysical realist, this metaphor clarifies how a distinction between the inevitable (noncontingent) true "content" and the contingent "form" of scientific knowledge can be understood.

By using this metaphor, Giere implicitly suggests that the possibility of distinguishing between the observable things we can point at (e.g., the Earth) and our contingent, sometimes incompatible representations of its features (e.g., maps) also applies to unobservable objects in science, which makes the way in which he explains the incompatibility of different accounts of reality acceptable for a metaphysical realist.

Conversely, the epistemological constructivist disagrees with the epistemological distinction between *access to* and *knowledge of* (the existence of) unobservable things, on the one hand, and *representations* of their features, on the other—which is why the notion of inevitable, noncontingent, true "content" of scientific knowledge is incomprehensible within this stance. Denying that this epistemological distinction can be maintained in the domain of science, in my view, is the point of van Fraassen's (1980) much disputed distinction between knowledge claims about *observable* and *nonobservable* things, and his claim that the attribution of truth only makes sense for claims about observable things, whereas knowledge claims about nonobservable things are empirically adequate at most (see also van Fraassen 2008; Boon 2012a). The epistemological constructivist accepts the incompatibility of different accounts of reality 'all the way down,' and explains it in terms of, for example, different paradigms or perspectives within which scientific knowledge must be constructed.

In conclusion, the question of whether "noncontingent knowledge" is an epistemologically meaningful notion with regard to scientific knowledge boils down to an epistemological issue, namely, whether the distinction can be maintained between knowledge of (the existence of) a thing (e.g., a theoretical entity or structure or law) and its representations. I do not believe that this issue can be decisively solved. In a metaphysical realist stance, the belief that this distinction is meaningful will be supported by the metaphysical presupposition that the world has a well-defined structure, which goes smoothly together with the view that scientific research has the character of discovering what there already is, similar to expeditions that discover what there is by exploring the world (and/or by intervening with the independently existing material objects, as Hacking [1983] suggests). My epistemological constructivist stance draws on another picture, which better suits the experiences of scientific researchers in chemistry and engineering. From this viewpoint, the amount of empirical information that can be gathered about the world is infinite and therefore requires scientists to impose different kinds of structures for organizing, representing, and interpreting this information (see also Massimi 2008). What is more, scientists usually aim to develop experimental setups and technological procedures such that they obtain manageable, well-ordered, and reproducible information. On this matter I have in mind Cartwright's (1999) notion of *nomological machines* (see Nomological Machines). Furthermore, the role of instruments and apparatus developed and employed for gathering this information should not be understood, metaphorically, as seeing the world through some kind of magnifying glass. In many cases we do not have independent access to what is "behind" or "in" the instrument or apparatus. Often, we cannot clearly distinguish between "world" (i.e., the purported phenomenon or the object under study) and the instruments used in our explorations (also see Harré's [2003] notion of apparatus-world complex, and Boon [2004]). In these cases, the epistemological distinction between knowledge of (the existence of) purported entities and representations of them cannot be maintained. Epistemological constructivism is a direct consequence of this, in the sense that, if we can no longer hold a sharp distinction between the entities and their representation, then both constructivism and realism are unable to grasp the complex relation between knowledge and reality, which instead is what epistemological constructivism is supposed to do.

Observing the World: Perspectives and Interfaces

Scientific Perspectivism as a Viable Alternative?

When adopting scientific perspectivism (Giere 2006), the concern is whether at least some knowledge of the world "behind our perspectives" is possible. Is some kind of "direct" knowledge possible, or must our knowledge be considered perspectival "all the way down"? The two ideas are sketched in figure 5.1.

The divide between the two stances already concerns the character of empirical knowledge. In a metaphysical realist stance our starting point in the production of empirical knowledge is some kind of direct or immediate knowledge of the independently existing object or phenomenon we point at or intervene with (e.g., an apple, the Earth, the brain, the Universe). Next, we acquire empirical knowledge of this thing through different perspectives. This view is more or less forced upon us through our common language: We say that we have "a representation *of* X," or "a perspective *on* X," which implies a clear epistemological distinction between our representations of the thing called X and our 'direct' knowledge of it. This idea can also be illustrated by Giere's descriptions of how images of the brain

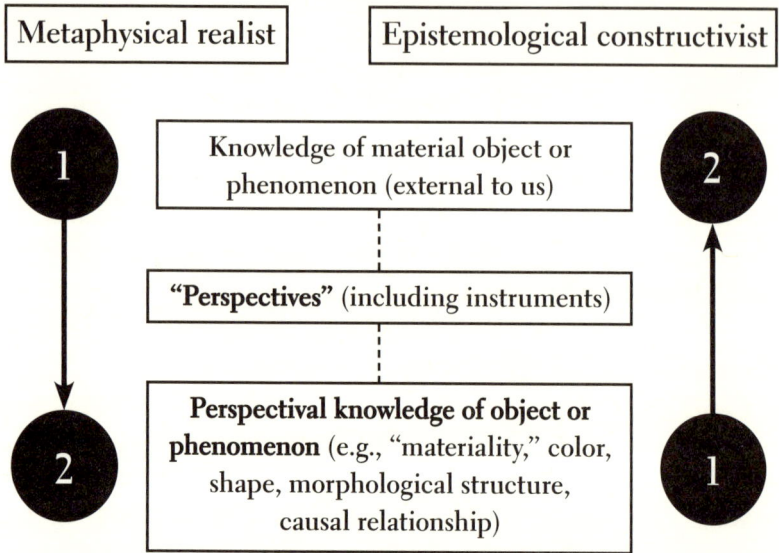

FIGURE 5.1. Philosophical presuppositions about the direction of empirical knowledge production, of the metaphysical realist (left: top-bottom) and the epistemological constructivist (right: bottom-top)

(i.e., representations of object X and its features) are produced by means of instruments such as computerized axial tomography (CAT), positron emission tomography (PET), or magnetic resonance imaging (MRI) (i.e., the perspectives), and the images of objects in the universe by means of optical or gamma ray telescopes. In both these cases computer programs do a lot of work in processing the data. These images do not simply present images of the brain or objects in the universe. Instead, according to Giere, one has images of the brain *as produced by* the process of CAT or *as produced by* MRI and so forth. This character is what makes them perspectival.

In an epistemological constructivist stance, the direction of empirical knowledge production is the other way around. Knowledge of (unobservable) objects and their features is already considered to be perspectival. Such perspectival knowledge results from an interaction between the external world and our perceptual apparatus and cognitive faculty. In the Kantian tradition, the ontology exemplified in perspectival knowledge and in the language that expresses empirical knowledge (e.g., that there are objects with properties and causal relations and interactions between things or events) is regarded as resulting from the role of so-called regulative principles (also see Chang [2009a], who calls them ontological principles).

Against this background, the significant controversy between meta-physical realism and epistemological constructivism is whether we believe either that: (1) there exists a "pregiven" ontology (the furniture of the world consisting of, e.g., definite material objects with definite properties, and/or causal structures, etc.), which can be known in a way that is independent of the ways that humans structure and interpret "information," and which is knowable because our perspectival knowledge is somehow *similar* to it; or (2) the ontology (the furniture of the world) we "point at" or "refer to" or "represent," and which consists of, for example, definite material objects with definite properties, and/or causal structures, and so on, *results from an interaction* between the external world and our perceptual apparatus and cognitive faculty and, in many cases, scientific apparatus and instruments. In accordance with the belief expressed by (1), a metaphysical realist defends the case that the "content" (rather than the "form") of our representations of the "pregiven," knowable ontology is inevitable. The constructivist disagrees on epistemological grounds that we could decide whether there is a pregiven ontology, let alone that it is knowable. Based on (2), he may turn the issue into the question of whether the interaction between the apparatus and the independent material world could be the locus of inevitability.

How Inevitable Are the Results of Science, or Could Science Have Been Different?

According to Giere (2006, 88), "Wholesale scepticism about the existence of so-called theoretical entities now seems to me almost quixotic." He agrees with the physicist Steven Weinberg, who states that "it is true that natural selection was working during the time of Lamarck, and the atom did exist in the days of Mach, and fast electrons behaved according to the laws of relativity even before Einstein" (Weinberg 2001, 120). This belief of the inevitabilist, which also seems to be held by Hacking (2000a), is supported by the metaphysical realist idea of an independently existing, inherent, and knowable structure in the world. Does an epistemological constructivist believe that atoms did not exist in the days of Mach? And does he believe today that they exist? Hacking (2000a) argues that a constructivist such as Pickering actually believes that the *idea* of the theoretical entities is constructed, but not the entity. In Distinguishing Inevitability/Contingency from Realism/Constructivism, I suggested that this divide between knowledge of the real existence of an object versus the *idea* of the object is intelligible for a scientific realist, but not for a constructivist. The epistemological constructivist does not build his view of science on the presupposition of an inherent and knowable structure. Instead, in his view, theoretical entities (and laws of nature) have been epistemologically and ontologically "carved out" through the interaction between the independent world and our perspectives. In addition, some theoretical entities such as atoms have also been "carved out" through the interaction between the independent world and technological instruments to such an extent that we can intervene with them. Hence, rather than claiming that the object is *constructed*, the picture of the constructivist is that the theoretical entity is *"carved out,"* both materially and epistemologically.

Metaphorically, we can think of how the sculptor carves out a statue. Regardless of our stance, most of us will agree that the statue exists and deny that the statue already existed as an inherent, material, and knowable structure in the marble. Most of us also agree that the piece of marble (in an interaction with the sculptor and the techniques he has at his disposal) sets limits to the statue that can be carved. What is meant by "carving out" theoretical entities can also be metaphorically understood by the example of constructing a route on a map. The independent world (and the map) sets limits to the route that can be "carved out" but does not determine it. Nevertheless, the route exists as an independent entity as soon as it has been constructed. The metaphysical realist may reply that in some cases, there will be only one possible route, and this is where our knowledge is determined by the world independent of us, hence, inevitable. The epistemological constructivist responds that the case is mistaken because there are many possible routes, but only this one happens to be practical given our criteria and needs. Therefore, he will claim that "reality seems capable of sustaining more than one account given of it" and "we are never allowed to claim that such and such scientific achievement could not have been otherwise and was inevitable because of the intrinsic nature of the world."

A Viable View of Science for Scientific Practices

Interfaces

We cannot attain mirror-like knowledge of the inherent structure of the world. Moreover, knowledge is contingent because the perceptual apparatus and cognitive faculty are human, and technological and scientific instruments are produced by humans. Furthermore, we only have limited instruments at our disposal (perceptive, cognitive, measurement, and theories), and thus a limited number of perspectives and a limited amount of information. Yet, this is not to say that every aspect of science is contingent.

Scientific perspectivism is extremely useful for developing a viable view of scientific practices. It presents us with a fruitful metaphor for understanding the similarity between the functional roles of the perceptual apparatus, cognitive faculty, and technological instruments as well as theories for the production of 'pictures of the world' (or, as Boon and Knuuttila [2009], and Boon [2012b] call it, "epistemic tools"). I suggest that pointing out the roles of apparatus and instruments in science, (implicitly) makes use of our commonsense understanding of the functioning and causal workings of *machines*, thus avoiding philosophically problematic notions of "world," "mind," (or mind-independent world and world-independent mind), and especially of how the interaction between these fundamentally different substances can be understood such that it makes intelligible the way that the mind gathers knowledge of the world. This "perspectival and instrumental" account of perception, cognition, and measurement makes use of our commonsense understanding of *materiality*. Materiality is "solid" and capable of exerting robust, stable, and reproducible causal processes. These characteristics of materiality (robust, stable, causal) also apply to our perceptual apparatus and cognitive faculty, and to scientific instruments, as they are supposed to perform robust, stable, and reproducible (causal) operations, thus transforming some kind of "input" into perception and cognition of data, images, properties, processes, and so forth.

As an additional metaphor for furthering the fruitfulness of this notion, I propose that scientific perspectives function as *interfaces* (e.g., as used in computer technology). An interface robustly, stably, and reproducibly transforms *input* of one sort into *output* of another sort. This metaphor may add to the machine metaphor in the sense that it makes the mind-world interaction more intelligible. The interface transforms something that is not color at all to color on a screen or color perception. It transforms one substance into another substance. Material or symbolic or electronic or whatever *input*, which cannot be directly perceived by us, is transformed to *output* that can be perceived, experienced, and/or conceived (e.g., numbers, tables, graphs, a picture or text on a screen, a melody, a substance; and also numerous physical properties such as scent, color, heat, fluidity, solidity, elasticity; as well as physical processes such as coloration, heating, breaking, flowing, solidifying, evaporating, and mixing). Conversely, an interface can transform something that can be perceived by us into something unperceivable (that can subsequently be processed further, e.g., by computer programs or technological instruments).

Similar to our perceptual apparatus, cognitive faculties, and scientific instruments that function as interfaces between us and the independent world, theories (e.g., the abstract principles of Newton's theory of motion) function as interfaces that enable us (e.g., someone like Newton) to transform observations and/or measured data of parts of the world into Newtonian models of those parts of the world, producing, for instance, the theoretical model of the moon orbiting around the earth or the theoretical model of the harmonically oscillating spring or pendulum. Similarly, scientists use theoretical knowledge of unobservable properties or phenomena to transform measured data into a picture or model of 'underlying' processes. In this case, observable input (data) is transformed into pictures or models of aspects of the world that are not observable in any direct manner.

Regarding the independent world, instruments and the like, which are considered to be interfaces, transform "inaccessible," "meaningless" input into "accessible," "meaningful" output through supposed stable causal interactions with aspects of the independent world (often in conjunction with stable data processing, which indeed involves theoretical presuppositions). We commonly believe that the stability of these transformations is warranted by the *materiality* of instruments, which is the sense in which these transformations are considered noncontingent. At the same time, an epistemological constructivist believes that these instruments do not provide us with inevitable knowledge of the independent world because, metaphorically speaking, the input remains without meaning. In other words, in spite of the many different scientific instruments and procedures we have developed for examining the world, we do not acquire knowledge of the world "behind the interface."

Nomological Machines

The epistemological constructivist stance proposed here involves the idea that, by means of different kinds of instruments, scientists ontologically, epistemologically, and technologically "carve out" aspects of the independent world, simultaneously generating reproducible physical phenomena (or, "regular behavior") as well as meaning and structure. Accordingly, descriptions of theoretical entities, properties, and/or laws of nature are not taken as representations of the inherent nature or structure of the world. An account of how entities, properties, and/or "laws of nature" are "carved out" may be clarified somewhat further by Nancy Cartwright's (1983, 1999) notion of *nomological machines*. When introducing this notion she writes:

> [I will reject the story that] laws of nature are basic. . . . Sometimes the arrangement of the components and the setting are appropriate for a law to occur naturally (e.g. the planetary system), [but usually] it takes what I call a *nomological machine* to get a law of nature. . . . [Here] a law of nature is a necessary regular association between properties. . . . [The kind of associations chosen] tend to be just the cases where we understand the arrangement of capacities that give rise to them. . . . Laws hold only ceteris paribus—they hold only relative to the successful repeated operation of a nomological machine. A nomological machine is a fixed arrangement of components, or factors, with stable capacities that in the right sort of stable environment will, with repeated operation, give rise to the kind of regular behaviour that we represent in our scientific laws. (Cartwright 1999, 49–50)

As Cartwright has argued time and again, there is little regularity in our world. Regularity really only exists in machines and in the laboratory. Finding regularities often requires creating "nomological machines"—devices that produce robust, reproducible physical behavior. Cartwright stresses the crucial role of "shielding," which means that, guided by their knowledge of conditions that "disturb" the effect they aim at, scientists build the machine such that it shields against "disturbing" conditions. Yet, the law-like behavior thus produced in our laboratories should not be interpreted wrongly— that is, we cannot infer from the observed regularity to knowledge of laws (and/or entities and properties) that operate independently in nature. The machine and the procedure of running it are part of the conditions that produce the observed regularity. Nevertheless, the system that has been constructed (i.e., the nomological machine) operates independent of us, and the knowledge that scientists have achieved about relevant conditions and effects of that system, is true about aspects of that system.

In scientific practices, it is important to understand how "laws of nature," or even mere "empirical knowledge" is produced, that is, to recognize the indispensable role of instruments in the production of this knowledge.

Such an understanding enables scientists to see why and how available scientific knowledge is applicable for the modeling of specific target systems, and/or how to make predictions about their behavior. At the same time, it makes them cautious, for example, with regard to the kind of certainty the application of a law of nature provides. Often their application is empirically successful as a result of "how the target system is, independent of us," but often it also turns out that predictions based on "laws of nature" are incorrect. In brief, in the case of failure, scientists understand that "the nomological machine" by means of which the law has been produced was most likely "shielded" against conditions that happen to be relevant for the target system at hand.

Cartwright's notion of nomological machines, therefore, illustrates why scientific knowledge resulting from experimentation is both contingent and inevitable. Expanding on Giere's scientific perspectivism, a similar functional role in the production of scientific knowledge can be attributed to our perceptual apparatus, our cognitive faculty, and our theories.

Properties and Interfaces

Inevitabilists such as Hacking consider theoretical *entities* as part of the furniture of the world. They believe that theoretical entities such as proteins, electrons, and quarks exist independent of us and make up the ontological structure of the world. Taking into consideration several of the ideas introduced in this chapter (i.e., scientific perspectivism, interfaces, nomological machines), we may wonder how these entities are discovered. When considering the role of instruments, inevitabilists and/or metaphysical realists may assume that theoretical entities are discovered, through their purported causal behavior—that is, by means of the properties or capacities they "exert" (also see Cartwright 1989). Epistemological constructivists, on the other hand, may agree with Giere (2006) who claimed that properties such as colors do not actually exist in external objects—rather, color is the result of an *interaction* between aspects of the world (e.g., physical light in the environment) and the human visual system. More generally, our experience and knowledge of properties result from interactions between a "target system" (e.g., aspects of a specific substance or material and/or aspects of an experimental setup) and measuring instruments in conjunction with other aspects such as human perception and cognition, theories, and technological and mathematical procedures. This also implies that new properties may result from technological developments. Indeed, ever more instruments and technological procedures have been developed by means of which all kinds of properties of a target system are measured. We only have to take a look at the famous *CRC Handbook of Chemistry and Physics* (Haynes and Lide 2014) to acknowledge how many properties of materials have been established. For example, S (a substance, material, or object) has a melting point, a specific density, a viscosity coefficient, an elasticity coefficient, a thermal and electrical conductivity coefficient, a diffusion coefficient, a hydrophobicity coefficient, a coefficient of surface tension, a friction coefficient, a thermal expansion coefficient, an elasticity coefficient, a specific heat coefficient, an absorption coefficient, an atomic number or weight, a wave length, magnetic permeability, magnetic and electrical field strength, a magnetic flux density, a magnetic moment, a crystallinity index, a refractivity index, a reflexivity coefficient, chemical concentration, potential and affinity, and solubility. Every physicist knows that the manifestation of these properties is often dependent on the technological procedure for measuring them. The color of gold, for instance, happens to be red at the nanoscale. The description of such properties usually has the character of an operational definition, which means that it encompasses aspects of the measuring instrument and procedure (also see Boon 2012b).

This is another insight that I find significant in teaching science students. The measuring instrument usually is not some kind of magnifying glass that makes the (properties of) S perceivable without somehow interacting with S. Instead, the technological procedures by means of which these physical properties *manifest* must be understood as *interfaces*. Hence, following up on Giere, we do not have "images" of (the properties of) S, but instead, "images" (e.g., the measured data pattern) of a property of S as produced by the interaction between (aspects of) the target system under study, on the one hand, and (aspects of) the instrument on the other. In other words, "carving out" a property involves a nomological machine, and the way in which the nomological machine provides access to aspects of the world is understood as the working of an interface. This understanding makes it hard to maintain that physical properties belong to the inherent structure of the world. At the same time, we believe that the manifestations we detect result from material workings, that is, from a causal interaction between an aspect of the independent material world and the measuring instrument. Hence, although the instrument may be contingent, we believe that, once it is in place, its results are inevitable (see also Hacking 1992).

An Overview of Contingency and Inevitability According to Epistemological Constructivism

In this chapter, I have analyzed the meaning of contingency/inevitability from the point of view of two different philosophical stances. Furthermore, I have aimed to develop a constructivist view of scientific practice that nevertheless takes a position as to which parts must be taken as inevitable in order to allow for a viable view of scientific practices that work in application contexts. Ideas about contingent and inevitable aspects of science within the proposed epistemological constructivist stance are summarized in column 2 of table 5.2, and column 1 summarizes my interpretation of Hacking's (2000a) position.

It has been argued that the inevitable content of our knowledge is determined by a stable interaction between instruments and apparatus that enable access to and/or interventions with the independent world. This idea is different from Hacking (2000a), who seems to assume that the "content" of our knowledge is determined by the independent world only. Even so, the apparatus and instruments (including our motor system, perceptual apparatus, and cognitive faculty) and the constellations in which they are used are still contingent in the manners explained by Hacking (2000a).

Saying that the results of science are *contingent* due to the contribution of these contingent aspects does not mean to say that they are *arbitrary*. In constructing scientific results, the role of general epistemic criteria such as logical consistency, coherence with relevant accepted knowledge, reproducibility, and empirical adequacy is an inevitable aspect of scientific practices. However, the way in which these criteria are employed is not through a deterministic procedure, and not in that sense inevitable. Nevertheless, although their *role* is inevitable, it may be disputed *which* epistemic criteria are regarded as inevitable. In my view, well-known epistemic criteria such as simplicity and generality do not belong to this list. Scientific practices that work in application contexts often aim at reliability and preciseness rather than generality and simplicity (also see Boon 2012a).

Different from what a constructivist position suggests, I have argued that in order to account for the possibility of constructing scientific knowledge, next to the *inevitable role of epistemic criteria*, some additional presuppositions in science must be considered inevitable. Yet, these presuppositions are not inevitable in a metaphysical sense—nor are they inevitable in a logical sense insofar as their negation does not imply a contradiction. In-

Table 5.2. Inevitabilist/contingentist aspects of metaphysical realism and epistemological constructivism

	1. Metaphysical realism: Metaphysics as a starting point for explaining science	2. Epistemological constructivism: Epistemology as a starting point for explaining scientific practice
a. Inevitable aspects	*Metaphysical presuppositions* about "how the world is": • The inevitable content of our knowledge is determined by the objects, properties, physical phenomena, laws of nature, and so on that *exist* independent of us.	*Regulative principles* of scientific research: • There exists an external (physical) world independent of us. • The world is stable in the sense that the same conditions will produce the same effects. • The inevitable content of our knowledge is determined by an interaction between (1) apparatus and instruments that enable access to the world and (2) the world. *Inevitable epistemic criteria:* • Logical consistency; coherence with relevant accepted knowledge; reproducibility; and empirical adequacy.
b. Contingent aspects	*Knowledge:* • Concepts or "forms" representing these objects and so on. *Means:* • Instruments that measure or isolate or bring about phenomena.	*Knowledge:* • Concepts representing these objects and so on, and methods of structuring and interpreting "data." *Means:* • Instruments that measure or isolate or bring about phenomena. • Interfaces that make the world accessible for us. *Epistemic criteria related to "epistemic purpose":* • For example, simplicity, generality, preciseness, reliability, efficiency.

stead, these inevitable presuppositions are epistemological in character. Kant called them *regulative principles*. One such principle is that we must presuppose an independent (physical) world that sets limits to our knowledge of it. This presupposition is regulative in the sense that scientists, in order to construct knowledge of the world, must investigate the world and take into account the possibilities it provides and the limits it sets. At the same time, constructing knowledge of the external world involves some autonomy of the scientist that results in the contingency of knowledge. Another regulative principle is the presupposition that the physical world is stable in the sense that the *same physical conditions will produce the same physical effects*. Again, this is not first and foremost a metaphysical belief about "how the world is," but a principle without which the production of scientific knowledge would not be possible. This principle "regulates" the way that scientists develop stable, robust, and reproducible instruments and procedures (i.e., nomological machines), and why they believe that the outcome of a causal interaction between an instrument and a target system is inevitable. In conclusion, from an epistemological constructivist stance "inevitable knowledge that reflects the inherent structure of the world" is a problematic notion. Nevertheless, in this manner, one arrives at discussing metaphysical issues, but via epistemology rather than the other way around. Some aspects of science are held to be inevitable: first, the notion that "the physical world is independent of us," which means that we presuppose an independent world that sets limits to our knowledge; second, the notion that "the physical world and technological instruments are material," which means that their physical functioning is independent of us and implies their *stable* and *robust* workings; and third, the indispensable *roles* of epistemic criteria and regulative principles, which are "inevitable" for the possibility of doing science anyway and are formed by the possibilities and limits of the human cognitive system as well as the possibilities and limits we experience of the independent world. Therefore, the epistemological constructivist has an understanding of inevitability in science that is different from that of the metaphysical realist. Accordingly, regarding the discussion in Contingency/Inevitability Disconnected from Realism and Constructivism, an epistemological constructivist admits that it may very well be the case, as Hacking (2000a) argues, that any investigation of roughly the same subject matter, if successful, would at least implicitly contain or imply the same results (see Inevitable versus Contingent Aspects), and that physics, at least in the long run, yields (a^1) more or less the same results or (a^2) different but reconcilable results, and (b) the same ontology as our own (Soler 2008b, 233; see Contingency/Inevitability Disconnected from Realism and Constructivism). However, the way in which an epistemological constructivist would explain this situation is in terms of a different notion of inevitability.

PART III

In Search of a Concrete and Empirically
Tractable Way of Framing the
Contingentist/Inevitabilist Issue

CHAPTER 6

Contingency and "The Art of the Soluble"

HARRY COLLINS

P eter Medawar (1967) defined science as "The Art of the Soluble" in a book by the same name. He suggested that the "trick" of science was to ignore questions that it could not solve and to concentrate on those that could be solved. I am going to approach the question of contingency versus inevitability in that spirit. I will make two moves in this direction. First, I will turn the philosophical topic into an empirical one or, at least, a quasi-empirical one, by asking not whether "this scientific position" (p), or "that scientific position" (not-p) is inevitable but whether, as a matter of fact, societies can maintain both p and not-p at the same time. The answer will be "yes." The most cursory observation shows that societies do in fact hold opposing views in respect of science at the same time.

To make the empirical problem a little closer to the philosophical question we can agree that little would be proved if the community of scientists endorsed p while some section of the general population endorsed not-p. To make the empirical solution philosophically interesting it must be that the same evidence and body of theory can support both some "p" and the equivalent "not-p" within the scientific community. Furthermore, the p's will not exist in isolation, they will be embedded in their own conceptual nets, experimental and observational ways-of-going-on, and bodies of data, with the equivalent "anticoncepts," "anti-actions," and "antidata" for the corresponding not-p's. Even when we elaborate the idea of holding p and not-p in this way, however, we still find communities of scientists supporting both at the same time.

Examples of p's include the existence of high fluxes of gravitational waves, the constancy of the speed of light, the existence of paranormal phenomena, and so forth. In each case, both p and not-p were held within the scientific community for a considerable time embedded within their corresponding bodies of evidence and experimental procedures—their mini- or not-so-mini-"paradigms" or "forms of life."[1] In the case of high fluxes of gravitational waves, the miniparadigm lasted for somewhere between five years, after which most of the scientific community had ceased to believe in them, and forty years, at which time scientific papers corresponding to the high-flux claim were still being published.[2] In the case of the constancy of the velocity of light, the controversy lasted from 1887 to at least 1933, when Dayton C. Miller published an experimental review paper proving that the velocity was not constant.[3] In the case of paranormal phenomena, the argument has been going for at least a hundred years and is still going on, depending on what one means by inside and outside the scientific community.

Unfortunately, if we ask whether societies can continue to hold both p and not-p *indefinitely* we stray back into the realm of the insoluble. First, as Popper pointed out, there is no secure, long-term theory of history so it is hard to predict the future in regard to any still ongoing scientific controversy. Second, it is hard to be sure that, in the case of controversies that did close down, the reason they closed down had to do with the "logic" of science. It is not hard to imagine scientific controversies closing down for all kinds of trivial reasons—for example, all the proponents of one unpopular miniparadigm might die of old age or be imprisoned by a hostile regime. Fashions in art or literature die out for reasons that have nothing to do with the logic of art or literature so why should fashions in science not die out for similar reasons? It should come as no surprise, then, that certain authors such as Hasok Chang (2011) and Malcolm Ashmore (1993) can mount cases for resurrecting dead scientific paradigms/ideas—the phlogiston theory and N-rays, respectively—without violating any scientific logic.[4]

A third reason it is hard to know whether p and not-p can be held indefinitely is that it is not easy to be sure when a scientific controversy is over. It was suggested above that contingency was supported only when the two opposing sides, p-ers and not-p-ers, belonged to the scientific community: it could not be that the scientific community endorsed p while a section of the general population endorsed not-p—that would not count as scientific opposition. But where are the boundaries of the scientific community? In the case of parapsychology, there are competing sets of mutually exclusive journals, one insisting that there are only four basic forces and another ready to accept the existence of a fifth force associated with paranormal effects and reporting empirical findings that support these effects. The empirical and statistical procedures reported in the parapsychology journals are similar to those used in every other science and they operate with peer review and produce publications that are similar in appearance to that of mainstream science. Are the parapsychology journals science and is the fifth force an example of not-p currently living alongside the p of the four forces? It is hard to say. Likewise, the electronic preprint server arXiv has had difficulty finding grounds to exclude a wealth of papers that deal with exotic interpretations of the foundations of quantum theory: is this a scientific dispute or not? At the outer edge of the question are the letters, and today e-mails, that physical scientists with a public profile receive regularly, promoting alternative theories of gravity and stressing the inconsistencies of relativity theory.[5] Do these constitute a scientific controversy? Are the alternative theories in the letters and e-mails examples of a continuing dispute about the soundness of relativity with alternative positions—p and not-p? To be sure that the dispute about relativity had finally closed one would have to find ways of excluding these letters from the body of science, and it is hard to imagine how to do this according to criteria that belong to the realm of philosophy. But whether or not we can find a way of answering the question of whether such things are continuing disputes, we still cannot be sure that the logic of science will not ensure that they come to an end at some time in the future.

The Theory of Contingency

Therefore, and this is the second move, the secure way to be sure that we are practicing the art of the soluble is to restrict the question to the short term. In that way the problem can be solved: the answer to the transformed question is clear, empirically demonstrable, and theoretically understood, it is "contingency not inevitability." It has to be contingency not inevitability unless, self-servingly, every single instance of short-term scientific disagreement is to be ruled out as a special case.

The "empirically demonstrable" part of the above claim has already been illustrated but what about the "theoretically understood"? Here we turn to the sociology of scientific knowledge (SSK). SSK's lasting legacy is to have revealed the mechanisms that get in the way when attempts are made to use "purely scientific procedures" to settle scientific controversies—at least, in the short term. One example of such a mechanism is the experimenter's regress, which turns on the idea of tacit knowledge. To do an experiment properly requires skill and expertise. A good part of that skill and expertise is located in the "tacit knowledge" of the experimenter—those things that the experimenter knows but cannot tell and that can only be transferred by the equivalents of socialization and apprenticeship. Unfortunately, the only way to demonstrate the successful completion of a socialization or apprentice-ship is with virtuoso performances. But in disputed sciences, no one can be sure what a virtuoso performance comprises—one party believes it will be the discovery of X whereas the other side believes it will be the nondiscovery of X. Thus experimenters cannot know when they have done an experiment successfully so they cannot use experiments in any straightforward way to settle their disputes. That is why there can be contingency in the short term even in the heart of the hardest sciences. It has been shown that there is a similar theoretician's regress and there are regresses of the same sort wherev-er one looks. Again, this should be unsurprising because it is a consequence of philosophical reasoning such as Wittgenstein's rules do not contain the rules for the own application (Wittgenstein 1953).[6]

What Is the Short Term?

It can, then, be empirically demonstrated and theoretically understood that, in the short term, science is contingent in the sense adopted here. But what is the short term? Given the examples of disputes over the constan-cy of the velocity of light and the existence of high fluxes of gravitational waves, a nominal "Fifty years" might make a good starting point. Fifty years also fits nicely with "Planck's dictum"—that science advances funeral by funeral. Though neither experiment nor theory can force disputing sci-entists to change their minds, death often silences the minority dissenters and effectively ends the controversy.

Of course, the short term is not always fifty years. The parapsychologists have been going a lot longer than that and, doubtless, some controversies die almost immediately. But the exact time does not matter—the logic is that there is always a short term. If you tell me that controversy "X" died out in only two years I will say, "I'll define short-term as two years in the case of X." So there is always a short term and therefore always a time when we can be sure that science is contingent in the sense that both p and not-p, with all their correlates, can be held by different groups within the scientific community.

New Questions Given Short-Term Contingency

This solution gives rise to a pressing question: given that many important decisions that turn on science (and technology, taken as understood from here on) have to be taken in the short term, and science is contingent in the short term—in the sense used here, that the community of scientists can support both p and not-p and their correlates—how can science be used to help make such decisions? The policymaker's question is different from the scientist's question. The scientist has a duty to try to work out the truth of the matter however long it takes—the scientist wants to escape from the short term. The policymaker lives in the short term, however. One might illustrate the relationship between the policymakers and scientists with Gary Larsen's well-known analysis of "the four basic personality types." In Larsen's cartoon each personality confronts a half empty glass. The first says: "The glass is half full." The second says: "The glass is half empty." The third says something like: "Half-full, half-empty—what was the question?" The fourth is a big ugly-looking fellow who insists, "Hey—I ordered a cheeseburger." The policymaker does not want an answer to a refined question but something that can be immediately consumed, even though, as far as the scientists are concerned, there is still doubt about what is there to be consumed.[7] What can those who study science offer policymakers in the way of a cheeseburger?

Over recent decades the most salient approach coming from the broad area of science and technology studies has been to let the public decide whenever uncertain science and technology are encountered. The call has been for ever more public participation in technological decisions—this is a move in the direction of what can be called "technological populism."[8] A full-blown technological populism would mean that wherever there is some scientific doubt about a matter the decision is made by the general public rather than the body of experts. It is, however, a very dangerous trend with potentially disastrous consequences. On the one hand, the public is in a poor position to understand the true balance of scientific doubt where there is an element of controversy. This is well illustrated by the recent scare over mumps, measles, and rubella vaccine (MMR) in the United Kingdom. A single doctor, on the basis of no scientific evidence, announced at a press conference that he considered MMR to be associated with autism. The newspapers took up his cause with the result that there was a popular revolt against MMR, the consequences of which are a minor measles epidemic in the United Kingdom that has caused unnecessary suffering.[9] A similar lesson about the problems of technological populism could be drawn from the popular sentiment in the United Kingdom supporting the return of capital punishment as a deterrent to murder.[10]

The Argument from Wisdom

To avoid technological populism one can adopt another kind of pragmatic argument that is based on the common sense that pertains in a society that places some value on expertise in general and scientific expertise in particular. One simply says that even though experts may not be able to come up with a correct and certain answer one would still prefer their answer to that of the general public simply because one prefers one's decisions to be made by those "who know what they are talking about." Their conclusion may be wrong but it affirms our way of living whereby we place the technical decision in the hands of those who know what they are talking about rather than, say, tossing a coin or asking the general public. This is not to say that in the realm of politics there is any obligation to act on policy recommendations that seem to follow from the conclusions of the scientists, the only obligation is to accept the technical judgments and use them in making the policy decision even if they are explicitly overturned.[11] If we are to reaffirm the role of science and technology in our society we should prefer the technical part of a political decision to be delivered by the technically "wise" rather than those who have had no opportunity to gain technical wisdom.

But what is the right kind of wisdom in any particular case? The question has three levels: the first level is what has become known as the "framing problem"—whose technological business is the point under dispute? I will not discuss it here except to say that some proposed "solutions" to the framing problem take us right back to technological populism because technical problems are framed entirely in terms of deeply held and widespread beliefs that are not subject to what would normally be thought of as technical input. An example is the disposal of the North Sea oil rig the *Brent Spar* at the end of its life. "Green" opposition to its being sunk in the North Sea led to it being disposed of on land, but all parties eventually agreed that this caused more pollution. The opposition seemed to have more to do with notions of purity and reverence for nature than pollution itself.[12]

The second level of what counts as wisdom is the old problem of demarcation. Assuming one has accepted that it is *scientific* wisdom that should be applied to the technical part of technological problems, one has to decide what science is in any particular case and how it is different from other kinds of wisdom-producing cultural activities. As is well known, the problem of demarcation has not been solved. Every attempt to produce a demarcation criterion has failed under close philosophical scrutiny. For example, the last and most promising was Popper's falsificationist criterion, but it failed when Lakatos showed that its foundation—the asymmetry between corroboration and falsification—did not hold up.

Under the sociological view, however, all the old failed demarcation criteria are resurrected—they are all successes not failures. This is because instead of looking for criteria with the purity of glass, which is entirely shattered by only one small crack, one is looking for the softer, and therefore more resistant to assault, properties of a certain form of life held together by the "family resemblance" (in Wittgenstein's sense) of its varying activities. Any one element that does not exhibit all the defining characteristics does not shatter the whole—the whole integrated network of ways-of-going-on is hardly touched by the occasional violation of what, under the old more philosophical approach, would be a rule with the status of logic. The form of life of science is characterized by much that is already familiar, such as logical positivism's preference for claims backed up by observation, Popper's insistence that scientists should try to state the means by which their claims might be falsified, Merton's scientific norms, and so on. Given the looseness of the notion of family resemblance, all these ideas remain standing even after it is shown that they are not logically perfect and that they are often violated in practice. It may be that new demarcation criteria can also be discovered—for example, we believe we have found a new one called "the locus of legitimate interpretation," which has to do with who is given a societal license to interpret the meaning of a cultural output. In the arts the locus of legitimate interpretation is more distant from the seat of production than it is in the sciences (see Collins and Evans 2007, ch. 5).[13]

The third level applies once the proper kind of *scientific expertise* to be brought to bear upon a technological problem has been established. The question is "who is it that has this kind of expertise?" Robert Evans and I have already done considerable work on this question, much of which is gathered together in our book, *Rethinking Expertise* (2007). The fulcrum of the book is what we call "The Periodic Table of Expertises," which classifies expertise according to the extent to which people have access to the body of "tacit knowledge" pertaining to a specialist technical domain. The Periodic Table is reproduced here (table 6.1) and all categories are explained in Collins and Evans (2007).

The second line of the table—the categorization of specialist expertises —can be used to exemplify what is at stake here. The first three categories in this line are more "information" than expertise because to acquire them depends on no more than "ubiquitous expertise"—that which we all possess by being full members of our native society. In this case, merely being able to read enables one to acquire the three left-hand categories of information. The two right-hand categories depend on the acquisition of specialist tacit knowledge.[14] Contributory expertise is the normal expertise of well-practiced specialists but the category that has caused most interest and given rise to most new work is interactional expertise. This appears to be a new concept. It is an expertise that is sufficiently rich to enable good technical judgments to be made—it encapsulates the wisdom of a specialist technical domain—but it consists of fluency in the linguistic discourse of the domain rather than the practical abilities. It has been argued, and to a good extent shown, that it is possible to acquire this expertise through deep immersion in the spoken discourse alone, without engaging in the domain's practices. This, of course, is a controversial claim given the salience of bodily practices in much contemporary philosophical discussion of the source of human abilities, but we have gone a long way toward establishing it. The arguments for its possibility are long, set out in several places, and continually being refreshed; the latest developments can be found on our Web site (www.cf.ac.uk/socsi/expertise).

One way to demonstrate the existence of this kind of expertise experimentally is by playing the "Imitation Game." Here one kind of expert tries to pass as another while a judge tries to determine who is who. We have tried the experiment in a number of domains but one set of experiments (among others) that touches nicely on the philosophical point concerns those who become blind in very early life. It turns out that they can engage in the discourse of sighted people so well that it is hard to tell they are blind even when a judge freely and interactively asks questions that turn on the

TABLE 6.1. The Periodic Table of Expertises

UBIQUITOUS EXPERTISES		
Dispositions	Interactive ability / Reflective ability	
	Ubiquitous tacit knowledge	Specialist tacit knowledge
Specialist expertises	Beer-mat knowledge Popular understanding Primary source knowledge	Interactional expertise Contributory expertise
		Polimorphic / Mimeomorphic
	External	Internal
Meta-expertises	Ubiquitous discrimination Local discrimination	Technical connosseurship Downward discrimination Referred expertise
Meta-Criteria	Credentials Experience Track records	

matters that one would imagine only someone who had been fully sighted in adult life would be able to understand. This is because the blind have been fully immersed in the discourse of the sighted all their lives. In the quasi-control condition we show that sighted people are quite unable to pass as blind people.[15]

The idea of interactional expertise gives new answers to the question of who is a technical expert (though they do not seem to be such new answers once one realizes that contributory experts learn their practical skills via interactional expertise and deliver their judgments via interactional expertise and, indeed, that there could be no division of technical labor without interactional expertise) (see Collins 2011). As the rest of the Periodic Table comes to be explored in similar depth we ought to be in a better position to understand what should be going on when we look for people who know what they are talking about to make the technical part of our technological judgments.

In some cases this approach will produce a clear answer in regard to a technical decision. For example, it produces a clear answer in the case of the administration of antiretroviral drugs to pregnant women in South Africa. Under Thabo Mbeki's regime the drugs were not administered be-

cause Mbeki said there was a scientific dispute about their safety and effi-
cacy. The approach described above shows that there was no dispute of the
kind that could justify such a policy.[16] In the majority of cases, however, the
approach does not produce a technical answer. It is to be expected that a
sociological approach could only occasionally produce an answer to a question pertaining to the natural sciences. Even in these cases, however, the
sociological approach still provides the cheeseburger because it defines the
community of people to whom the technical part of the decision should
be handed over. It says, "let *this* group of people argue it out and reach
a conclusion and then make the political decision taking that technical
decision into account." It accepts that given a high level of disagreement
and uncertainty, and knowing the consequences of "this" decision rather
than "that" decision, it may be that the technical is a minor element in
the policy decision but, crucially, it says, however minor the element, it
is still the prerogative of specialist scientific and technological experts, as
can now be defined by clear criteria. It says that their conclusions might be
wrong, and often will be wrong, but that is the best conclusion that can be
had in regard to technical matters. It is best in two ways: at worst, it may
have a better chance of being right than a decision made by the toss of a
coin or by the general public, and, perhaps much more important, it is a
decision that reaffirms the importance of scientific expertise in our society
and holds back the rush toward technological populism.

CHAPTER 7

Contingency, Conditional Realism, and the Evolution of the Sciences

RONALD N. GIERE

A little over a decade ago, Ian Hacking posed the question: "How inevitable are the results of successful science?" "This is," he claimed, "one of the few significant philosophical issues that arises in constructionism debates about science" (Hacking 2000a, 61). Although perhaps not attaining the status it deserves, this question has attracted sufficient interest so that it is becoming increasingly difficult to add something new to the debate.[1] Indeed, Hacking himself, albeit often only briefly, raised most of the issues subsequently discussed. In particular, he pointed out that each of the key terms in his question, "inevitable," "results," and "successful science," needs clarification lest the answer to the question be too easily "not at all inevitable" or "completely inevitable." Hacking also named a proper contrary to a thesis of inevitability, a "contingency thesis" according to which any existing scientific field might now exhibit results that are in no significant way equivalent to existing results, although the field as a whole would be judged as successful as the existing field now is.

Of these two notions, inevitability and contingency, I find contingency to be the more transparent. Something is absolutely contingent if its description is neither a logical truth nor a contradiction. That is a clear but not very useful notion. On the other hand, no scientific result could be contingent on nothing whatsoever because it is possible that no sentient beings ever evolved anywhere in the universe. So we are left with the relativized notion, "contingent on. . . ." Thus, as Hacking already noted, any scientific results are contingent on science being done at all. More to the point, we can fairly well understand the sociological claim that an existing "consensus" in a given scientific field (which defines existing "results") is contingent on various social arrangements and interactions. This means that, keeping the data and methods (and a whole lot else) constant, the consensus would have been significantly different had just the social context been different in plausible ways.[2] Of course, we might also want to say that, for the same reasons, the actual consensus at that time was "not inevitable."

Here we have a way of understanding "inevitability" that I shall employ in this chapter. It is simply that a result (or "consensus") becomes "inevitable" when there are no remaining plausible contingencies to divert the ensuing consensus. Of course, everything hinges on what counts as "plausible." Here there is no general formula, but in specific contexts it can be reliably determined that there are simply no remaining alternative avenues of investigation that the relevant scientific community as a whole will take seriously. This way of understanding inevitability has the consequence that a result that was once inevitable might later be overturned. That is, later developments may open up new avenues of inquiry, and thus new contingencies. Inevitability is not forever. On reflection, this should not be surprising. Newtonian mechanics became the inevitable consensus sometime in the eighteenth century but was overthrown at the beginning of the twentieth.

As Hacking also noted, the feeling on the part of many scientists and some philosophers of science that our current results were in some sense "inevitable," and thus not "contingent," is often based on a particularly strong sense of scientific realism. Since I shall be arguing for a mild form of contingentism, I will begin by outlining some conceptual problems with this "absolute objectivist" or "metaphysical" realism. In its place I will propose a more modest "conditional" form of realism compatible with a reasonable contingency. I will then introduce two forms of inquiry that, to the best of my knowledge, have not been considered in the literature on inevitability in science. The first is an evolutionary understanding of the course of science. The second is the practice of "counterfactual history" among general historians. Using this framework, I will review one case in the actual history of science that exhibits both contingencies and, in my restricted sense, inevitability.

Conditional Realism

As I see it, the main conceptual problem with standard understandings of scientific realism is that they incorporate a metaphysical account of truth according to which there are truths about the world that exist independently of human existence.[3] The aim of science to discover these truths. This view may also include the view that there are "laws of nature" that are literally true descriptions of the world itself (Weinberg 2001, 123). This account requires that the world itself contain something like "facts" that mirror the linguistic structure of statements describing them. But this leaves us with the problem of understanding how the world itself could have independently acquired a structure corresponding to human languages, which are humanly created artifacts. Indeed, given the vast variety of human languages (think of Chinese), it seems necessary to invent something more abstract than actual linguistic expressions ("propositions"?) to be the counterparts of supposed facts. But the same problems arise for these invented abstractions.

There is a way of understanding metaphysical realism reminiscent of Nietzsche's many reflections on truth. Recall the opening lines of the Gospel of John: "In the beginning was the word." This suggests that there is a language in which the Christian God spoke the universe into existence. So there is a language that perfectly mirrors the structure of the world. At the time of the Scientific Revolution, this idea could be understood quite literally. Indeed, Newton was supposed to have understood the mind of God, thinking the thoughts of God after him, presumably in God's own language of mathematics. Newton's laws are then among God's laws for the natural world. But this way of thinking is unavailable to most contemporary students of science.[4]

Of course our actual concepts must bear some important relationships to the world or else they would not be useful. But there are many such relationships exhibited in many different ways. One such relationship, which I regard as especially important for understanding scientific representations of the world, is *similarity* between models and selected aspects of the world. Models range from actual scale models, through pictures and diagrams, to symbolically constructed imaginary objects. Thus, in the sciences, symbolic representation is not a matter of a direct language–world relationship, but of a relationship mediated by models.

Rejecting a metaphysical account of truth, we can still have a form of realism, but it must be conditional on the concepts we deploy as well as on known empirical results. Instead of declaring categorically "It is true that the world is such and such," we more modestly say "Given our symbolic resources and current evidence, the world appears to be similar to our models in such and such various respects." In doing so, we lose nothing but an inflated metaphysics. The more modest conditional claim requires as much grounding in experience as the inflated categorical claim. To the typical accusation that such conditionalization amounts to an unacceptable relativism, the reply is that the categorical claim is in fact no less relativistic. It only avoids the appearance of relativism, pretending to a metaphysical grasp on the world.

Nor need we avoid ordinary uses of the notion of truth. We merely adopt a deflationary ("disquotational," "redundancy") understanding of this notion, invoking the logical commonplace that to say "It is true that snow is white" is to say no more than "Snow is white." This is just another way of saying the same thing, a way that refers indirectly to the world by referring directly to a statement about the world.[5] And, indeed, within a framework, most ordinary intuitions about truth apply. Of two contradictory claims, only one can be true. And one can still raise all the usual questions about the extent to which claims to truth are socially constructed.

Finally, and most importantly for the present topic, conditional realism is compatible with contingency. A claim conditional on one set of concepts and empirical facts leaves open the possibility of there being another very different claim about the same general subject matter that is, however, conditional on different concepts and empirical facts. Nor need there be anything in the initial conditional claim to conflict with the later conditional claim being judged to be better grounded. We are not faced with a choice between metaphysical realism and a debilitating relativism.

The Evolution of Scientific Fields

The best scientific example of historical contingency is organic evolution as described by evolutionary theory. The late Stephen Jay Gould emphasized the importance of contingency in evolution with the following dramatic thought experiment: "I call this experiment 'replaying life's tape.' You press the rewind button and, making sure you thoroughly erase everything that actually happened, go back to any time and place in the past. . . . Then let the tape run again and see if the repetition looks at all like the original. . . .

Any replay of the tape would lead evolution down a pathway radically different from the road actually taken" (Gould 1989, 48–51). This suggests a dramatic excursion into counterfactual history of science. Rewind human history back to 1400, say, and hit the play button. The suggestion is that the history of science would travel down a road radically different from the road actually taken. This provides a dramatic statement of a contingency thesis for science. Unfortunately, unlike the case for organic evolution, we lack a substantial evidential basis for affirming such a claim. But neither can we rule it out. Nevertheless, the example of organic evolution provides a useful conceptual framework (a "perspective"?) for thinking about the role of contingency in the historical development of the sciences.

According to Gould's own account, organic evolution has three basic elements: (1) *variation* in traits among members of a population; (2) *selection* by the environment of members possessing traits making them more likely to leave relatively greater numbers of offspring; and (3) *transmission* of selected traits to the next generation. This schema can be applied to sciences. In the scientific case our concern is with the evolving consensus regarding the principle theories of a field. (1) *Variation* in suggested concepts and hypotheses is provided by contingent differences among scientists such as education and scientific experience. Where one received graduate training and with whom one studied is a major source of variation. (2) *Selection* of candidate hypotheses for consensus ideally is provided by experiment and observation in the context of an effective methodology. In fact, of course, many other contingent forces are at work. (3) *Transmission* of consensus views is provided by education and apprenticeship. This process is subject to many contingencies that provide the variation needed to keep the field evolving.

This is a prescription for a minimal evolutionary understanding of theoretical progress in a scientific field. It provides a prominent place for contingencies as part of a larger, evolutionary process. It is emphatically not an "evolutionary epistemology." The epistemology is in the details of the selective processes. This account does propose a strong analogy between the evolution of traits among members of a species and the evolution of consensus views in a scientific field. Both processes involve an interaction between organisms and their natural environment.

There is also a disanalogy between organic and scientific processes, in that major evolutionary changes among organisms are typically due to major changes in the natural environment, such as ice ages. In the scientific case, we presume the nature of the natural world is fixed, for example, the nature of mechanical motion or the properties of chemicals do not change during the history of science. The natural sciences (as opposed to the social sciences) have a fixed target.[6] This favors, though of course does not ensure, convergence in the natural sciences. On the other hand, both the social and the technological environment have changed dramatically over the few centuries since the Scientific Revolution. It is hard to argue that these latter changes were always inevitable. This favors contingent changes in consensus views even in the natural sciences. Nevertheless, an evolutionary perspective by itself provides no general resolution to the conflict between contingency and inevitability in science. The real usefulness of evolutionary notions is to be found in their application to the study of developments in particular scientific fields. Working within an evolutionary framework, one cannot neglect the possibility of significant historical contingencies.

Counterfactual History

Although Gould's dramatic counterfactual applied to the history of science is insufficiently specified for it to be seriously considered, the general possibility of doing what historians call "counterfactual history" is very relevant to any consideration of contingency in science. Among historians, it seems, counterfactual history has many more detractors than defenders. This general suspicion of counterfactual history by historians is unfortunate because, from a theoretical point of view, counterfactual reasoning is implicit in any attempt to give causal explanations. To understand a causal system is to know at least some counterfactuals about that system (Woodward 2005). And few would deny that history is a causal process.[7]

Several recent works, however, provide some guidelines for thinking counterfactually about the history of science. One is *Unmaking the West: "What-If" Scenarios That Rewrite World History* (Tetlock, Lebow, and Parker 2006). This series of essays by diverse authors is devoted to the general question: Was it inevitable that "the West" should come to dominate the world by the end of the nineteenth century? What makes this an interesting question is that "the West" (i.e., Europe) occupies a relatively small proportion of the world's land mass and includes a relatively small part of the total world population. And it developed much later than civilizations

such as Egypt, India, and China. Here I am not concerned with the particular counterfactual arguments of the various authors but with the general
methodology advocated by the editors and with the nature of some of their
conclusions.[8]

Their primary methodological strategy was to keep the counterfactual assumptions both as minimal and as plausible as possible, consistent
with the possibility of there eventually being a large historical impact.
This required a very careful choice of counterfactual assumptions, but it
also meant that they were left with most of their historical knowledge of
what actually happened intact. So their projections of what would have
happened under the counterfactual assumptions were as well justified as
possible. Their main conclusion is: Up until about 1800, it is possible to
imagine plausible counterfactual conditions that would not have led to
the hegemony of "the West." Later it becomes increasingly inevitable that
a hundred years hence "the West" will dominate the rest of world. This
conclusion is an instance of an obvious general principle: For any event,
the farther back in history one goes, the more contingent (less inevitable)
that event will be. Some alternative paths that once were possible get cut
off. The detailed historical question in any particular counterfactual study
is at what point in time this happens.

A second recent counterfactual study is *Vietnam if Kennedy Had Lived*
(Blight, Lang, and Welch 2009). These authors insist that what they are
doing is not "counterfactual history" but "virtual history." They denigrate
counterfactual history as asking unanswerable questions such as what
would have happened if Hitler had stuck to painting.[9] Almost as bad, they
claim, is asking whether Kennedy would have introduced United States
combat troops into Vietnam in 1965 when Johnson did. Asking the question in this way ignores the fourteen months between Kennedy's assassi-
nation and Johnson's decision. The situation in 1965, they insist would
surely have been different if Kennedy had been president throughout this
period. The only way sensibly to ask whether Kennedy would have escalat-
ed the Vietnam war, they argue, is to learn as much as possible about Kennedy's views and feelings regarding military interventions in small countries prior to his death. Here they emphasize his decision not to commit
American troops to an invasion of Cuba during the Bay of Pigs disaster.
The relevant fact is that Kennedy was willing to accept defeat rather than
intervene militarily. In general, the authors construct a relevant model of
Kennedy at the time of his death and use this model to run a conceptual

simulation of the subsequent months with Kennedy rather than Johnson as president. They conclude that Kennedy would have made do with U.S. "advisers" until after the election in 1964, and then sought a face-saving way to withdraw rather than introduce U.S. combat troops. I will now apply these ideas to a specific example, the geological study of the Earth's surface from roughly the middle of the nineteenth century to the middle of the twentieth.

A Hundred Years of Earth Science

Here I wish briefly to survey the history of theories of the surface of the Earth for the one-hundred-year period from roughly the middle of the nineteenth century to the middle of the twentieth.[10] In particular, I will be concerned with the question of whether the distribution of land and sea on the Earth has always been more or less as it now is (stabilism) or, on the contrary, whether there have been major changes in this distribution (mobilism). My main concern, of course, is to uncover both contingencies and inevitabilities in this history and to show how these changed over time.[11]

    I begin with the 1858 publication of a pair of engravings of the globe of the Earth centered on the Atlantic Ocean.[12] One of the pair shows the configuration of the continents as we find them today, with North and South America separated from Europe and Africa by the Atlantic Ocean. The other shows North and South America joined with Europe and Africa forming one land mass, with open ocean to the west. The thesis of the work in which these engravings appeared is that the original single land mass was split and the Atlantic Ocean created by the biblical flood at the time of Noah. There are two separable hypotheses here: (1) that there was one land mass later split into two (mobilism), and (2) that the split was caused by the Noachian flood. I maintain that the emergence of both hypotheses around this time was, given conditions at the time, inevitable.[13]

    The conditions that made explicit formulation of the hypothesis of mobilism inevitable were the discovery of the New World and the subsequent detailed mapping of the coastlines on both sides of the Atlantic. Anyone with an understanding of mapping and an interest in global geography cannot help but notice the congruity of the respective coastlines.[14] We can even bolster this claim with our current scientific knowledge that the human visual system is particularly well designed to notice spatial similarities and that the visual system is integrated with systems for memory and thought. As emphasized in recent cognitive studies of science (e.g., Nersessian 2008), the idea that the "discovery" of significant scientific concepts is a matter of mysterious intuitions and "creativity" is simply not true.

Something similar holds for the Noachian hypothesis. In the nineteenth century, Christianity in the Western World was strong but increasingly threatened by the growing influence of the sciences. Christian thinkers were eager to find support through association with science. In this climate, it was inevitable that some would link the idea of mobilism with the biblical Noachian flood. This is a case where the cultural origins of a hypothesis could not be clearer. The hypothesis could not even be formulated in the absence of the biblical story. Overall, from the viewpoint of 1858, it was hardly inevitable that a century hence mobilism would become the reigning scientific theory. The idea of attaching a probability to this eventuality hardly makes sense. Any claim of the inevitable triumph of mobilism could be nothing more than an expression of metaphysical realist faith.

Its association with "flood geology" was one reason among many that professional scientists paid little attention to the mobilist hypothesis. In, say, 1900, one could not predict when, if ever, mobilism would be given serious scientific consideration. That this should happen fifteen years later was highly contingent and, in fact, due primarily to the efforts of just one man, an initially obscure German astronomer turned meteorologist, Alfred Wegener. Like so many others, Wegener was originally attracted to mobilism by the congruencies in the coastlines across the Atlantic. These play a prominent role in his presentation of mobilism.[15] He set about searching for physical evidence from many sources including geology, geophysics, paleontology, paleobotany, and paleoclimatology. His best evidence came from comparisons of several striking geological features that seem clearly to leave off in Africa and pick up again at exactly the expected location in South America.

It is hardly surprising that a book published in 1915 by an obscure German scientist advocating a suspect theory should initially have attracted little professional attention. In spite of these discouraging contingencies, however, Wegener's book gradually became sufficiently better known to merit translations into English, French, and Spanish in 1924, Russian in 1925, and Swedish in 1926.[16] A high point in his career, and in that of his mobilist views, came in 1926 in New York City at a "Symposium on the Origin and Movement of Land Masses Both Inter-Continental and Intra-Continental, as Proposed by Alfred Wegener." Wegener himself was a participant. The proceedings were published two years later (van der Gracht, van Water-schoot, et al. 1928). The fourteen participants were roughly equally divided among (1) those generally sympathetic to mobilism, (2) those critical but thinking it worth discussing, and (3) those opposed or outright hostile.[17]

Two contingencies in Wegener's life are worth noting. In 1906, he accepted a position as a meteorologist on a two-year national expedition to Greenland, at the time a Danish territory. As it happens, Greenland is a place where physical evidence for mobilism is relatively plentiful. Absent this experience, it is doubtful Wegner would have pursued his interest in mobilism to the degree he did. Wegener returned to Greenland in 1930 on his own expedition to gather further evidence for mobilism. He died from an apparent heart attack traveling alone on skis between two observation posts. If he had lived, he would have kept interest in mobilism alive some years longer. As it was, it retained interest only a short while longer, mainly in South Africa. Thus, as late as 1930, the eventual triumph of mobilism did not seem in the least inevitable. There were many contingencies still to be overcome.

The strongest arguments against mobilism came from geophysics, the most prestigious of the earth sciences. Assuming, following Pierre-Simon Laplace, that the Earth was initially formed by the gravitational attraction of small bits of matter surrounding the sun, one could make estimates of how hot the center might become and how long it would take for this heat to dissipate through the surface. On any such model, the Earth would have started cooling soon after formation and begun shrinking and its surface crinkling like a drying apple. In such a model, there is no possibility of large lateral movements at the surface. So mobilism, rather than being considered inevitable, seemed simply impossible.

An answer to this objection was inspired by a scientific development that, from the standpoint of the earth sciences at the time, was totally contingent: the discovery of natural radioactivity around 1900. The possible relevance of this discovery to the possibility of mobilism, however, was not realized until around the time of Wegener's death. A British geologist, Arthur Holmes, suggested that natural radiation within the Earth might produce enough heat to create what are in effect convection currents that rise to the surface displacing existing crust that is then drawn back down as much as several thousand miles away. Such a current rising under a land mass could tear it apart, moving the pieces away from each other. The continents would ride on top of the crust as on a "conveyer belt." This could explain how the Americas got separated from Africa and Europe.[18] The suggestion lay fallow for thirty years, leaving the expected fate of mobilism little changed. Holmes himself thought it would be many generations before there might be direct empirical evidence relevant to his hypothesis.[19]

Following the Great Depression, the Second World War, the development of nuclear weapons, and the early years of the Cold War, the world of the 1950s and 1960s was vastly different from the world of the 1920s and 1930s. Few would argue that these changes were inevitable. Yet they made possible various scientific developments, none initially connected with the issue of mobilism, which, in retrospect, made the triumph of mobilism within the decade of the 1960s inevitable. That is, during that decade, all plausible contingencies that might have prevented a consensus on mobilism were removed.

One of these postwar developments was the discovery of a worldwide system of massive ridges on the ocean floors. These ridges typically have two high walls with a deep depression between. Notably, one such ridge runs down the middle of the Atlantic Ocean from north to south (Giere 1988, figs. 8.2, 8.3). Funding for this research came primarily from the United States Office of Naval Research, the Navy then developing the capacity to launch ballistic missiles from submarines as a deterrent to the Soviet Union. If one is going to hide submarines in the oceans, one had better know the territory. Absent this motivation, it is difficult to imagine when the details of this ridge system might have been discovered. The existence of mid-ocean ridges inspired a Princeton geologist, Harry Hess, to revive Holmes's idea of convection currents rising from the Earth's core. But Hess pictured the currents rising in the middle of the oceans rather than under a supercontinent, thus producing the ridges and providing a mechanism for an ocean floor to spread out in both directions from the ridges, descending some place far away (Giere 1988, fig. 8.8; 1999, fig. 7.8). The idea, nevertheless, remained in the realm of speculation, Hess himself describing his 1962 paper as "an essay in geopoetry."

About the same time, but initially having nothing to do with mobilism, a group at Berkeley was investigating scattered claims going back a half century that the Earth's magnetic poles had reversed direction as recently (in geological time) as one million years ago. This research required two newly developed technologies: potassium-argon techniques for dating minerals in the one- to five-million-year range, and sensitive magnetometers for measuring low levels of magnetism in minerals. The Berkeley group focused especially on material taken at various depths from lava flows. Iron particles in molten lava tend to line up with the Earth's magnetic field and then get locked into place as the lava cools, thus indicating the direction of the magnetic field at the time of cooling. Combining samples from North America, Hawaii, Europe, and Africa, they concluded the field was reversed beginning about a million years ago, but "normal" beginning around two and a half million years ago, and reversed again beginning about three and a half million years ago (Giere 1999, fig. 7.9). Refining their measurements from 1963 to 1966, they discovered that each of these three "epochs" included shorter (about one hundred thousand years) "events" of corresponding reversed magnetism (Giere 1988, fig. 8.5; Giere 1999, fig. 7.15). The result was a quite distinctive "signature" of reversals found around the world. There could be little doubt that the Earth's magnetic field had indeed reversed in a distinctive pattern over the past four million years. But so far there was no connection with mobilism.

The connection was forged by a young member of the Department of Geodesy and Geophysics at the University of Cambridge, Drummond Matthews, and his new graduate student Fred Vine. Matthews was involved in a project mapping the magnetic field produced by minerals making up the ocean floor, particularly regions across ocean ridges. His focus at this point was on the Carlsberg Ridge in the Indian Ocean. This research was made possible by the development of a sensitive "proton precession magnetometer" that could be put into a torpedo-like container and dragged along the ocean floor behind a research vessel, all the while sending up data on the local magnetic field. Upon returning, Matthews assigned Vine the task of analyzing the data. Using then new computer technologies being developed at Cambridge, Vine found that the magnetic profile across the ridge exhibited "anomalies," alternating positive and negative 5 percent differences from the current Earth's magnetic field. Having himself personally heard Hess present his idea of seafloor spreading, and knowing about the work on geomagnetic reversals at Berkeley, Vine quickly surmised that what he was seeing spread out on the ocean floor in bands parallel to the ridge was a record of geomagnetic reversals. As in the case of lava flows, iron in a convection current deep under the earth would be free to line up with the Earth's magnetic field, then get locked into place when the current cooled as it got closer to the crust and spread out in both directions. The so-called Vine–Matthews hypothesis was initially mostly ignored. But then the data got refined for other ridges in the North Atlantic below Iceland, the North Pacific west of Vancouver, and the South Pacific west of South America. A profile of the "Pacific-Antarctic Ridge" extending ten million years back in time was particularly impressive, especially because of its striking symmetry on both sides of the ridge (Giere 1988, fig. 8.14; Giere 1999, fig 7.13). With the realization that the same distinctive signature pattern existed around the world both on land and in the seafloor, mobilism, and what we now know as plate tectonics, was undeniable. No one could think of any plausible alternative explanation for these data. In other words, there were no remaining major contingencies that could divert the general acceptance of plate tectonics, which now plays the same sort of overarching role in the earth sciences that evolutionary theory plays in biology.

The above is little more than a sketch of the main developments over a century in the Earth sciences. The more deeply one goes into historical details, the more contingencies one finds, particularly in the later period for which the historical record is much richer. Nevertheless, one can draw some more general conclusions regarding the contingency or inevitability of the triumph of mobilism. At the end of the Second World War, no one could argue that the triumph of mobilism in the Earth sciences was inevitable. The mid-ocean ridges had yet to be discovered, the concept of seafloor spreading had not been invented, and good evidence for the existence of geomagnetic reversals did not exist. By 1960, these crucial elements were beginning to come together. Although no one at the time was in a position to proclaim the inevitability of mobilism's victory, in retrospect we can see that it was inevitable, and most likely within a decade. There were no remaining plausible contingent paths leading to a different consensus. Of course, many details remained contingent. For example, it did not have to be Fred Vine who first saw the connection between magnetic anomalies across ocean ridges and geomagnetic reversals. Others without his unique experiences could soon have taken that step, and someone surely would have.

It has been a staple in the sociology of science that results of observations and experiments need to be interpreted, and that this can always be done in more than one way. All it takes is sufficient imagination.[20] Taken in full generality, this view seems to assume that what is involved is just logical possibility. But logical possibilities are far too cheap. What is involved is more like real historical possibility. In this case there were just no such contingencies remaining. One might say that there was just no conceptual space for a stabilist (or other) alternative in the overall empirical and theoretical context. The best evidence for this claim is that prominent longtime critics of mobilism, and there were many, very quickly gave up and joined the revolution.

Conclusion

The question of whether some conclusions of the sciences are inevitable or remain forever contingent is unanswerable in the abstract. In particular, it is impossible to give an answer that does not reduce to vacuity because the background conditions that might support a nontrivial answer are too many and too various. In the case of the Earth sciences, for example, it seems imaginable that a stabilist approach, which began with Charles Lyell in the 1830s, could have gone along quite nicely through the twentieth century. But one would have to imagine, for example, that the matching signatures in geomagnetic reversals and magnetic anomalies across ocean ridges remained undiscovered. It is difficult to say what that would have required. That the Second World War or the later Cold War did not happen? That submarine warfare and the detailed study of the ocean floors was not pursued? That the necessary sensitive instruments were not invented? Here we may have a case where methodologically responsible counterfactual history is impossible. More troubling still is the realization that all scientific research could have come to an indefinite halt in 1963. That was the year of the Cuban Missile Crisis, the time when the United States and the Soviet Union came closest to all-out nuclear war. This possibility reminds us that any consideration of contingency and inevitability in the sciences presumes a vast amount of stability in the broader physical and cultural environment.

It might be objected that Earth science is an atypical science. The 1960s mobilist revolution was highly data driven and the theories were stated in quite general terms such as "convection currents," "seafloor spreading," and "geomagnetic reversals." Little was said about underlying mechanisms originating in the core of the Earth. The situation might be quite different for a science such as cosmology, which is much more theory driven. I agree.[21] In my terms, it may be more difficult in an abstract science to eliminate plausible alternative paths. Nevertheless, one detailed study of contingency in an area of fundamental physics, James Cushing's (1994) history of the development of quantum theory, follows the type of historical pattern I suggest. One can describe his thesis as being that a choice between a deterministic and indeterministic microphysics was contingent until roughly 1927, after which time the indeterministic Copenhagen interpretation took on the mantle of inevitability. For whatever reasons, a deterministic path was no longer regarded as plausible.

Focusing on contingency in science supports thinking of the history of science as an evolutionary process in the specific sense of proceeding in part by random variation and selective retention. Approaching the history of science from this perspective is a good antidote to thinking of history as an inevitable trajectory from the past to the present. Finally, an evolutionary perspective nevertheless allows us to view a stable consensus in a way that is both realistic and not metaphysical. The realism is tempered by making it conditional (which is to say "contingent") on existing evidence and on the concepts used to express it. So we can say without metaphysical pretensions that, given a mobilist framework and the lack of a plausible alternative, it really does appear that the continents have moved significantly over geological time.

CHAPTER 8

Necessity and Contingency in the Discovery of Electron Diffraction

YVES GINGRAS

T he French philosopher of science Gaston Bachelard tells us that from the perspective of its rationality a discovery is never really contingent. Though it may look so at first sight, that is, from the point of view of its singular history, a discovery loses its contingent appearance once viewed from the rational point of view of the theory that explains it. In his unique style he goes as far as saying that "the contingency of scientific discoveries is often the view of the ignorant. Discoveries come as a surprise only to those who do not make the effort to understand, those who do not benefit from the research tension that animates the scientific city" (Bachelard 1953, 7). As usual with Bachelard, this polemic sentence contains an interesting intuition on the question of contingency and inevitability that he does not develop in any detail but that can serve as our departure point.

The opposition between contingency and necessity could thus be related to the capacity to link scientific facts to theories or paradigms that give them meaning by placing them in a deductive structure. For a rationalist *and* constructivist philosopher like Bachelard, the more science is structured with abstract concepts the more its consequences are necessary, not contingent.[1] For behind the partially contingent order of historical development there is a coherent meaning constructed by a rational agent. In that context, "contingent" can be defined, following Antoine Augustin Cournot, as "what cannot be explained by theoretical knowledge" (Cournot 1956, 450).

Beyond Bachelard's brief mention of contingency, which already poses the problem of what is meant exactly by "contingent," this question has been discussed mostly in the context of realism and relativism, a debate strongly influenced by the constructivist sociology of science, which, since the middle of the 1970s, has always insisted that "things could have been different." Social constructivists often presented this contingency (implicitly or explicitly) as a direct consequence of the Duhem thesis of the underdetermination of theory by observation (Gingras and Schweber 1986; Franklin 2008). Following the contributions of Ian Hacking (2000a), who reformulated that question in the language of "inevitability," Léna Soler (2008b) and Howard Sankey (2008) have suggested making finer distinctions between the different meanings one can give to the words "contingent" and "inevitable." Despite these efforts at clarification, there is still, I think, much loose talk in the discussion of contingency. Hence, Hacking talks about the "results" of science, whereas Sankey considers that the question of the inevitability of science is about "whether the methods and practices of science might have led to a different set of scientific *claims about the world*" (Sankey 2008, 260). Soler points out that there are different "ingredients" in science like "speed of light," "Maxwell's equations," and "electrons." She has also tried to separate "benign" forms of contingency, which are compatible with the inevitability of scientific results (like following a different historical route leading to the same discovery), from "harmful" ones in which we would be confronted with two incompatible representations of a given piece of science (Soler 2008b). Finally, in his study of the role of contingency in the history of the interpretation of quantum mechanics, James Cushing also insisted on the existence of "trivial and philosophically uninteresting historical contingency of who did what when" (Cushing 1994, xii). One should note, however, that there may, in fact, be important reasons why it is X and not Y who made a given discovery, so that even this kind of apparently "benign" contingency may not be as trivial as it seems from the point of view of a historical sociology (Raman and Forman 1969). I thus think one must take a closer look at the nature of "historical events" when they are about science. To do that, I will use the historical case of the discovery of the wave properties of electrons to discuss the many constituents of science that could be said to be contingent or inevitable, since it is important to distinguish science from its history, the role of contingency being different in the former than in the latter.

Levels of Contingency

As the French philosopher Georges Canguilhem insisted, the *object of science* is not the *object of the history of science* (Canguilhem 1989, 9–23). So, history of science, often used as a resource in philosophical discussions about science, must first be characterized in relation to contingency, and it is important to distinguish between contingency in history and contingency in science. Though one can write the history of a science, there is a difference between history and science. As the French mathematician, economist, and philosopher Antoine Augustin Cournot noted in his 1851 *Essay on the Foundations of Our Knowledge*: "What makes the essential distinction between history and science is that the former includes the succession of events in time, while the latter is concerned with the systematizing of phenomena *without reference to the time when they occur*." For him, "there is properly speaking, no history where all events necessarily and regularly derive from one another" (Cournot 1956, 351; emphasis added).

From that point of view, the best way to define *historical contingency* is to use Cournot's famous definition of chance as "events brought about by the combination or conjunction of other events which belong to independent series" (Cournot 1956, 41). In short, a contingent event, which introduces the truly "historical element," is one that emerges at the meeting point of independent causal series. Though within a given causal series one can have deterministic behavior, the independency of each series makes it impossible to predict their meeting point, if any. A causal series thus creates a trajectory in which one can have a regularity that makes change predictable to a large extent: the growth of computer power (Moore's law) or the energy consumption of large cities of industrial countries for a period of time, to give only two examples from different domains. But as the latter case has shown, external perturbation like the 1973 political decision of the Organization of the Petroleum Exporting Countries (OPEC) to radically raise the price of petroleum, can indeed disturb a causal series and, in this case, change the consumption habits and trends observed during the "Glorious Thirty" postwar years.

So, history having few (if any) intrinsic necessities has a large place for contingency in the history of science, as in history in general, and it is obvious that many phenomena could have been—and in fact have been—discovered at different times and in different contexts and for different reasons. This kind of contingency could fall into the "benign" category, though a detailed historical analysis could also show that what did happen was not really contingent given the context of the time. In other words, any *explanation* tends, by construction, to *necessitate* events and thus diminish the sense of contingency.

This brings us to the meaning of terms like "results," "phenomena," "theory," and the like. Simply saying, as is often the case in relativist accounts of science, that things "could have been otherwise" without saying what exactly that "otherwise" could be, implicitly suggests that anything could have been possible and that there are few if any constraints on events and phenomena. Given that no (or very few) *empirical* phenomena can be considered "necessary" in the logical sense, they are thus contingent in the sense of "not necessary." As Emile Boutroux noted in his thesis, written in 1874, on *The Contingency of the Laws of Nature*, "from the analytical point of view, the only proposition wholly necessary in itself is that which has for its formula $A = A$" (Boutroux 1902, 8). Though the contrary of "contingent" is "necessary," it still makes sense to talk about events that are "contingently necessary" in cases where a *contingent* decision (to study something or not) entails *necessary* consequences. If one decides, for some contingent reason, to measure the speed of light, then it is, by necessity, either finite or infinite (as Descartes thought). Necessary elements can therefore be introduced once a given question is posed: Ole Rømer measured the speed of light and found it finite, though he had no exact measure of this finite value. Finally, one must also distinguish contingent from *arbitrary*: choosing on which side to walk or drive a car—right in North America and left in the United Kingdom—is arbitrary. One can always (in principle) change such an arbitrary decision (though usually at a cost, economic, social, or otherwise). By contrast, a contingent fact is not arbitrary because it cannot be changed once it is known (excluding errors and mistaken identities). That an electron is a particle or a wave (or both) is thus contingent, in the logical sense of not necessary, but it is not arbitrary because once it is found to have a given property, that property cannot be changed.

Talking about contingency can thus be slippery and degenerate into deaf dialogues if one does not use well-defined terms or does not identify the kind of contingency one is talking about. The kind of contingency that may arise also depends on the level at which we analyze science. Depending on the type of object we are talking about, it is plain that the kind of contingency of a historical *event* involving humans, a particular scientific *concept*, or a formal *theory*, is not the same as the kind of contingency of an *entity* (like an electron or neutron) or of an *effect* (like the Zeeman and Stark effects), all of which have a different *mode of existence*. Already in his 1951 book on the rationalist activity of contemporary physics, Bachelard had noted the importance of *effects* in physics as opposed to *entities* and commented on the role of reason in making these effects real (Bachelard 1951). Likewise, saying that "Maxwell equations" are contingent is a statement that remains vague as it can refer to an abstract set of laws relating electric and magnetic fields propagating in a vacuum or to a specific mathematical formulation using specific notations. Therefore, one cannot consider the contingency of "finding Maxwell's equations" and, say, the contingency of the statement the "speed of light is finite" as being of the same kind or on the same level insofar as the former is a complicated set of mathematical equations involving abstract concepts whose understanding may depend on the notation used, whereas the other is a simple ratio of distance and time that characterizes the movement of light, independently of its ontological nature (photon or wave).

From what we have said above, asking whether or not "science is contingent" is thus much too vague. One must also be careful not to move from one level to another as if there were no difference and that an argument applied at one level (theory, say) is also valid at another level (entity).

In summary, the degree of contingency of science can vary according to the kind of elements we are analyzing: phenomena, entities, laws, mathematical formulations of theories, and so on. In order to make sense, any discussion of inevitability of "results" must always specify the element being discussed. While theories usually have many contingent formulations as well as different interpretations, the existence of given phenomena (what appears) is more inevitable when one meets them (like supraconductivity) even when one may differ about their meaning and interpretation or explanation. For example, Cushing's book on quantum mechanics focused on the historical contingency of the dominant *interpretation* of the mathematical formalism and not on the contingency of the *empirical* consequences of two different formalisms or of the Schrödinger equation itself.

Real History versus Imaginary Worlds

As a historian and sociologist of science, I prefer looking at reality—already quite rich and diverse—to inventing imaginary cases. I am not denying the interest of completely imaginary universes with supposedly different sciences but suggest that we should use that tool of possible worlds and thought experiments only when no cases can readily be found in historical or contemporary situations. More than thirty years of work by historians and sociologists of science have provided an immense amount of knowledge on real cases of all sorts. Note that such a methodological choice would not completely exclude discussions of counterfactuals but would certainly constrain them to plausible inferences. Cushing's book provides a nice example of this approach. He does raise counterfactual questions about what would have happened if the Copenhagen interpretation had not become dominant by the end of the 1920s, but he stays within plausible sequences consistent with the context of the time (Cushing 1994, xii–xiii).

Even talks of two successful but incompatible sciences can be discussed on the basis of historical cases using, for example, the well-known controversy on the age of the Earth. For about fifty years we did have in this very case two totally incompatible theories on the age of the Earth: that of physicists claiming a young Earth and that of geologists and naturalists promoting a very old Earth (Burchfield 1990). There is no real need here to imagine two planets with two kinds of scientists creating two incompatible theories. The advantage of using real cases is that we can then analyze how real scientists do in fact react when faced with very different theories that pretend to cover the same phenomena. Such an approach also raises an interesting question: can scientists really accept being faced with two successful but incompatible sciences covering the same object? What we know about the history of science suggests that they will in fact do everything they can to make these two theories compatible even when they involve clearly incompatible ontologies like those of particles and waves.

The history of the two main formulations of quantum mechanics is interesting here in that it took less than a year for Wolfgang Pauli and Erwin Schrödinger to independently show the equivalence and intertranslatability of the two formalisms.[2] And only a few years later, Dirac showed that one could formally unify under a single abstract mathematical structure two ontologically incompatible models of reality, namely, waves and particles (Dirac 1930). Their work on the formal equivalence of different formulations suggests that they implicitly considered that the two theories had the same object domain and hence had to be equivalent, at least at the formal level if not at the ontological one. This case helps to refine what exactly "mutually incompatible" may mean: ontological incompatibility or formal incompatibility? Similarly, the history of the early uses of Schrödinger's equation and Heisenberg's matrix mechanics, shed light on how physicists react when confronted with apparently different theories of the same object domain. Given that the two theories were considered empirically and formally equivalent, both formalisms had a priori equal chances of being used. But despite such a logically contingent choice, a sociological analysis suggests that, given the "initial conditions" of doing physics in the 1920s, the chances were strongly biased in favor of Schrödinger's formulation: differential equations were well-entrenched tools for all physicists whereas matrix theory was an esoteric field of mathematics completely unknown to physicists. So, depending on the weight one puts on the word "contingent," one could say that, given the historical context of the time, it was not really contingent that the Schrödinger formalism came to dominate practice and marginalize the uses of matrix mechanics. There was a kind of "lock-in," as experts in technological innovation call it (Arthur 1989), where the accumulated expertise with differential equations, incorporated in the habitus of physicists (Bourdieu 1998), as well as its intuitive appeal were such that changing that training completely just for the sake of a matrix mechanics was too high a price to pay for most physics departments. Moreover, Shrödinger's version was based on the very classical theory of optics (Hamilton–Jacobi equations) learned by most students in their physics training. They could thus directly transfer their knowledge and know-how (and intuitions) of optics to apply them to the new quantum phenomena through a formal analogy. Physicists will in fact rapidly develop a complete "Optics of electrons" (Thomson 1931). Add to that the fact that techniques of solutions seemed more at hand for differential equations and their panoply of transformation techniques as compared to the awkwardness of the matrix machinery, and the case is foreclosed and quite inevitable.

In summary, the inertia of the training system as well as of the existing habitus of physicists are important constraints on the "contingent" trajectories that can be taken at a given time. One can call these "sociological accidents" (Cushing 1994, 175), but sociological considerations must certainly be part of the picture if counterfactual *history* is supposed to be plausible and thus consistent with the way physics is practiced and not limited to purely logical necessities. One could also say that all this was indeed "contingent" because anything *could* have happened if the initial conditions had been different. That is true but tautological: different conditions would indeed cause different histories. It is obvious that a completely different history of mathematics and its relation to physics could have affected the training of physicists making matrix the dominant tool in the 1920s. Finally, such historical examples—which I will not develop further here—also show that any attempt at a historical explanation of a series of events tends to *diminish* the sense of contingency and raise the sense of inevitability, as Bachelard suggested in the quotation given at the beginning of this chapter.

The Necessary Consequences of Contingent Existence

Beyond the question of levels, another important aspect of contingency is related to the fact that even when the basic properties of the world are contingent in the sense of not logically necessary, a great many *consequences* of a contingent existence are necessary once the existence of a given property of the world is admitted.

It is obvious that the properties of what we call "electrons" are contingent and only an empirical analysis can tell us what their properties are. For one can make predictions, but they will not necessarily always be confirmed by observation. In the rest of this chapter, I will discuss this question on the basis of a particular case study, that of the discovery of the wave properties of the electron in the mid-1920s. In presenting the basic elements of that case, I will focus on the segments of that story pertinent to the question of contingency, pointing out which constraints limited the spectrum of possible occurrences, including our very thought processes. By the latter, I mean the fact that all humans trained to enter the dynamic of the scientific field consciously or unconsciously apply schemes of thinking such as analogy, the principle of sufficient reason, and the principle of contradiction.[3] Also implicit in all schemes of thinking are the Kantian antinomies: things are either continuous or discontinuous, finite or infinite, created or eternal, and so on.

This historical case seems particularly relevant for a discussion of contingency since it involves at least three levels: (1) the contingency of who could discover the phenomena of electron diffraction, (2) the contingency of the phenomenon itself, and finally (3) the contingency of the ontological inferences made about the nature of the object "electron" that can produce the observed phenomena of diffraction and interference. Though philosophers may consider the contingency of who discovered a new phenomenon as "benign" if not even trivial, things are not so simple. After all, conditions had to be met to discover electron diffraction, and few physicists were in a position to observe the phenomenon. Moreover, as we will see in the case of George

Paget Thomson, the interpretation offered by physicists depended on their prior training. From a historical and sociological point of view it is thus of interest to begin with the context that led to the observation of the diffraction of electrons in order to evaluate the degree of contingency of that discovery.

Clinton Davisson as a Contingent (Lucky) Discoverer of Electron Diffraction

At the time our story begins, all physicists took for granted that the electron was a particle (Arabatzis 2006). They knew how to apply Lorentz's equations to the motion of electrons in electric or magnetic fields and deduce what would happen to its future movement in a given experimental configuration. They could do those calculations in advance of any experiment and thus make a prediction or do them after the fact and explain them. In the case of electron diffraction, we have, to a large extent, both of these cases in action.

The series of events that led to the observation of diffraction patterns produced by electrons interacting with a crystal followed two independent routes related to two distinct research programs. The first was that of Clinton Joseph Davisson and his collaborators in the United States and the second was that of George Paget Thomson in England. Comparing the two cases helps us to identify the necessary and the contingent aspects of the discovery of the wave behavior of the electron.

Going back to the beginning of the 1920s, Davisson's research program was based on the paradigm of scattering of particles in direct analogy with Ernest Rutherford's previous study of the scattering of alpha particles by different targets of matter: analyzing the speed and angular distribution of recoiling electrons would give, it was thought, information on the internal structure and distribution of electrons and ions inside the metal (Gehrenbeck 1978). Their major observation was that, as was the case in Rutherford's experiments, a small fraction of the emitted electrons have the same energy as the incident ones and are elastically scattered back from the atomic structure of the target. It is important to note that, in these experiments, the electron, taken to be a simple particle, is not being tested for its properties but used as a tool to investigate the structure of metals like nickel, platinum, and magnesium (Davisson and Kunsman 1921).

In 1923, Davisson and Kunsman published the details of their apparatus and results for platinum and magnesium in *Physical Review*, noting that one reason why they could see the elastic scattering at high potentials, where others did not observe anything, was that they used "a tube exhaust-ed to 10^{-7} mm of mercury" whereas "much of the experiments in this field were done prior to the development of present day vacuum technique" (Davisson and Kunsman 1923, 244). In other words, a technical feat made possible their success where others had failed. Using a simple and classical model of the atom, they could account for "the main features of the distri-bution curves so far observed for nickel" on the supposition that "a small fraction of the bombarding electrons actually do penetrate one or more of the shells of electrons which are supposed to constitute the outer structure of the nickel atom" (Davisson and Kunsman 1921, 524).

At this point, we have typical Kuhnian normal science going on, but using a high quality apparatus with the best vacuum possible at the time, a detail that explains the existence of few competitors in the race to test Lou-is de Broglie's ideas between 1925 and 1927. Before the summer of 1927 then, Davisson was not *studying* electrons and thus was not looking for their wave behavior because this clearly fell outside the space of what was thinkable at that time; he was simply using the electron as a tool to under-stand atoms. So what did happen to change Davisson's research program?

In their December 1927 *Physical Review* paper presenting the details of their experiment showing the existence of the wave properties of the elec-trons along the lines predicted in 1924 by the French physicist de Broglie, Davisson, now working with Lester H. Germer, noted that, "the investiga-tion reported in this paper was begun as the result of an accident which occurred in this laboratory in April 1925. At that time we were continuing an investigation, first reported in 1921, of the distribution-in-angle of elec-trons scattered by a target of ordinary (poly-crystalline) nickel" (Davisson and Germer 1927b, 705).

It is worth noting that neither the 1921 *Science* paper nor the 1923 *Phys-ical Review* sequel mentions the polycrystalline or amorphous structure of the targets, which were simply presented as "metals" without further details. Nothing then suggested that it made a difference to take note of this detail in the context of their research program. Only when the idea of a wave behavior enters the picture does it become necessary to take that detail into account and be sensitive to the crystal structure of the target. Hence the *retrospective* description of their first experiments has having been done with "poly-crystalline nickel."

The accident had altered the nature of the nickel target in such a way that, when they used it for new experiments, the results obtained were very different from what they usually observed. The new results, they note, "were the first of their sort to be observed" (Davisson and Germer 1927b, 706) and they traced the difference to the recrystallization of the target due to the accident that generated a prolonged heating.

This accident provides a perfect case of historical contingency, followed by a striking instance of serendipity (Merton and Barber 2006; Catellin 2014), to which a second historical contingency must be added: Davisson's trip to England in the summer of 1926 when he assisted, in August, at the Oxford meeting of the British Association for the Advancement of Science meeting. He then learned, during a talk given by Max Born, about de Broglie's theoretical work on matter waves and Walter Elsasser's prediction of electron diffraction based on it (Born 1927; Gehrenbeck 1978, 37–38). These two historically contingent events completely reoriented Davisson's research activities and *reversed* the relation between the tool (electrons) and the object of research (the metal target): the original "target" became a kind of detector/analyzer and the electrons became the objects investigated instead of being simply a tool to study the target. Davisson's research program then moved from analyzing the distribution of scattered electrons from metals to testing de Broglie's theory of matter waves. Since he already had the instrument, it did not take long to get the measurements done and the first results were published in *Nature* on April 16, 1927 (Davisson and Germer 1927a). Explicitly referring to x-ray diffraction, Davisson and Germer wrote in their detailed *Physical Review* paper of December 1927:

> Because of these similarities between the scattering of electrons by the crystal and the scattering of waves by three- and two-dimensional gratings, a description of the occurrence and behavior of the electron diffraction beams in terms of the scattering of an equivalent wave radiation by the atoms of the crystal, and its subsequent interference, is not only possible but most simple and natural. This involves the association of a wave-length with the incident electron beam and this wave-length turns out to be in acceptable agreement with the value h/mv of the ondulatory mechanics, Planck's action constant divided by the momentum of the electron. (Davisson and Germer 1927b, 707)

From an epistemological point of view, this paragraph is rich. First, the authors interpret the observed phenomena in analogy with radiation waves. They insist that such an analogy is "most simple and natural." Here "natural" implicitly refers to the fact that physicists are used to interpreting all physical phenomena either as waves or particles. Whether or not that dichotomy—like that between continuous and discontinuous, finite and infinite—is related to incontrovertible dichotomies of our "mental wiring" or simply due to "bad" human habits, which we should do everything possible to transcend, does not change the fact that we always construct models based on these dichotomies. Second, the authors never say that the electron *is* a wave but that a wave *is associated with* the electron, which is the view de Broglie himself held at the time.

From the point of view of the scientific habitus of the typical physicist, the way to proceed to test de Broglie's ideas, once they had been expressed, were thus obvious and in fact many physicists knew what experiment to perform, even when they had no resources or expertise to conduct them. What is at work here is the analogical schema frequently used by scientists to transpose the methods and results of one field to another once it is clear that the attributes of the objects are the same (Gingras 2015). Thus, if electrons behave as waves then they *must* diffract when interacting with a crystal. Not surprisingly, after diffraction was shown to exist, Davisson studied the other properties of waves, namely, reflection, refraction, and polarization. In doing so, he was just making the automatic inference any physicist (even average) would have made. Likewise, the notion of a refraction index for matter waves was suggested to explain the different positions of the peaks for x-ray diffraction compared to electron diffraction.[4] This kind of thinking based on analogical transfer explains the occurrence of simultaneous discoveries in the cases where the inferences are direct and obvious consequences of standard paradigms.

Whereas the events that led Davisson to investigate the behavior of electrons are historically contingent, the deductions made from the hypothesis that there exists a wave property somehow associated to the electron are not contingent at all but, rather, necessary. As we have just said, they are even quite automatic given the habitus of a physicist trained in the dominant paradigm. As Arturo Russo observed in his detailed study of the events leading to the discovery: "It is likely that electron diffraction would never have been discovered at Bell laboratories had not Davisson decided to spend holidays in England with his wife in the summer of 1926, and to seize the opportunity to attend a meeting of the BAAS in Oxford during August" (Russo 1981, 141). We have here what is, I think, a typical combination of *historical contingency* and *conceptual and logical necessity* that corresponds to a conditional truth: if Davisson looks for a wave behavior of electrons he will, or will not, observe it.

It is thus quite clear that there is no reason to doubt that without Davisson, the discovery would have happened within a year or two because, following de Broglie's prediction of matter waves associated with electrons, at least three other groups were working on the problem at different levels (Russo 1981, 141–44). In his Oxford talk in August 1926, Max Born, probably eager to promote the new quantum mechanics, considered that "indications of such an effect [electron diffraction] are given by the experiments of Davisson and Kunsman" and even that "a complete verification of this radical hypothesis is furnished by Dymond's experiments on the collision of electrons in helium" (Born 1927, 356). For him, a detailed experimental proof providing exact numbers to test the validity of de Broglie's equation did not matter as much as they did for the experimentalist Davisson, who could not accept the idea that his first results were real proofs of the wave character of electrons.

George Paget Thomson and the Necessity of Discovery

In analogy with the study of x-rays, the two basic methods for observing electron diffraction are (1) reflection on and (2) transmission through a crystal. The first method was used by Davisson, whereas the second would be used by the British physicist George Paget Thomson, the son of Joseph John Thomson, then professor at the University of Aberdeen. Like Davisson, he too was already well equipped to observe the phenomena once he learned (also at the Oxford BAAS meeting) it was predicted by de Broglie. As was the case with Davisson, Thomson's apparatus was already in use for the study of the scattering of positive rays. As he later recalled, "it was an extremely easy experiment to do" with "little more than reversing the current in the gaseous discharge which formed the rays" (Russo 1981, 153; Navarro 2010, 264; Thomson 1968, 7). A few weeks after the BAAS meeting, he altered the configuration of the apparatus to study electron diffraction (Navarro 2010, 262). Using photographic plates recording the transmitted electrons through a thin film of gold, platinum, and aluminum, he obtained the familiar rings observed in x-ray diffraction. His results were published in *Nature* just two months after those of Davisson and Germer (Thomson and Reid 1927).

For a brief moment, it was suggested that the diffraction pattern was not due to the electrons themselves but to a kind of Bremssrahlung, that is, x-rays produced by the collision of the electrons with the atoms. Noting that x-rays are not influenced by magnets, that possibility was easily excluded by applying a magnetic field and observing that "the electrons which formed the rings . . . were equally bent with the quite numerous electrons that had gone through holes of the film" (Thomson 1961, 824).

In summary, Thomson followed a more classical route to discovery by testing the theory once he learned of it during Born's talk in Oxford. What is also important, however, is the fact that, as was the case with Davisson, his particular situation equipped him with the tools needed to provide measures precise enough to test not only the qualitative aspects of diffraction but also the quantitative ones. In that sense, that Davisson and Thomson were the discovers of electron diffraction was not so "contingent" after all.

Conflicting Interpretations of the Nature of the Electron

Once known, electron diffraction was rapidly taken for granted and transformed into a tool of analysis, thus defining a new technological trajectory that ultimately led to the electron microscope. But all these practical applications left the theoretical question of the nature of the electron open and physicists were then faced with the same dilemma they had met twenty years before with Einstein's particle conception of light: was the electron a wave or a particle? Accepting that there was a kind of wave phenomena associated with the motion of electrons did not of course mean that all physicists knew the exact kind of wave it was. Asking such questions forced physicists to "leave the sure foothold of experiment for the dangerous but fascinating paths traced by the mathematicians among the quicksands of metaphysics" to use the words of G. P. Thomson (Thomson 1928, 281). Thomson himself sided with de Broglie and talked about a wave being "associated with the electron, guiding it in its trajectory" (Thomson 1928, 281–82). Also in 1928, Davisson suggested a pragmatic approach, admitting that electrons "are behaving as if they were waves" and that physicists can describe what they "observe by pretending that they are waves" (Davisson 1928, 606). After all, the only hard facts were that electrons are subject to diffraction in a crystal just as x-rays are and that the equivalent wavelength obeys de Broglie's equation (Navarro 2010, 269). It seems that only J. J. Thomson tried to provide a detailed model of the electron based on "tubes of force" and Maxwell's equations, but the ad hoc features of its hypothesis, needed to derive de Broglie's equation, do not seem to have convinced many physicists and was not further developed.[5] As is well known, the idea of a wave guiding the electron will be marginalized after the Copenhagen interpretation imposes itself at the end of the 1920s (Cushing 1994).

A second possibility was that the electron is simply a particle and that there is thus no need of any wave to guide it. Though never really debated in the 1920s, this approach will, as we will see below, be revived in the 1950s by the physicist Alfred Landé. The theoretical basis for a purely corpuscular explanation of diffraction were provided in 1923 by William Duane for the case of x-ray diffraction in a crystal. By a simple application of the laws of energy and momentum conservation applied to photons (particles) in a crystal lattice, he showed that one can deduce Bragg's law of reflection, which was usually obtained through an analysis of the interference of waves (Duane 1923). What Duane showed on the basis of a dimensional analysis was that the "so-called interference" is the consequence of a length parameter "a" that defines the periodicity of the crystal and not the effect of a wave, thus reversing the usual reasoning. Though this was particularly useful for solving the apparent contradiction between the existence of photons and the phenomena of light interference and diffraction, by showing that one could indeed deduce the formula without attributing wave properties to photons, that solution never seems to have been applied to the case of the electron, and most physicists instead adopted the idea of a wave behavior associated with the electron, which culminated in the idea of "wave-particle duality" promoted by Bohr as part of the Copenhagen interpretation of quantum mechanics.[6]

Though Duane's ideas were briefly mentioned by David Bohm in his 1951 textbook (Bohm 1951, 71–72, 134), it was Alfred Landé who revived and developed them to promote a coherent particle view of quantum mechanics against the then dominant Copenhagen interpretation based on wave-particle duality (Landé 1955, 74–76). Since then, many papers by Landé and others have tried to further develop Duane's particle view of interference and diffraction, but their approaches remained marginal (Rosa 1979; Olavo 1999; van Vliet 2010). Many reasons can explain this situation.

According to Landé, Duane's solution long remained, "a closely guarded secret" (Landé 1965, 11). The fact that Duane's paper appeared a few years before the discovery of electron diffraction and that its analysis was applied only to x-rays may have hidden, Landé suggests, the more general nature of its results (Hickey 2013, 14–22). In fact, contrary to what Landé and others have suggested,[7] Duane's results have not been neglected and the explanation for the lack of a purely corpuscular explanation of electron diffraction lies elsewhere. Only a few months after the publication of Duane's paper, Arthur Compton showed that "the general statement of the quantum postulate leads directly to the result that the momentum of the crystal changes by integral multiples of h/a as Duane assumes." He thus showed that the basis of Duane's results are profoundly linked to the quantum aspect of nature (Compton 1923, 359). Building on this work, Paul S. Epstein and Paul Ehrenfest (1924) developed the quantum theory of the Fraunhofer diffraction. Interestingly they came back to this problem in 1927, saying that "the recent discovery by Davisson and Germer gives to the problem of corpuscular diffraction a new interest and importance." However, what all their calculations show is that the classical theory and the quantum theory, which incorporate Duane's rule for the change in linear momentum, "give identical results" (Epstein and Ehrenfest 1927, 407). In his series of lectures on quantum mechanics given in Chicago in 1929, Heisenberg also presented Duane's explanation, saying that he had "given an interesting treatment of diffraction phenomena from the quantum theory of the corpuscular picture" (Heisenberg 1949, 77–78).

What all these treatments suggest is not that Duane's work has been neglected but, rather, that it was rapidly incorporated into the quantum formalism in such a way that made the opposition between waves and particles obsolete. In the introduction to the book based on his Chicago lectures, Heisenberg notes that in his approach "a particular emphasis has been placed on the complete equivalence of the corpuscular and wave concepts." This formal equivalence of both points of view (waves and particles) is such that Compton can conclude his paper by saying that "even from the quantum viewpoint electromagnetic radiation is seen to consist of waves" (Compton 1923, 362), whereas Duane concluded to the contrary that one could see x-rays as particles (photons) and not waves. Even Landé admits that "it certainly is remarkable that the result of the quantum mechanical reaction of particles can also be obtained in the roundabout way of wave calculations" (Landé 1965, 11). It is thus plausible that the lack of interest in the 1920s in choosing between a particle or a wave interpretation of electron diffraction was due to the perceived *formal* equivalence of the two points of view, despite their *ontological* incompatibility. This formal approach, which unified, or one could better say dissolved, the different ontological points of view under an abstract mathematical formalism,[8] was already chosen by Dirac in 1930 in his *Principles of Quantum Mechanics* and justified on the basis that the symbolic method "seems to go more deeply into the nature of things" (Dirac 1930, viii). Similarly, Heisenberg insisted that the "symmetry with respect to the words 'particle' and 'wave' shows that nothing is gained by discussing fundamental problems (such as causality) in terms of one rather than the other" (Heisenberg 1949, iv). So, even when they disagreed on the nature of the wave to be "associated" with moving matter, physicists admitted that neither the experiments nor the formalism could force a decision on the ontological nature of electrons, and most left the "quicksands of metaphysics," as G. P. Thomson called it, under the rug of an abstract mathematical apparatus that provided a symmetrical treatment of matter and radiation.

Conclusion

Once de Broglie's idea of matter waves became known, it was inevitable that it would at some point be put to the test and that a conclusion would be reached—positive or negative. There is of course an asymmetry between positive and negative results, insofar as the latter can be criticized as being an experimental system not good enough to really show the effect, as was the case of Davisson in 1923 when he invoked a better vacuum to explain the new kind of elastic scattering he observed. But even then there are constraints such that a negative result cannot indefinitely be rejected because all arguments are not equally acceptable given the state of knowledge.

Having shown the many constraining elements that explain why it was Davisson and Thomson who showed the existence of electron diffraction and why the wave properties of matter became accepted, we could start all over again and inquire into the contingency of de Broglie's own peculiar theory. After all, absolutely no empirical phenomena suggested in 1923 the bizarre idea that electrons could be waves or had waves associated with them. To do this we would have to inquire into de Broglie's scientific trajectory and analyze the role of formal analogies in his work, which are at the basis of his new conception. Like Einstein, he took seriously the heuristic of mathematical analogies and inferred that behind them there was in fact a corresponding reality (Gingras 2015). Of course, not all formal analogies work and official history usually retains only those that did succeed. Though we can reconstruct the rationality of de Broglie's choices, there remains a fundamental and noneliminable element of empirical contingency in his concepts insofar as the predictions could have been wrong had the electron been of a different nature than it happens to be. In other words, we could have lived in a world of electrons as simple small billiard balls. In that world there would not have been a Nobel Prize for de Broglie or for Davisson and Thomson who would simply have shown de Broglie's theory to be wrong.

Now, could the electron have been conceived *first* as a wave and only *later on* as a particle? On the one hand, given that the technical instruments and theory of x-ray diffraction did not exist at the time J. J. Thomson did his work on electrical discharges in an evacuated tube, and the need for a high-quality vacuum to detect the scattering of electrons by matter, the chances are low that the waves properties could have been known *before* the particle properties. On the other hand, it is plausible that the empirical investigations of electron scattering in gas, metals, and crystals during the 1920s, would have led to the idea that the electron does behave as a kind of wave as suggested by Davisson's new results observed after the 1925 accident in his laboratory, which led him in that direction. Reminiscing about that time, his collaborator Lester Germer himself believed that "If de Broglie's theory had been developed three years later, [Davisson and Germer] would have recognized [their] scattering patterns as diffraction, but it would have been slower and there would have been considerable stumbling around before the correct interpretation was found" (Germer 1964, 21).

The conclusion of this chapter thus lacks any revolutionary or radical character as it simply suggests that since the world is not the pure result of our will and its properties cannot be completely deduced from thought— as Descartes dreamed of doing—there will always be a dialectic relation between contingency and necessity. Or to put it in the words of the well-known psychologist but less well-known *epistemologist* Jean Piaget, all knowledge is the result of the interaction between the subject and the object, an interaction that leads to (partial) equilibrium through the assimilation of the objects into existing schemes (practical and conceptual) and the accommodation (that is, adaptation through change) of these schemes to new objects. For the meaning of an object is only given through its assimilation into a scheme (Piaget 1985). And it is through this coordination that discoveries acquire, as Bachelard noted, their necessity after having, for a while, appeared contingent.

PART IV

Contingency and Mathematics

CHAPTER 9

Contingency in Mathematics

Two Case Studies

JEAN PAUL VAN BENDEGEM

A mong the sciences, mathematics occupies a special place. For a start we should leave open the question of whether it is to be considered a science among the others or something else, for example, a language, that relates to the sciences. This, of course, has an immediate bearing on the problem of what the referents for mathematical objects could be and hence on any discussion concerning their reality. But, in addition, mathematics seems to have been rather immune to all developments that took place in the philosophy of science. For plenty of reasons, mathematics seems to have maintained its status of "discoverer" of "eternal" truths. This implies that notions such as alternativity and contingency are not obvious at all in the mathematical context and, instead, that uniqueness and inevitability are the main characteristics of the mathematical enterprise.

That being said, there have been occasional attempts to show that mathematics is not as "unshakable" as it is often presented. The first example that comes to mind—and one that has inspired me in my as yet unfinished search for (an) alternative mathematics (and hence to show its contingency)—is a chapter in David Bloor's *Knowledge and Social Imagery* titled "Can There Be an Alternative Mathematics?" ([1976] 1991, 107–30). There he raises the important question of the possibility of an alternative mathematics. The fact that this question needs to be dealt with is rather obvious to him: if (the production of) knowledge is a social process, then this must also apply to the part of human knowledge that seems to resist this sociological turn most strongly, namely, mathematics. And he stresses the point, with which I agree as this chapter will try to show, that the answer to this question should be in the form of concrete examples: "To decide whether there can be an alternative mathematics it is important to ask: what would such things look like? By what signs could they be recognised, and what is to count as an alternative mathematics?" (Bloor [1976] 1991, 107). It is worthwhile, I think, to take a brief look at the specific examples Bloor presents in that chapter because it will help in understanding the nature of the two examples that I will present here. He presents four cases, here grouped into three:[1]

(a) *The nature of numbers.* Here Bloor argues that in different historical periods numbers such as one and zero were interpreted in quite different ways compared to today's practices and theories. To give one small illustration: there is indeed a world of difference between saying that "the equation $x^2 + x - 6 = 0$ has two solutions, $x = 2$ and $x = -3$, but I use only the positive solution" and saying that "the equation $x^2 + x - 6 = 0$ has exactly one solution, namely, $x = 2$," simply because negative entities can be anything you like, but they are not numbers.

(b) *The metaphysics of numbers.* Here Bloor deals with the Pythagorean view of numbers as part of a larger metaphysical framework. These metaphysical considerations are part and parcel of mathematical activities and the resulting theories. In support of this claim, he discusses the great classic of all (mathematical) times: the irrationality of $\sqrt{2}$. Actually by formulating the theorem in this way, one is already subscribing to a particular metaphysics, namely, where there are other numbers besides the rational numbers, so the assumption is accepted that the rational numbers do not exhaust all numbers. But a quite different conclusion can be drawn: $\sqrt{2}$ is not a number.

(c) *The case of the infinitesimals.* This, of course, is the best-known example of an alternative approach in differential and integral calculus. Bloor here quotes the work, among others, of John Wallis, more specifically the case of the surface of a triangle. Let me briefly present this case. Wallis slices up the triangle into an infinite series of very thin layers, parallel to the base of the triangle. If the height of the triangle is h, and the number of layers is ∞, then a layer has height h/∞. The length of each layer varies from b, the base of the triangle to 0 at the top. This forms an arithmetical series, the sum of which is equal to the product of the average of the terms and the number of terms or, in this case, $b/2 \cdot \infty$. Hence the surface of the triangle is $h/\infty \cdot b/2 \cdot \infty = h \cdot b/2$. It is obvious that different standards of rigor are at work here, but this is precisely what is at stake: the standards of rigor are not fixed once and for all but are susceptible to (deep) changes.

These cases show at least two things: first, that there have been periods in the history of mathematics where mathematical concepts were interpreted otherwise, compared to today, and second, that these different interpretations can indeed be seen as alternatives. The way we look at numbers today is certainly different as compared with the past and infinitesimals have disappeared from mathematics altogether. Although perhaps this last statement needs a qualification. A staunch defender of the unique-inevitable view will probably argue as follows:[2] yes, true, once we "messed" around with infinitesimals and we did not get our concepts right, but then came the well-known ε-δ approach and they were no longer needed, but today there is such a thing as nonstandard analysis and that captures the idea of an infinitesimal in an exact way. Because it is not clear at all how this approach could be further improved, we may rightfully claim that we now have the correct notion of infinitesimal. To answer such a critic is not an easy matter, although I would argue that the burden of proof rests with the critic because the full claim is what needs to be defended, namely, that for all mathematical concepts and theories, there will be a moment either now or in the (near?) future when these will get a "final" interpretation (which can include their disappearance).

The main difficulty with Bloor's examples is that the alternative corresponds to a particular historical moment or period and thus these alternatives are 'real' and not on the level of possibilities, whether historical or logical. A particular weakness of these examples is therefore that, even if one accepts Bloor's cases as good examples of alternative mathematics, nevertheless, for each historical moment there is only one kind of mathematics around and, if a certain stability is reached, this can be counted as the 'final' stage. The challenge therefore remains as to whether other forms of alternativity and/or contingency are possible.

Alternative and Contingent Mathematics: A Wide Range to Explore

A first important observation to make is that if we accept logical possibilities as genuine alternatives, then it is obvious that alternative mathematics exists. Here is a straightforward example: if we had an arithmetic wherein $2 + 2 = 5$, surely this must count as an alternative. Perhaps, but definitely not if that statement was merely the result of a different way of counting, using the series 1, 2, 3, 5, 4, 6, 7, . . . And, if one were to claim that $2 + 2 = 0$, then this might perhaps be seen as an alternative to classical arithmetic but it is a perfectly acceptable statement in an arithmetic modulo 4. And a modulo arithmetic is surely not seen as an alternative to classical arithmetic but, rather, as a derivative. Phrased in a different way, if these were the only examples we could come up with, that is, *purely* logical possibilities, then it would be acceptable to conclude that, even if one were willing to accept the alternativity of these logical possibilities, they nevertheless do not show the contingency. I will therefore not focus any further on these possibilities in this chapter.

Ethnomathematics

That being said, the seemingly simple phrase "existing mathematics" already generates an interesting discussion. Does the phrase include all forms of mathematics occurring in different cultures on earth? If so, then at least in the weak sense, it must be clear that alternatives do exist that, in addition, support the contingency thesis in the form that different cultures generate different forms of mathematics. This is indeed a weak sense because it is still compatible with the view that a particular culture will develop the unique sort of mathematics that fits in it. Nevertheless, if it is indeed the case, this is a first important step in the direction of contingency. So do different cultures generate different mathematics? Fortunately, there is a scientific discipline that gives us the immediate answer, namely, ethnomathematics, and the answer is a clear yes. Any introduction to or handbook on ethnomathematics (see, e.g., Ascher 1994; or D'Ambrosio 2006) will show that non-Western cultures have developed all kinds of notation systems for (particular sets of) numbers, where large numbers can have a certain vagueness, where infinity is usually lacking, and so on. More often than not, there is no notion of mathematical proof in an axiomatic setting. In short, it is clear that all these mathematical systems rightly deserve the label "alternative." However one should take care of a number of issues that need to be dealt with:

- Very often the form of mathematics that is found in different cultures comes in the form of a set of practices, techniques, procedures, and the like, and rarely in the form of a (more or less) clearly formulated theory. So, if one insists that a mathematical theory is required, then whatever these practices are, they cease to be mathematics.

- Even if one does not have a problem in attributing the label "mathematics," the question remains as to how these practices can be compared with Western mathematics. At first sight, it seems nearly impossible. One might argue that the degree of "sameness" that is required is too low to warrant a claim of alternativity. On the other hand, quite obviously, the practitioners of ethnomathematics do believe that it is mathematics. In other words, there is a genuine danger here of a "rescue by definition" strategy: if something could be a genuine alternative, redefine the subject such that this something ceases to be part of the subject.

- Contrary to the previous remark, there is another view that suggests full comparability along the following lines. Non-Western mathematics should be seen as an approximation of (often rather trivial) parts of mathematics as we know it today in our culture. The caricature is well known: starting from "one, two, many," "many" will gradually change from a finite number into infinity and there we are! Apart from the fact that this picture is truly a caricature, it has unfortunately not lost its force.

The picture that emerges is that of sailing between the Scylla of too great a difference to make comparison possible and the Charybdis of all too easy comparability, making any differences disappear. In the first case, we cannot talk about alternatives, and in the second the alternatives cease to be such as they become incorporated and integrated. To further complicate matters normative issues are at play. To avoid Scylla, one can enlarge the definition of what mathematics is and, to avoid Charybdis, one can point out how much force is needed to obtain the integration. In summary, it is clear that if the extremes can be avoided, the possibility of alternatives is a genuine one and thus, as mentioned, supports at least the weak contingency view.

Apart from the fact that we should see whether stronger forms of contingency can be found in mathematics, the ethnomathematical approach still leaves us with a major open question. To what should one compare advanced topics in Western mathematics, say group theory, complex numbers, quotient fields, or partial differential equations, in non-Western mathematics? This needs to be examined as well.[3]

Mathematics Here and Now

When we look at present-day mathematics in our culture, at first sight things look quite promising. As it happens, mathematicians and philosophers of mathematics do occasionally use the term "alternative" themselves. A perfect example is the domain of foundational studies: (neo-)formalism, (neo-)logicism, intuitionism, all sorts of constructivisms, including extreme cases such as strict finitism, paraconsistent and inconsistent mathematics, and so on are often presented as alternatives by their practitioners. That of course does not (and cannot) mean that thereby the claim has been substantiated. For it still needs to be shown that these alternatives are indeed alternatives that support the idea of the contingency of mathematics.

Let me first deal with an important counterargument for the contingency of mathematics. Actually, it seems to me that this is the main objection, so once we have dealt with it, we can at least conclude that contingency is not to be excluded.

The argument runs as follows. All modern mathematical theories tend to be or are already formalized, they even try to be axiomatic, they hardly discuss the idea of a mathematical proof—even "severe" constructivists accept the idea that there are such things as proofs—they accept the idea that one counterexample to a universal statement is sufficient to refute that statement, and so on. If one were to insist that, say, an intuitionist does have a different notion of proof, the reply will be that, even so, it remains comparable to other theories. When we say, for example, that intuitionists do not accept the excluded middle, does this not presuppose the comparability? So what we see is that other mathematical theories can be reconstructed as subtheories of other theories and hence they cease to be "genuine" alternatives or rivals but become extensions of one another. In addition, precisely because we have a high degree of comparability this counterargument goes together with the idea or, rather, the belief that, in the end, mathematics will prove to be unique and singular—that is, in the long run one of these alternatives will prove to be the right one. In summary, it seems that formalization gets us even more quickly into Charybdis.

Sometimes one finds a direct expression of this belief of unicity within mathematics itself. Here is such an example. Many handbooks dealing with number theory and its extensions present the sequence starting with the natural numbers, then the integers, then the rationals, then the reals, then the complex numbers, and, finally, in some cases, the quaternions. The story is often told in terms of completion or closure in the sense that an inherent teleological idea seems to be guiding the development of mathematics, where a curious game is played to try to identify the historical development with an internal logical development. A number of examples in the present-day mathematics of theorems express the idea that in a particular domain all has been said that could be said and that hence this domain is completed. A famous case is the theorem that, given a division ring D that contains the real numbers as a subring, if D is of finite dimension over the real numbers, then it must be isomorphic either to the quaternions, or to the complex numbers, or to the reals themselves (see, e.g., Mac Lane 1986, 121), there are no other possibilities.

Concepts and Their Meaning

What pleads against the counterargument? The core objection is, concisely formulated, that "translation," as it is used in the counterargument, corresponds to "syntactic translation." Suppose, to remain within the example of intuitionism, I say that "p or q" for the intuitionist means "$\Box p$ or $\Box q$" ("$\Box$" standing for necessity) for the classical mathematician, then I have said nothing about the meaning of "or" in the statement "p or q," or, in other words, what does "or" mean for the intuitionist? It would be plainly wrong to assume that the "or" occurring in both sentences is the same "or" with the well-known classical truth table. Compare with physics. A "translation" from Einsteinian physics into Newtonian physics by replacing, where it is allowed, v/c, when v is sufficiently small, by 0, does not show that the latter is now *part* of the former.

The above means that the answer must be found not in the syntactical properties of the concepts we use but in their semantical properties and their meanings. Or, put differently, the argument that, if two theories share the same labels and expressions for concepts (e.g., the same names are used such as number, point, function, topology, etc.) they are translatable and hence cannot be alternatives anymore, is inconclusive. What we have to look at is the meaning of the concepts and, together with it, their use. I do not want to enter here into a Wittgensteinian discussion concerning the relations between meaning and use—all I need is the idea that part of the meaning of a concept is shown through its use. If so, then it is clear that not only mathematical theories but mathematical practices become equally important.

Take (mathematical) proof as an example. A classical mathematician, with a Hilbertian inspiration, sees a proof as a labeled list of statements, starting with the premises, ending with the conclusion, and such that every other statement in the list can be justified either as an axiom or as the result of the application of one of the logical rules. An intuitionist, with a Brouwerian inspiration, will look at "real" mathematical proofs, where what the statement means is crucial in order to be convinced by it. A logical reconstruction can, if necessary, be done afterward. Can we claim that they have the same notion of proof? If meanings have to play their part, the answer must be no.

To pave the way toward the two examples that are presented in the next two sections, let us look in more detail at what alternatives can be. There seem to be (at least) two possible routes: (1) a theory T_1 is an alternative of a theory T_2, if both share the same labels or names for concepts but these have different meanings, or (2) a theory T_1 is an alternative of a theory T_2, if there are concepts occurring in one theory and not in the other and the domains of both theories are sufficiently similar—that is, they talk (more or less) *about* the same things. Bloor's examples that deal with the "nature" or the metaphysics of numbers, clearly fall under (1), as the same names for numbers are used throughout, whereas the infinitesimal example falls under (2) because, before the nineteenth century, infinitesimals occurred in theories about integral and differential calculus, whereas in present-day formulations of the same calculus infinitesimals have disappeared. The examples I present here fall squarely under (2) with this additional property: the similarity of the domains that is referred to in the stipulation of (2) is total. In other words, we end up with two theories talking about the same things, in the sense that the missing concepts of the one theory can be *syntactically* defined in the other theory and vice versa. The first example deals with complex numbers and describes an alternative route that historically could have been taken. The alternative that I propose introduces concepts such as holes in a plane, surfaces that are missing and can be added, in short, a list of concepts that do not occur in any of the descriptions that we have today of complex numbers. The second example shows that arithmetic (and in principle any mathematical theory) can be developed without the notion of proof as we know it. If such a crucial concept (or, at least, so it is believed to be) can be dispensed with, then surely mathematics here and now could have been quite different.

Mathematics of Holes and Complex Numbers

Elementary Concepts

The alternative mathematical theory for the complex numbers will be presented in a geometrical informal presentation. It starts with the Euclidean plane that acts as a *reference plane* (RP). To this plane geometrical objects, such as points, lines, and surfaces can be *added*. They can also be *removed*. The special feature is that parts of RP can also be removed, thus producing *holes*. Note: it is this element that turns the theory into an alternative because to my knowledge I have not encountered it anywhere else.

Take now the special case of an added square with side a. Its surface A will be a^2, no mystery there. If we now ask what the size of a surface of a missing square, that is, a hole in RP, must be, then a simple argument leads to the answer $-A$. For, if a square of surface size A is added, we obtain RP and all surfaces have thus disappeared, hence the surface of the missing square must be $-A$. The inevitable next question must be what is the length of the side of a missing square? And the answer seems equally inevitable: $\sqrt{-A}$.

This means that given a square and its surface A (positive, zero, or negative), three scenarios are possible:

1. If $A > 0$, then A is the surface of a square with side $\sqrt{A}$,
2. If $A = 0$, then both surface and side are equal to 0,
3. If $A < 0$, then A is the surface of a *missing* square, with side $\sqrt{A}$.

A special notation, namely, the symbol m, is now introduced to indicate present and missing surfaces and lines in a slightly different, but equivalent way. The notation is of the form $\alpha/m\beta$, where α is the side of a square that is present (case 1 above) and β refers to the side of a missing square, but due to the presence of the operator m, it will now be indicated by the positive number, $\sqrt{-A}$, if $A < 0$ is the surface of the missing square. So, the three cases will now be represented by the following notations:

1. If $A > 0$, then the side of A is $\sqrt{A}/m0$,
2. If $A = 0$, then the side of A is $0/m0$,
3. If $A < 0$, then the side of A is $0/m \sqrt{-A}$.

The reason for the introduction of this notation is that it allows us to write down mixed expressions, such as a/mb. This allows us to talk simultaneously, as it were, about the side of a square that is present and the side of a square that is missing.

How to Calculate with Sides of Surfaces Present and Missing

The first question to deal with is addition. Consider first two special cases:

(a) Both numbers are of the form $a/m0$, that is, nothing is missing. Then it is straightforward to assume that $a/m0 + c/m0 = (a + c)/m0$,
(b) Both numbers are of the form $0/mc$, that is, nothing is present. Then it is equally straightforward to assume that $0/mc + 0/md = 0/m(c + d)$.

What is present remains present, what is missing remains missing, hence for the general case it seems obvious to simply combine the two special cases and to arrive at the following rule for addition:

$$(a/mb) + (c/md) = (a + c)/(m(b + d)).$$

Another way to justify this rule is to rewrite (a/mb) as the sum of $(a/m0)$ and $(0/mb)$.

Multiplication is somewhat trickier and needs a few preparatory steps. Suppose we have a missing square with surface A, $A < 0$. Then its side is $0/m \sqrt{-A}$. So what we do know is that $(0/m \sqrt{-A}) \times (0/m \sqrt{-A}) = A$. Obviously for a square that is present, with surface A and side $\sqrt{A}$, $(\sqrt{A}/m0) \times (\sqrt{A}/m0) = A$. The question is now what to do for the general case: given two numbers a/mb and c/md, what could their product $(a/mb) \times (c/md)$ be?

To get started, let us consider an intermediate case, where the two numbers to be multiplied, are the same: $(a/mb) \times (a/mb)$ or $(a/mb)^2$. A possible answer to this question could be formulated in the form of table 9.1.

The left-upper and right-bottom corners are clear enough: they determine surfaces present and surfaces missing. It is not immediately clear

TABLE 9.1. Multiplication table—incomplete version

	a	mb
a	surface present a^2	?
mb	?	surface missing $-b^2$

however what the meaning of $a.mb$ could be. The following argument might provide an answer: $a.mb$ indicates either a length or a surface. It clearly cannot be a surface because it is neither a positive nor a negative number but still involves the operator m. So it has to be a length. It cannot be a length that is present, so therefore it has to be the length of a missing side of a surface. In other words $a.mb$ is to be read as $m.ab$. In modern terms, a acts here as a scalar. One might remark that a certain ambiguity is present here but this seems to be unavoidable, we believe. In the very same way that, given a line with length L, and given the expression $a.L$, then this either can be read as the length L a times over, so that we obtain a length, or a can be read as the other side of a rectangle and in that case $a.L$ refers to the surface of a rectangle. This argument suggests that "missing length of a line" is a correct replacement for the question marks in table 9.1. This leads to table 9.2.

TABLE 9.2. Multiplication table—completed version

	a	mb
a	surface present a^2	missing length
mb	missing length	surface missing $-b^2$

If we allow ourselves an extension of the notation we have introduced, then we could write:

$$(a/mb)^2 = (a^2 - b^2)/m(2ab).$$

The extension comes down to the idea that an expression of the form a/mb indicates on the left all the expressions that do not contain the operator m and on the right the expressions that do contain m.

The intermediate case makes clear how the original question can be answered. It produces table 9.3.

TABLE 9.3. Multiplication table—general version

	a	mb
c	surface present a.c	missing length b.c
md	missing length a.d	surface missing −b.d

Using the same notation, this gives:

$$(a/mb)(c/md) = (ac - bd)/m(ad + bc).$$

The set of numbers of the form a/mb, equipped with the operations of addition and multiplication corresponds quite neatly to the structure of the complex numbers via the simple (syntactical) translation $T(a/mb) = a + i.b$. The "work" done by the imaginary unit i has obviously now been shifted to the missing operator, the major difference being of course the direct interpretability of the operator in contrast to the mysterious square root of -1. However, it is more important that this approach uses a concept that is totally lacking in classical geometry, namely, *holes* (in a *reference plane*). Of course, other solutions have been found to get rid of the imaginary unit, such as the translation into couples of real numbers (and hence a geometric interpretation for free in the Euclidean plane). So have I done nothing more than add just another interpretation to the already existing ones? In a sense yes, but it is an interpretation that is sufficiently different semantically to be called an alternative.

Potential Historical Relevance

The inspiration for this particular alternative for the theory of complex numbers (that, to my knowledge, has not been formulated before) comes directly from one of the obvious sources where complex numbers are discussed, Girolamo Cardano's *The Great Art* ([1545] 1968). In chapter 37, "On the Rule for Postulating a Negative," he deals with the following problem: "Divide 10 into two parts the product of which is 30 or 40; it is clear that this case is impossible." Having written that, he immediately continues to provide a solution. Suppose the product has to be 40. Then, if 10 is divided in 2, we have 5 and the square of 5 is 25. This leaves us $40 - 25 = 15$ "short." The square root of that number, $\sqrt{-15}$ is to be added

CONTINGENCY IN MATHEMATICS 235

to and subtracted from 5 and that will be the answer. So the solution is $5 + \sqrt{-15}$ and $5 - \sqrt{-15}$. The demonstration that follows is not easy to understand at all and small wonder that Cardano himself ends the demonstration with the statement: "So progresses arithmetic subtlety the end of which, as is said, is as refined as it is useless" (Cardano 1968, 219–21). However, in less useless terms of the alternative presented here, the geometrical answer would be that 10 is divided in $5/m\sqrt{15}$ and $5/m(-\sqrt{15})$, such that the sum is of course $10/m0$, that is, 10 and the product $(5/m \sqrt{15}) \times (5/m(-\sqrt{15})) = (25 + 15)/m0 = 40$.

I have further elaborated this alternative to show that polynomial equations of third degree can be dealt with as well (Van Bendegem 2008). Of course, my claim here is a more modest one, namely, to show that genuine alternatives can exist. It is a quite different matter to estimate whether a particular historical development could actually have happened, were circumstances slightly different. Then the answer has to be negative because Cardano was using the geometry of his day, which should have been different to start with. It thus seems nearly inevitable to end up with the question: what are the chances that a geometry of holes could have been developed in the first place? That question, for sure, I consider as good as impossible to answer. On the one hand, if one equates geometry with geometrical writings, diagrams, and pictures, then it seems not obvious at all to come to the idea of holes and gaps. On the other hand, if one thinks of the environment one inhabits, then holes and gaps are plentiful. And let us not forget that today we do have a mathematical theory that deals, among other things, with holes—topology, where the distinction between a sphere and a torus is precisely one hole, expressed (roughly) by the *genus* of these surfaces.

No Proof Required in Random Mathematics

The alternative to be presented here is a radical departure from mathematics as we know it presently. I will try to show in outline that it is possible to arrive at a mathematical theory, taken in the sense of a set of true statements, *without the notion of proof.* As in the previous case the presentation will be semi-informal in the form of a plausible scenario (although, in comparison to the previous example, there is no historical inspiration to be found anywhere).

The Basis: Addition and Multiplication

Imagine a culture where arithmetic is developed in the following way. Let us assume that the members of that culture have some notions of numbers in the sense that they can generate names for numbers that are locally ordered. By this I mean that they know 3 follows 2 and comes before 4, but 3 compared to 1,000 does not necessarily make (any) sense. They also have some idea of addition in the following restricted sense. When they are presented with an equation of the form $n + m = k$, then they can also generate the equations $(n - 1) + m = (k - 1)$, $n + (m - 1) = (k - 1)$, $n + (m + 1) = (k + 1)$ and $(n + 1) + m = (k + 1)$. Call these four equations the *neighbors* of the given equation. Do note that the numbers in parentheses are meant to be names for those numbers.

Imagine further that empirically they discover (e.g., through the manipulation of certain objects) that $2 + 3 = 5$. They then accept this equation as correct and the neighbors as well. Such equations are put on a list, titled "Things we know for sure." In the course of time, people ask questions of the type "What is the outcome of $n + m$ (for some specified n and m)?" If the answer is on the list, that is indeed the answer. If not, nothing is added to the list (and the answer will be held under consideration until relevant information is available to add it to the list). In the former case, note that all the neighbors are added to the list as well. In the course of time, this list will therefore grow.

What will happen is that, under the assumption that, given sufficient time, any equation is likely to turn up,[4] at the end of time the list will contain all true arithmetical statements involving addition. A simple argument to see why this must be the case is to consider a two-dimensional lattice, where each square corresponds to a couple (n, m). One starts (as in the example presented here) at $(2, 3)$ and the neighbors are $(1, 3)$, $(3, 3)$, $(2, 2)$, and $(2, 4)$. This corresponds neatly to a random walk in the plane and, as is well-known, given sufficient time, all squares will be visited. Hence the label *random* mathematics.[5] Formulated in those terms, if we call T the set of all true arithmetical statements involving addition, then in the limit T will be reached. It is perfectly acceptable therefore to say that people in this culture know how to add.[6]

A similar approach can easily be written down for multiplication. If $n \times m$, for specific n and m, can be empirically established, then the neighbors can also be calculated, thus $(n - 1) \times m$, $(n + 1) \times m$, $n \times (m - 1)$ and $n \times (m + 1)$.

Full Arithmetic

There is an obvious objection to this approach. So far, I have only been discussing specific cases, dealing with constants. Proof does not amount to much in these cases, so small wonder they can do without. What if variables are introduced for numbers and a language such as first-order predicate logic is available? Surely there is no way to arrive along this route at universal statements such as $(\forall x)(\forall y)(x + y = y + x)$. If the rules of classical logic are to be followed, then the answer, as far as I can see, is indeed negative. But a different procedure could be followed, which has a surprising side effect.

Consider formulas of the form $(\forall x)A(x)$ and $(\exists x)A(x)$, where $A(x)$ itself does not contain any quantifiers and x stands for a series of variable x_1, x_2, $\ldots, x_n$. How to deal with more complex formulas follows in a rather straightforward fashion from this ground case. The procedure is quite simple:

> (a) If $A(x)$ is the case for all cases listed in "Things we know for sure," accept $(\forall x)A(x)$, if not reject $(\forall x)A(x)$.
> (b) If $A(x)$ is the case for a case listed in "Things we know for sure," accept $(\exists x)A(x)$, if not reject $(\exists x)A(x)$.
> (c) Every time the list is extended, repeat procedures (a) and (b).

This procedure has some quite interesting consequences:

> (a) As the list grows longer, some statements will change status, some will not. The commutativity of addition or multiplication will obviously remain true right from the start. But for existential statements, the situation can change at every moment. If the number n that makes a statement $A(x)$ true (for a single variable x) is not yet on the list, the statement $(\exists x)A(x)$ will be rejected, until that number turns up, and then it will be accepted.
> (b) As stated, the underlying logic here cannot be classical logic because statements can change truth values along the line. Rather, we are dealing here with a nonmonotonic, default logic, or more generally with an adaptive logic in the sense of the Ghent logic group.[7] This is in itself an interesting result because if the question is whether mathematics has to be necessarily monotonic, then, if this example holds, the answer is no.
> (c) Some statements will remain true or false all the way through the

procedure and only at the end of time do they change truth value. A fine example is this. If we assume, which is quite reasonable to do, that in due time the mathematicians will have the idea of division and also the notion of a prime number and the associated predicate $Pr(x)$, expressing that x is a prime number, then at every step of the procedure there will obviously be a largest prime number. So the statement $(\exists x)(\forall y)(Pr(x)\ \&\ (y > x \supset \sim Pr(y))$ will be true at all stages of the procedure but false in the limit. What this means is that at every stage of the procedure we have a mathematical theory that is inconsistent with the classical theory, yet in the limit they are the same. This is a rather surprising result.

This approach is not restricted to arithmetic but can be easily generalized. Take any classical mathematical theory T. As the set of sentences is countable, it is always possible in principle to write down a list of all sentences. The only tricky part is to define a relation of neighborhood that can be connected to either empirical data or concrete practices (with justifications, if necessary). On the other hand, given a theory T, if it has models, then surely it must have local models and such models are easy to link to certain practices. In addition, from a formal point of view, one could argue that any mathematical theory can be coded into arithmetic and it is clear that for addition and multiplication such a connection can be established. Of course, the weak point of this argument is that, according to the code used, the corresponding arithmetical statements might be difficult if not impossible to connect to a practice.

In conclusion, this perhaps somewhat bizarre example shows that it is absolutely not necessary to have a notion of proof (as weak or as strong as one would like to have it) to arrive at a set of true arithmetical statements that corresponds neatly, that is, perfectly with what holds in classical arithmetic. What happens here is that local bits and pieces are glued together as time goes on and the only requirements are (a) some local knowledge of local bits and pieces (but this can be learned quite easily in an empirical fashion) and (b) a "willingness" to change one's views whenever necessary.

I end this chapter with a comment and a question. The comment concerns Bloor's examples compared to ours. One might remark that perhaps my case studies are indeed stronger than Bloor's, but, as noted in the introduction, at least what he presents is as close as possible to the actual historical development of mathematics, whereas my examples seem to be nothing but logical possibilities and thus far removed from mathematics as we know it. I cannot deny that for the complex-numbers alternative the "distance" from the histor-ical case is indeed large but not so large that this alternative should be labeled a "mere" logical possibility. The strongest argument in favor of that claim is that it is a geometrical theory and at least fits in perfectly with the time frame. As to the proof-free arithmetic, I would claim that this example too is not a 'mere' logical possibility because I think it fits in well within the ethno-mathematical view. Few cultures produced a proof concept as we know it, so "proof-free" mathematics seems to be the norm rather than the exception.

Finally, the question I would like to raise is quite simply this: Did I or did I not succeed in showing that mathematics is indeed contingent? If I take one of the definitions in Soler (2008b, 233), adapted to mathematics instead of physics as in her example, then I think the answer has to be no (at least for the time being). For Soler's definition stipulates the following (and one should keep in mind that this definition has been put forward to address the question of what a contingentist has to do to convince the inevitabilist, a concern that I share and that is at the very heart of this chapter): a theory T is contingent if an alternative theory T' can be developed such that:

1. more or less the same initial conditions obtain as those which have occurred in the history of our own science,
2. nevertheless, the possibility, as 'final' (subsequent or later) condi-tions, at least in the long run, of an alternative mathematics
   2.1. as successful and progressive as ours, and
   2.2. that yields results irreducibly different from ours (notably that involves an ontology incompatible with ours).

Requirement 1 is not really satisfied for the case of the complex numbers (although I have tried to defend its historical relevance in the section on complex numbers) and not satisfied at all in the case of proofless mathemat-ics, although of course it does exclude all the far-fetched scenarios involv-ing twin earth and the like so this could be a matter of debate. Subclause 2.1. is, I believe, satisfied in both cases, but subclause 2.2, not at all. This means that I cannot claim to have shown that mathematics is contingent in this sense on the basis of a maybe (1), a yes (2.1.), and a no (2.2). However, I do believe that the case for mathematics is a special one compared to the empirical sciences, and that I have shown here that genuine alternatives do exist, and this is the first and major ingredient needed to achieve the final goal, namely, a defense of the full contingency thesis.

CHAPTER 10

Freedom of Framework

JEAN-MICHEL SALANSKIS

A s I finally put it in explicit words in the last section of this chapter, I cannot answer the question about inevitability or contingency of our scientific results, at least if I stick to my philosophical options. Still, I wish to take the issue seriously, so I have adopted the following plan:

1. In the first part of the chapter, The Case of Mathematics, I have honestly tried to analyze how and why some people consider mathematical progress as inevitable, and what kinds of objections suggest on the contrary that each significant step is contingent: I play the game of the contingentist/inevitabilist (C/I) debate on the basis of mathematics. My main result in this part is that it brings to the fore the function of framework. I begin the section by commenting on some past discussions of the philosophy of mathematics (involving Jean Cavaillès and his reception), and then I outline some possible arguments favoring contingency of the emergence of Lebesgue integration theory (this belongs to the philosophy of mathematics, although it requires some history of mathematics). Later I focus on the contingency of mathematical frameworks (nonstandard frameworks, constructive frameworks), which can in some sense be inherited from results proven inside such frameworks (and this again

appears to belong to the philosophy of mathematics). In this part, I try to accept the inevitability/contingency issue—at least I do not criticize it.

2. In the last two parts (Transcendentalism and Physics and Back to the C/I Debate), I explain why I cannot accept the issue. Transcendentalism and Physics shows that the framework function underlined in the mathematical context also operates in physics and appears as the essential resource of renewal and change in theoretical physics. Back to the C/I Debate shows that we have to take change of framework as an expression of our freedom, but that we would not be keeping our philosophical line in labeling it as contingent (thus my title).

The Case of Mathematics

First I choose a historical approach. I comment on Cavaillès's inevitabilism, as he formulated it in a talk he gave in 1939, and on how it was received.

Cavaillès's Thesis and Its Reception

Jean Cavaillès was an important voice in French philosophy of the twentieth century, not only as a contributor to the field of philosophy of mathematics (Worms 2009). He happens to have sustained a rather surprising and strong thesis about the necessity of each progress in mathematical science. This sounds like a case of inevitabilism, but we should look at it more carefully.

Useful for that purpose is a reading of the paper titled "La pensée mathématique," a document resulting from a special session of the Société française de philosophie to which Cavaillès was invited with his colleague Albert Lautman so that both could reveal their ideas concerning the philosophy of mathematics (Cavaillès 1994, 593–630). The session took place in February 1939. Importantly, some famous mathematicians attended: Élie Cartan, the great name in differential geometry and father of the recently deceased Henri Cartan, founding member of the Bourbaki team, whose mathematical contribution is too vast to be summarized here; Paul Levy, who did important work in probability theory and stochastic mechanics; Maurice René Fréchet, who contributed much to point set topology; and Charles Ehresmann, well-known as a promoter of category theory, who also pioneered in differential topology.

Cavaillès's thesis concerns what he calls "mathematical becoming," and may be articulated in three points:

1. Mathematical becoming is autonomous.
2. Mathematical becoming is necessary.
3. Mathematical becoming is unpredictable.

The first two points are connected in Cavaillès's mind: we recognize the autonomy of mathematical becoming when we understand that it arose from mathematics' local and own necessity. But this autonomy is conceived of as a source of the alleged necessity. Cavaillès also writes "Autonomy, therefore necessity" (Cavaillès 1994, 601). Here Cavaillès is thinking in a Spinozian way: autonomy, for some thing, means obeying its own and proper necessity. That mathematical development is not triggered by external influences is manifested in the operation of its internal necessity.

This necessity itself is described as the necessity belonging to some kind of problem solving, and illustrated in the case of the invention of set theory. According to Cavaillès: "but, precisely in the development of this theory which might look like a perfect example of inspired theory, composed of a series of radically unpredictable interventions, I thought that I could perceive an internal necessity: the essential ideas arose from certain problems in analysis that gave rise to methods already foreseen by Bolzano or Lejeune-Dirichlet and that became the fundamental techniques established by Cantor" (Cavaillès 1994, 600–601).[1] It is important to concede, here, that Cavaillès acknowledges the impossibility of being certain of that necessity: "It is clear that the word 'necessity' cannot be specified here in any other way. One notices problems, and one sees that these problems required some new idea to arise; that's the best one can do, and to be sure it is too easy for us to use the word 'require,' because we are already on the other side, we can see what has been achieved. Nonetheless, we can say that the ideas that appeared really brought a solution to problems that were actually faced" (Cavaillès 1994, 600).

Cavaillès asserts that the new notions provide solutions for problems that had been stated, or at least faced. But what he cannot give evidence for is precisely the inevitability of such notions for solving such problems, as he recognizes. He says that these notions were "required," but he confesses that using the verb "require" is too easy for someone coming after the battle.

The same hesitation or the same difficulty shows up in the third point of Cavaillès's conception: mathematical becoming is unpredictable. Cavaillès wants to think it necessary, but at the same time he also wants to think it unpredictable. As I understand it, this means that no method or algorithm leading from the problems to the notions can be defined. The new notions are not computed from the problems. They are not necessary as the automatic output of some objective process that we would be in a position to explicate. They are more like clothes, which appear to be "made for" the person who is wearing them. It is because they appear to be so perfectly suited that one is tempted to take them as necessary. In the French philosophical legacy of Cavaillès's work, his necessity thesis joins the political debate about freedom and history. It is well-known that Cavaillès fought in the French Resistance, and that he was even caught and shot by the Nazis. Georges Canguilhem, in a small text paying homage to Cavaillès (cf. Cavaillès 1994, 674), used Cavaillès's name and story to argue in favor of what Foucault after him called "philosophy of concept, knowledge and rationality," contrasting it with "philosophy of subject, sense and experience," in which everybody recognized French phenomenology (cf. Foucault 1985). Canguilhem's main point is that in order to make the right ethical and political choice in France between 1939 and 1945, you did not have to follow a philosophy underlining contingency of human choice (Canguilhem is thinking above all of Sartre's existentialism, a typical example of "philosophy of subject, sense and experience"). Ironically, Canguilhem declares: "Let the supporters of phenomenology and existentialism do better than Cavaillès did, next time, if they can!" (Cavaillès 1994, 678). But more important in connection with our discussion, Canguilhem describes Cavaillès's commitment as being that of "resistance by logic" (the exact quote in French says that Cavaillès was "résistant par logique" [Cavaillès 1994, 677]): it is as if the fight against Nazism had emerged as the necessary behavior in view of the political problem, as did set theory for Cantor.

What could support Canguilhem's reading is the fact that Cavaillès's conception of necessity in the field of philosophy of mathematics was in no way a lazy one. He never meant to minimize Cantor's contribution, for example. The necessity of mathematical becoming, for him, arises from mathematical practice and mathematical experience (he identifies both, by the way). Such necessity requires our contribution: what happens takes place through our acting and behaving, even if this acting and behaving is required by some conceptual or logical features that build the situation's structure.

I now comment on Paul Levy's reaction to Cavaillès's talk. Paul Levy says that he disagrees with Cavaillès and argues that mathematical results are inevitable. But does Cavaillès not precisely say that they are? Levy has been more impressed by Cavaillès's third point, asserting the unpredictability of mathematical becoming. For him, unpredictability means contingency, or at least that the results were evitable. It is not very clear how Levy's intervention dialogically connects with Cavaillès's talk, but it is interesting in its own right for us. I will quote the way Levy argues that mathematical results are inevitable:

> Of course it would have been impossible to predict that a given theorem had to appear at a given point in history, but internal necessities still play a very big role, and there are some theorems about which I can say: if such and such a scholar had not found such and such a theory at such and such a time, and if such and such theorem had not been proved in such and such a year, it would have been discovered in the five or ten years that followed. As proof of this, a great number of theorems have been discovered separately by different scholars within a short time, because they were responding to a necessity in the development of mathematical thought at that time. (Cavaillès 1994, 612)

And Levy even gives what he sees as a very good example, with the theory of Lebesgue integration:

> It is to Mr. Lebesgue that we owe the definitive form of the idea of the integral, and, as you all know, it is now an essential tool for mathematicians. It is so indispensable that, without any doubt, even if Mr. Lebesgue had not lived, his integral would by now have been discovered a long time ago. In saying this, I don't mean to diminish Mr. Lebesgue in any way. On the contrary, I believe that I only strengthen his reputation by declaring that he brought to light an idea that was necessary for the further development of science. Would Mr. Emile Borel, who was already working in this field, have finalized this theory? Or would another of his students have done it? This I don't know. But, after the work of Jordan and Borel, in view of the level reached by humanity as a whole and the number of specialist researchers in mathematics, I believe it was necessary and

inevitable that, within ten or fifteen years, the Lebesgue integration theory would have been established. And in this respect I believe that, to a certain extent, the development of mathematics may be predicted. (Cavaillès 1994, 612–13)

Well, to begin with, Cavaillès would easily object to the reasoning of the last two sentences: Levy argues on the basis of the given Lebesgue theory. Knowing the terminal point (Lebesgue theory), he feels certain that it would have been invented by someone. But Cavaillès does not disagree with that: he simply says that we only know after the battle what it was necessary to invent, as Levy says in the quotation. That such an invention may appear necessary in retrospect does not make it predictable.

But let us come to the point itself. How convincing are Levy's words in favor of inevitability? I think he is making two arguments: (1) that in many cases the same theorem has been proved by several mathematicians around the world in the same years; and (2) that some results appear more or less summoned by the state of the art at some moment.

The first argument looks like a good one. Still we can remark that Levy does not go so far as claiming that "simultaneous discovery" is always the case. Ultimately, the argument does not seem to me absolutely conclusive: it remains conceivable that a plurality of mathematicians in the same period took the same avoidable step. The "plural move" only excludes that, for some psychological or cultural reasons (related to the discoverer as an individual), the step was not avoided. Here, I am tempted to quote the case of Gustave Choquet who, in some text for which I can no longer find the reference, affirms that he "nearly" arrived at the concept of distribution. As is well known, it was in fact his fellow student Laurent Schwartz who invented the idea. Choquet and Schwartz both entered the École normale supérieure in 1934; Choquet was ranked first at the *agrégation* in 1937 and Schwartz second.[2] A kind of friendly rivalry marked their whole careers and paths. Schwartz was awarded the Fields Medal for inventing and developing the distribution notion. Choquet tells us that he stopped just on the threshold of this major discovery and seems to regret that he did not dare to take one more step. Levy would read this story as proof that if Schwartz had not done the job, certainly Choquet would have. But could we not also understand that Schwartz, during his entire life as a researcher having been similar to Choquet, could very well have missed the theory of distribution as Choquet did?

I think the important argument is very clearly the second one, because it formulates in crude and direct scientist's language something similar to Cavaillès's necessity thesis. Only Levy refers to a less "framework-like" result: not to the invention of set theory but to the invention by Lebesgue of Lebesgue integration.

I will try to jump to my point. As Levy tells the story, we have the feeling that a finite number of relevant attempts were to be made, a finite number of given searchers, and even a definable finite range of time (fifteen years): he seems to mean that the arithmetical product of the number of searchers by the number of attempts that each one had time to carry on in the defined temporal window was greater than the number of relevant attempts, so that the "four-leaf clover" was to be discovered. Maybe we should add that researchers communicate, so that they avoid wasting time in duplicating the same attempts: this increases the number of attempts they are collectively capable of in some defined period. I do not attribute such reasoning only to Levy: I think the inevitability thesis very often rests on the assumption of reading the situation according to such a finite reduction scheme.

But it is not so clear, at least to me, that in any interesting case, the situation really fits the scheme. Levy was a far better mathematician than I can ever dream of being, but I feel compelled to say why I am surprised by the way he describes the case of Lebesgue integration.

I immediately have the feeling that indeed there is much in the Lebesgue integration theory—as I have learned it—that *does not* look like a predictable move inside some combinatory repertoire. I have in mind the content of a typical chapter in our textbooks:

- —It does not seem in any way straightforward to think of defining some integral on the basis of cutting into pieces the range of functions rather than their domain.
- —Introducing the concepts of σ-algebra and measure appears as a great new idea.
- —And there is still, on the way toward defining integrable functions and their integral, the move of managing to extend the obvious definition given for step maps to some large enough class (which can be done, from what I remember, in at least two ways, either working first with bounded positive or negative measurable functions or thinking directly in terms of the L_1-norm and completion).

Well, maybe it is possible to describe all these steps in such a way that the associated thought path appears necessary, each conceptual move being construable as the only relevant one: but at first sight it is far from being so. In other words, what this important scientific contribution is made of does not at all resemble the output of one among a finite set of predescribable moves. From what I know of mathematics, its advances and its history, the situation is similar in all interesting cases.

Continuing with Levy's example of integration, more must be said. First, we can at least open a discussion of whether Lebesgue should be integrally credited with the classical chapter of analysis that we now read, learn, or teach. According to Raymond Jean, Lebesgue defined his integral through a formula very reminiscent of Riemann, and the main thing he did was to prove on this basis the dominated convergence theorem (Jean 1975). According to Alain Michel on the contrary, at least if I read him well, Lebesgue's work introduced measure theory and the new integral in the same manner and spirit as we now do (or, rather, used to do when I was a student!) (Michel 1992). According to Jean again, even today we encounter diverse conceptions of the theory: some see integration as the key notion of measure theory, whereas some consider measure theory only an auxiliary topic inside integration theory. According to Jean: "The theory of measurement was sometimes regarded as a secondary topic, and sometimes as a more freestanding topic in integration theory. Stone, Riesz-Nagy, and Bourbaki epitomize the first trend, whereas Carathéodory and Halmos regard integration as the most important concept in the theory of measurement" (Jean 1975, 90). On the one hand, this seems to relativize my spontaneous contention that measure theory was part of the invention of the Lebesgue integral. But on the other hand, it points to an important issue that should be raised, which concerns the very identity of the result. For example, what do we call Lebesgue's result?

Moreover, as in all interesting cases about which I have ever obtained information, the historical event of obtaining the result is quite difficult to locate and capture. Historians of mathematics dispute when and by whom calculus was invented (the choices include Cavalieri, Newton, Leibniz, or perhaps even Cauchy or Weierstrass), who really deserves to be considered the founding father of probability theory (possible choices include Pascal and Komogoroff), or of differential manifolds (e.g., Riemann, Veblen?), and so on. It always happens that the big notions and their best exposition only arise through some unclear progressive story, where one can detect a posteriori an apparently contingent disguise of the "goal," the language, the content.

Beyond the historical problematic birth of the result, we have the technical debate about its content, which is conducted by later mathematics. Once one knows Lebesgue integral as it is exposed in the textbook currently in use, it still may be asked what the Lebesgue integral is really used for, what it allows and what it gives, in an attempt to grasp the core of the true and deep result behind the term "Lebesgue integral." I outline this debate to make my point clear. Some people say that the whole point of the Lebesgue integral is that we get the associated Hilbert spaces $L_p(X, R)$ and $L_p(X, C)$, where X is a measurable space. Some people say that the whole point of the Lebesgue integral is that we get the dominated convergence theorem, which in turn makes it easier to prove that some integrals classically called "improper" exist. Some people could also say that the Riesz theorem shows in a different way the essence of the Lebesgue integral, because it proves that to have a positive bounded functional on the space of continuous functions with compact support on some locally compact topological space and to have a Lebesgue integral associated with some measure on the Borelian subsets of this locally compact space are the same or equivalent.

Whether my hypotheses are the relevant ones or not, I do not know: in any case, I am sure that someone with greater mathematical knowledge would be able to formulate some accurate readings of what is important in the Lebesgue integral from the point of view of contemporary expertise. I mean that each of these considerations sets anew the problem of inevitability: we can ask whether it was possible to get the Hilbert spaces in some significantly different way or to get the dominated convergence theorem differently; or we can ask how much of what is made with the Lebesgue integral could be made directly with some positive bounded functional, by a kind of short circuit. Questions of this kind sometimes receive a mathematical answer, eventually to the effect that our theory appears indeed inevitable for such and such a consequence; in other cases, it leads to the exhibition of alternatives. The question of the identity of the result is a real one. It is not only that we have to distinguish robust and nonrobust results (in the field of mathematics, every result is in principle robust) but also that mathematical knowledge consists partly in always questioning what was really reached in the result.

Therefore, we understand that the part of the formula defining the inevitabilist position that states "on the same subject matter" cannot be handled easily in the case of mathematics. For someone who has physics in mind, or more generally any empirical science, "on the same subject matter" is naturally understood as a way of referring to some region of reality. It is argued that any science covering the same region of reality would in some way overlap with our actual accepted scientific knowledge. But in the case of mathematics, this "on the same subject matter" becomes much more problematic, because nothing else seems to exist but the theoretical meaning (most of the time, what mathematical "realists" themselves identify as mathematical reality is precisely this theoretical content). Therefore, when we decide on mathematical grounds that the result was not what we first considered as the result, do we have another science of the same subject matter or science of another subject matter? In order to make the situation more analogous with that of empirical sciences, we should be able to distinguish some part within mathematics playing the role of reality and some part playing the role of theoretical settings designed in order to cope with this reality. This leads us to the concept of mathematical framework, or more generally the "framework function" in mathematics, as we will see below.

Nonstandard or Classical?

I think a good starting point for our reflection and discussion is given in the following quotation from Abraham Robinson: "There are many fields in Mathematics where compactness arguments can be replaced by the use of Non-standard Analysis. . . . It is a matter of taste whether we wish to regard our present method as a remote reformulation of such argument or whether we wish to assert rather that compactness arguments (e.g. selection principles) were introduced in Analysis in order to fill a gap due to the historical breakdown of the method of infinitesimals" (Robinson 1979, 185).

Some explanations are necessary here.

Robinson invented a way of reintroducing talk, in contemporary formal set-theoretic mathematics, about infinitesimals as well as about infinitely great numbers. He used model theory to do so: his crucial concept of enlargement, providing the classical structures with the lacking infinitesimals, was based on the Gödel–Mal'cev compactness theorem (if every finite subset of some set of statements can be interpreted in some model, then the whole set also receives a satisfying interpretation). Robinson explains why and how we can solve classical problems by going through some enlargement: it is because the enlargements are elementarily equivalent to the structure they enlarge, which yields the so-called transfer principle.

In his 1966 book *Non-Standard Analysis* (Robinson 1966), Robinson showed how one could give a new exposition of a lot of contemporary analysis in infinitesimal terms: very often, the proofs of classical results using some adapted enlargement are much simpler and more straightforward than the classical ones. But in the paper from which my quotation is extracted, Robinson says something more subtle. He says it just after having gone through a nonstandard proof of a result that, classically, requires some logical compactness argument. And this leads him to his comment: it is as if nonstandard settings (enlargements), originating from the compactness theorem, had retained the strength of this theorem; thus they allow us to prove results that usually involve compactness only with the aid of these infinitesimals that we won thanks to compactness.

Still, the quotation says that neither compactness arguments nor infinitesimal ones are unavoidable: they exempt each other. His judgment seems to highlight a locus of freedom ("a matter of taste") concerning the way of exposing and proving some mathematical results. But then, after all, one can contest this point by remarking that the compactness theorem is in some way unavoidable: we have to use it at least once, in order to show in general the existence of enlargements, and only then are we able to forget about it in diverse situations. Robinson rather thinks, at least in my reading, that infinitesimal practice as a whole enfolds the scientific content that the Gödel–Mal'cev theorem articulates in the model-theoretic framework. He does not seem to take the dependency of his construction of infinitesimals on the compactness theorem as the ultimate truth of the story.

One could say that such Robinsonian insight has been justified by subsequent research. Work has indeed been done to disconnect the "nonstandard effect" from model-theoretic presuppositions.

First we have to mention the alternative point of view expressed by Edward Nelson in 1977, in his notorious "Internal Set Theory" (IST) paper (cf. Nelson 1977). Here, the possibility of infinitesimal talk is settled at the level of set theory: all we have to do is to consider some extension of the theory of Zermelo-Fraenkel with Choice ZFC,[3] equipped with a new one-place predicate *st*, and three axiom schemes governing the use of "external formulas" (formulas involving the new predicate). The new IST keeps all axioms of ZFC for internal formulas and the framework of first-order logic, to the effect that all of classical mathematics still holds. Moreover, IST can be shown to be a conservative extension. Therefore, we again get a degree of freedom: we may as well prove some internal formula inside classical ZFC or using the extended language of IST.

For the record, the statement of IST's conservative character is proved semantically, using some ultrapower, which means that something like the compactness theorem is still involved. However, Nelson offers something very near from a syntactic direct proof, which bestows a by-product that he calls the lexicon: an algorithm that translates external formulas in order to express their "mathematical content" at the level of the classical discourse. After Nelson's 1977 paper, people like Jacques Harthong, Robert Lutz, Pierre Cartier, and Edward Nelson himself worked on designing foundations for infinitesimal discourse logically independent of the compactness theorem, at least understood as a model-theoretic tool. If we take the example of the book *Predicative Arithmetic* (Nelson 1986), we do have an exposition that does not proceed through set theory, choice axiom, and the compactness theorem. For sure, one can always say that in some sense the idea remains there informally in the background. But even granting that, it has to be admitted that we are provided with two unequivalent ways of framing the same subject matter, as Robinson describes in the quotation cite above, no one being more inevitable than the other.

We can make similar observations outside the field of "nonstandard" research. After all, Errett Bishop and Douglas S. Bridges have shown that a lot of classical analysis and functional analysis was likely to be proved in Bishop's constructive framework, which differs from the classical one. So, what we more generally come to recognize is that, in many cases, mathematics itself shows some of its ways as avoidable. What we are considering here is the study of frameworks: part of the job of mathematics is to study alternative frameworks, and thus, making what seemed to depend on some definite framework free from it up to a certain point. This is not a tricky aspect of mathematics, already philosophical and betraying mathematics for relativism: it is in a way the mainstream, one of the biggest orientations of formal and set-theoretic contemporary mathematics. For what does it mean to generalize a result, if not to make some premises on which it was supposed to rest avoidable? And what does it mean to classify such and such a type of structure, if not to study the degree of freedom associated with some configuration, making explicit what can be avoided granted we are in the case of the mentioned configuration (in favor of some other avoidable case)? Contemporary mathematics systematically studies how situations and statements can depend or not depend on such and such premises. The case where it considers alternative frameworks is just the limiting case of a set of encompassing premises.

But as we said, one could ask whether there is in mathematics something not belonging to framework, something that does not participate in the framework function. This mathematical content would be inevitable "behind" all framework variations. The most natural answer would be, I guess: constructive facts and constructive mathematics. As a matter of fact, the results of constructive arithmetic—to begin with, the famous $7 + 5 = 12$—appear to be absolutely inevitable in whichever way mathematical research is going to configure the landscape of mathematical results. They are so, arguably, not based on some inexorable decree of nature but, rather, on belonging to a very intimate core of rationality, involving our use of symbolic systems and the structure of our grasping of multiplicities at the same time. Probably one can say that dealing with the subject matter of constructive arithmetic is inevitable (not only for the mathematician but also for the rational subject), and that addressing this topic cannot but lead to our constructive arithmetical knowledge.

We would then be tempted to reason in the following way: the contingent and avoidable part of mathematics is limited to the framework function. We have only alternative and equivalent ways or pictures in mathematics insofar as we are working at framing constructive facts.

But this conclusion does not look plausible, first of all because framing cannot be isolated inside mathematics in any easy and convincing way. Framing not only happens at the very general level of framing mathematical objectivity, as in our examples of set theory, nonstandard theories, or constructive theories. We also encounter ways of framing intermediate results, standing in the middle of the sea of mathematical theoretic content, as in the case of integration theory, which we discussed before, discovering that there was more than one way of framing what deserved to be considered as the main content delivered by Lebesgue integration theory.

But we now give another example, showing that the evitable character of the framework can be inherited by some specific result that appears to be in some sense contingent on the framework. I will refer here to the rather recent story of the theory of "ducks" (cf. Shubin and Zvonkin 1984).

Allow me to quickly tell the story. In the field of dynamical systems, one knows well the so-called Andronov or Andronov–Hopf theorem (or, in some other contexts, Andronov–Poincaré–Hopf theorem), which describes what is usually called "Hopf bifurcation": some dynamical system $X' = f_a(X)$ depends on a parameter a, and shows some fixed point M_a when a is below some crucial value a_0; the differential of f_a in M_a transforms in an important way while a moves from $a < a_0$ to $a > a_0$: two proper values in the complex plane that had a negative real part cross the imaginary axis and reach the $R(z) > 0$ part of the plane. The theorem states that in such circumstances, which of course have to be made a little more precise, the stable fixed point that the dynamical system enjoys in M_a for $a < a_0$ disappears for $a > a_0$ and in a way is "replaced" by a limit cycle (existing for $a > a_0$ and near enough to a_0); furthermore, this limit cycle's metric diameter is equivalent to $\sqrt{a - a_0}$ when a approaches a_0). So Andronov–Hopf's theorem stages a local transforming event of the system around the a_0 value of the parameter, called a Hopf bifurcation.

Nonstandard theorists have applied this theorem in the case of a dynamical system with two parameters a and ε, such that there is a Hopf bifurcation, for small enough ε at least, when a crosses the value 1. In any nonstandard framework, we are allowed to consider values of a infinitely near 1 and at the same time values of ε infinitely near 0 (while nonzero). For such ε, general Hopf theory applies, but it appears that the diameter of the cycle will not be for any choice of a infinitely near 1 what one could expect in view of the theorem (some value infinitely near $\sqrt{a - 1}$). Limit cycles of the size predicted by Hopf theory arise only in the end, when a gets sufficiently infinitely near 1, but we first see some other surprising forms and sizes of the limit cycle, called "ducks" by the searchers: they can very well explain and describe these "ducks" when studying the dynamical system for ε infinitely small. By making $a_0 = 1$ in the previous explanation, I was implicitly referring to the case of the Van der Pol differential equation (just in this paragraph). For more explanations of ducks' theory, see Salanskis (1999, 201–22), which includes references to the original papers.

My point concerns these ducks and their theory (which tells under what conditions they exist and what properties they have). In the first approximation, the whole package constitutes a result, which seems to depend on the nonstandard framework. It is, as a matter of fact, translatable in the classical context (as Nelson's general considerations allow us to anticipate), but the result becomes awkward: it has to be stated in terms of two convergence processes (toward 0 for ε and toward 1 for a), of which we have to compare the speed. Not only does the result become heavy and unpleasant for mathematical thought, it also stops appearing as a truly local result (only Andronov–Hopf's theorem is). We owe to the nonstandard framework the ability to see as local features everything that happens for values of ε infinitely near 0 and values of a infinitely near 1 (values that do not exist in the standard framework).

Thus, with the theory of "ducks," we have the case of a result that does not survive the framework change, at least in some sense. If we stick to the classical ZFC framework, we lose ducks' theory, either because we cannot state it anymore as a local result (it stops being mathematically interesting), or because, as a translated result (e.g., through Nelson's lexicon), it is awkward. Of course, in another sense, the result is quite robust: it counts once and for all as a deductive consequence of IST's axioms, for example; and we may even find a deducible statement in ZFC that translates it. But this does not maintain the "result" in some important sense of *result*: in the sense of what belongs to vivid knowledge, what gets reasserted, reproved, contemplated again and anew, communicated, used, and so on.

Again, I shall try to resume the section The Case of Mathematics. As I stated earlier, I wanted to take seriously the C/I issue for mathematics, to treat it as a sound acceptable problem. I have based my examination on two historical claims: Cavaillès's claim, which appears to support the inevitability of mathematical advances, and Robinson's claim, which instead appears to assert contingency of the choice of some language.

The discussion of these two claims shows:

> 1. On the one hand that Cavaillès's thesis, though metaphysically inspired (arising from some Spinozist mood), still rests on an epistemological conception of mathematics as problem solving, and Cavaillès confesses the inevitability he asserts to be unprovable; also mathematicians who adhere to the inevitability thesis seem to believe that research always faces some finitely reduced situation. In any case, what deserves to be called the result is part of what mathematics investigates, and what was or was not avoidable for our result changes when we interpret the result as such and such. Ultimately, the issue of inevitability or contingency, taken in this latter perspective, is a mathematical one rather than a philosophical one.
> 2. On the other hand, Robinson's contingency claim seems to be about the frame inside of which we work, rather than about our results. But are frames not part of what mathematics exhibits as out-

comes? This leads us back to the issue of the exceptional character of constructive mathematics. If one still takes for granted that our frameworks are always contingent, we have shown that certain results can be in some sense contingent on the chosen framework, which deepens the impact of the alleged contingency.

The most important point in this section is our emphasis on the framework function as an essential part of what mathematics is about and what it results in. This function manifests itself in various guises and calls instead for assessment as contingent. Still we do not know exactly what this contingency means or what kind of contingency it is: the meaning of that contingency seems to be in the last instance always mathematical (mathematics reveals all the time that we could have framed something differently, at any possible level).

The question is now to determine whether our case study can help us in tackling the general problem, or, in a more limited way, the C/I issue for physics.

Transcendentalism and Physics

One can look at the inevitability issue in a quite different way: one can think of inevitability as a subjective notion. What has been called transcendental in the continental tradition was defined as in some sense inevitable: when I think of external phenomena appearing to me, I cannot but put them in some framework, which I call space. I cannot avoid that move: thinking the framework is inevitable for me (which means, inevitable for anyone). When I gather representations in some unifying representation, I cannot but do it using one of the twelve forms enumerated in Kant's *Critique of Pure Reason*. So science is committed to something inevitable, but not of the "result" kind: the inevitable in science, according to the Kantian view, is its form. To be honest, there is one case in which the transcendental is also form and content: constructive mathematics, as we have outlined in previous paragraphs, has this character of "what we cannot but think, act, and consider as such": we can term it a transcendental body of knowledge as well as a transcendental form of knowledge. This inevitabilism is quite different from the one we have been examining until now: instead of objective inevitablism concerning the content, we have subjective inevitabilism concerning the form.

The big antitranscendentalist move undergone by international epistemology since the work of the Vienna Circle was in a way anti-inevitabilist: it was argued that nothing could be considered inescapable at the subjective level and that anything belonging to scientific knowledge was revisable under the pressure of experience. Therefore, nothing like a priori knowledge could exist. Contemporary scientific realism may have a preference for objective inevitabilism about the content, but it very much dislikes transcendental inevitabilism. And, at least in my youth, people often chose scientific analytical realism because they wanted to preserve science's freedom: they wanted to forbid Kant from imprisoning physics in the jail of Euclidian geometry. But does the present discussion about inevitabilism not show that analytical realism is also able to imprison? The new jail is the jail of necessity, arising from inexorable nature: a good jail, in the inevitabilist epistemologist's eyes.

I think it is important to understand that, after all, the transcendental conception does not have to be understood as restricting science's freedom. We could argue this roughly along the following lines, where we try to show that scientific audacities and novelties find their place within a Kantian subjective inevitable form.

This subjective inevitable form, indeed, takes science as always accommodating logical discourse with sense data under the conditions of some framework, receiving at each step of history some mathematical interpretation. On the one hand, we have the feeling that such a game is simply what we call science or at least exact science, and on the other hand, previous description fits the tradition of physics. For sure, our epistemological experience is that, whereas this picture looks like a good one, it is very difficult to identify the inevitable skeleton. Something like a "logic of experience" seems to be given, but it is dubious that we can identify it in a definitive way with the twelve forms of the table of judgment or the twelve concepts of the table of categories, or with first-order logic, to refer to a more recent enthusiasm. We may acknowledge a persistent "framework function" in physics, but we cannot associate it with some fixed mathematical structure, be it pseudo-Riemannian manifolds or Hilbert spaces. The "inevitable form" that Kant was exhibiting in his transcendental setting reveals itself epistemologically through its various content assignments. There is something common to classical Euclidian spaces, pseudo-Riemannian manifolds, and Hilbert spaces, which corresponds to their common position of mathematically interpreting the framework for mathematical objects extracted from experience (basically through measurement). But even if we try to pinpoint it by remarking that in all cases we have a locally complete uniform space or something like that, we are not sure that we have found the ultimate definition of what a mathematical interpretation of the framework has to be. The transcendental, which in Kant's language corresponds to the place where thought feels as if held by something inevitable, is at the same time par excellence the place of revision, change, and revolution. And the inevitable comes only to be guessed, felt, and anticipated in some way along the path of such revolutions: what is inevitable is, at least, that there has to be some imposed structure corresponding to the framework function, a structure that experience does not spell out. But again, we assert such an inevitability only from the subjective point of view: not in the name of some science saying what physics cannot but be, or in the name of some metaphysics spelling out the absolute law of knowledge.

To tell what precedes in another way, my point is that Kant's concept of pure intuition was a way of designating the framework function as it works in physics. That physics goes through pure intuition or remains ordained to pure intuition means, in Kant, that the theory of nature occurs as a mathematical theory making use of a mathematical structure playing the role of framework for data and determining the way that physics, at any specific time or in any specific case, conceives of things and change. Pure intuition does not designate any fixed certainty or Bible, but a place or a function, which is part of the physical setting since Galileo. This is the place of framework because it corresponds to the (subjectively and traditionally) inescapable experience that sense data have to unfold in some mathematical space: physics imagines the world, or imagines the real by reading it through a framework for phenomena, which technically will be a framework for phenomena translated into mathematical objects by measurement. We understand physics as a discipline that assumes the game of reconstructing the world by mathematically imagining its presentation: this settles some inevitable form for us—the one Kant exhibited—but this form is a very liberal one, as history shows.

The other very important point is that framework function in physics is strongly connected with framework function in mathematics: here resides the mystery of the deep interrelation between mathematics and physics. Frameworks that physics chooses for its theories are always mathematical frameworks in a sense that I wish to make more precise. For sure, one could argue, here, that Euclidian spaces, Hilbert spaces, and pseudo-

Riemannian manifolds are all objects defined in the mathematical framework of set theory, whereas each counts as "the" framework for some physical theory: the word *framework* seems thus to have two different meanings in the previous sentence. This much is true, but it does not create an impassable barrier. As a matter of fact, each of the aforementioned mathematical items corresponds to some level of axiomatization, introducing some kind of structure, purported to govern some multiplicity. The ZFC axioms govern the largest multiplicity to be considered in mathematics, and they intervene for that reason at the foundational level. But we still speak of the "axioms for Hilbert spaces," even though we do not have in mind some first-order theory, because the properties something has to possess in order to be a Hilbert space generally qualify a type of multiplicity (a type of space). Although this subsequent axiomatization moment is not equivalent to the foundational one, it stills bears the same function of defining and ruling some kind of multiplicity (this base multiplicity being, in general, given with some other objects or multiplicities providing it with structure—as the set of open sets for a topological space). The framework physics needs, in order to mathematically translate the intuition framework as conceived of by Kant, is always a structured multiplicity: a "space," as we say in contemporary mathematics. This means that such a framework brings the same kind of consequence: it changes the world of objects and relations we are able to consider and ultimately alters the kind of problems we are able to meet.

Back to the C/I Debate

If one acknowledges the importance of this framework function in physics, then one is tempted to connect it with the C/I debate.

I think it is impossible to claim that introductions of frameworks are dictated by experience or directly imposed by any kind of data. That is the point Kant was trying to make by associating framework introduction with what he called a priori, naming the level of possibility conditions in general. There is no algorithmic procedure for moving from one framework to another: Einstein had no way to compute his reframing from the Michelson–Morley experience, for example. The level of a priori is connected with a specific mode of revision: the relevant mode for changing the way we frame experience data, the way we mathematically identify them. One could call that revision mode the mathematical imagination of the world.

Cavaillès, for sure, would say that such revisions are necessary, even if we are unable to describe a derivation process for them. On the other hand, someone wanting to emphasize that science is a practice, and thinking that human practice enfolds freedom and contingency, will be prone, in opposition to Cavaillès's approach, to describe framework introduction, in physics as in mathematics, as contingent.

But in my opinion, this would not be better: it would still participate in what I understand here as a naturalist fallacy. Because we cannot decide whether historical moves of science are necessary or contingent without referring to an encompassing theoretical knowledge, including the subject of science as well as the world studied and described by science. We cannot have any hope of proving or disproving that science could have been different without mastering the variety of what can happen and what cannot happen in the "interaction" between scientific practice and world.

There are two classical ways of conceiving this totalizing setting, accounting for science and world—a setting that is both subjective and objective, if one prefers to say it that way.

The first is the scientific one. Here, we know two candidates. Today the most popular one is probably cognitive science. A naturalist account for the relation going on between human knowledge and the world could be recognized as the very definition of cognitive science, or of a cognitive science project. But we also have the other candidate: social science. Considering not only science as social practice but also world as a social construction, social science may claim to be able to account for the adventures of the relation between scientific practice and the world, which it pictures as social historical interaction between practice (as theoretical collective practice) and practice (as collective activity building and shaping shared reality). And, I guess, some positions in the C/I debate are formulated in terms of social science, or under the authority of social science.

Beyond these two contemporary scientific candidates theorizing the science/world relation, we have a very old discipline, which always claimed to have competence for dealing with that relation: metaphysics. Indeed, the second way we had in mind is the metaphysical one. Metaphysics offers general theories concerning every being, without relying on any specific hypothesis about how we access beings. It is quite possible to take a position about the C/I debate by relying on some metaphysical insight of this kind. For example, in one way or another one can reactivate Heraclitean metaphysics, according to which nothing exists but movement and change.

If this perspective is taken, then the relation between human subjective knowledge and the world it attempts to decipher is exposed to perpetual and unpredictable distortion. Everything at the level of that relation is so unstable and capable of bifurcation that we could never affirm any necessity: science could have followed an alternative path concerning some reality region, because the coevolution of objects of that region and of human investigation could have been perturbated (as we say in the theory of dynamical systems) in innumerable ways, at many levels.

Here I want to insist on my Kantian approach. I follow Kant in his rejection of dogmatic metaphysics, or better, in the limitation he assigns to any metaphysics of that sort. For Kant, any theorizing about any being whatsoever cannot bring interesting results: it will only deliver the universally valid forms of logic. When science manages to say more about things, to unravel laws of nature, it is because it considers objects not logically but as related to phenomena that we include/translate in some mathematical structure.

What we would need in order to decide the C/I debate is the knowledge of laws governing the relation of scientific discourse and world, whatever the circumstances of that relation may be. Only in reference to such laws could we evaluate whether the course of science, under such and such circumstances, would have led to results substantially the same or not. The examination of history never gives us any clue of that kind, it shows only what has been the case and not what could have been the case (unless we inadvertently use some laws of history that are then taken as the required laws of the knowledge–world relation).

Kant teaches both that dogmatic metaphysics cannot be anything but trivial and that there is no encompassing knowledge of the knowledge–world relation. What we know always belongs as such to the world, and the knowledge agency always escapes the world (at the juridical or epistemological level, which Kant called the transcendental level). We cannot build any encompassing logic, be it scientific or metaphysical, without forgetting about rationality as ours, as performed under our responsibility, and as having to be justified by us and for us: a facet of rational enterprise that refuses any objectivation.

Then we will not claim that physics could have been radically different about the same subject due to another mathematical imagination of the world, as may look tempting after having denied that the framework gestures were computable from data and situation. Because in order to make such a contingency claim, we would need the same kind of encompassing knowledge that we just recognized as philosophically meaningless (which does not mean that cognitive science or social sciences are impossible—just that they assert necessity or contingency only in a relative manner, from within their scientific framing).

Such refraining is part of the price we have to pay if we want to maintain an attitude of nonnaturalist epistemology. Since the beginning of the chapter we have faced the implicit naturalism of both assertive conclusions connected to the C/I debate, and at the same time, we have found reasons to doubt that inevitability or contingency is really advocated in such an absolute way by such and such in such and such context. Realists are prone to support inevitability in the name of their conception of science, but can they really convince us that the modal operator "it is a scientific result that" entails the metaphysical "it is necessary that"? Philosophers of mathematics and mathematicians may be inclined to think of mathematical advances as inevitable, but only insofar as they focus on an alleged finite reduction situation or because they neglect the internal mathematical reflection about inevitability, which perpetually asks what the result is and whether our ways toward it were inevitable. When one takes into account such a reflection, then one meets the framework function inside mathematics, which seems always to enjoy the freedom of defining differently what mathematical research aims at and what counts as a result for it. As far as we should recognize physics as inheriting the framework function from mathematics, after Kant, we are tempted to diagnose the same kind of freedom playing inside physics—a freedom that plays inside the (subjectively, not metaphysically) inevitable form of mathematical imagination of the world. But we should not conclude that the results of our science are contingent because this would ruin the benefit of having escaped the objective stance, in order to detect the "subjective" facet of science. The crucial point is that we do not need to prove contingency in order to ground freedom. What is important in freedom is that we may consider what we do as ours, that we allow ourselves a subjective picture of our acting. Good modest philosophy, not believing itself science or absolute knowledge, should know that freedom is only a descriptive notion belonging to the subjective perspective and does not have to be more than that. Again, Kant made this point with respect to morality, when he explained that we could not but understand our behavior as morally qualified as free, whereas we have to concede that it was determined in the naturalist perspective.

CHAPTER 11

On the Contingency of What Counts as "Mathematics"

IAN HACKING

Les mathématiques (cour.): ensemble des sciences qui ont pour objet la quantité et l'ordre.

—*Le petit Robert*

Au plur. Les mathématiques. Ensemble des disciplines qui procèdent selon la méthode déductive et qui étudient les propriétés des êtres abstraits comme les nombres, les figures géométriques ainsi que les relations qui existent entre eux.

—*Trésor de la langue française*

mathematics. Now (treated as singular) the abstract deductive science of space, number, quantity and arrangement, including geometry, arithmetic, algebra etc., studied in its own right (more fully *pure mathematics*) or as applied to various branches of physics and other sciences (more fully *applied mathematics*).

—*Shorter Oxford Dictionary*

mathematics. . . . Originally the collective name for geometry, arithmetic, and certain physical sciences (as astronomy and optics) involving geometrical reasoning. In modern use applied (a) in a strict sense, to the abstract science which investigates deductively the conclusions implicit in the elementary conceptions of spatial and numerical relations, and which includes as its main divisions geometry, arithmetic, and algebra; and (b) in a wider sense, so as to include those branches of physical or other research which consist in the application of this abstract science to concrete data. When the word is used in its wider sense, the abstract science is distinguished as *pure mathematics*, and its concrete applications (e.g. in astronomy, various branches of physics, the theory of probabilities) as *applied* or *mixed* mathematics.

—*Oxford English Dictionary*

math-e-mat-ics. 1. a science that deals with the relationship and symbolism of numbers and magnitudes and that includes quantitative operations and the solution of quantitative problems.

—*Merriam Webster's Unabridged Dictionary*

math·e·mat·ics *n.* (used with a sing. verb). The study of the measurement, properties, and relationships of quantities, using numbers and symbols.

—*American Heritage Dictionary*

Für *Mathematik* gibt es keine allgemein anerkannte Definition.

—*German Wikipedia*

The Contingency of "Mathematics," Not of Mathematical Truths

This chapter does not consider whether theorems of mathematics are themselves contingent and could have been otherwise. That is not as unthinkable as some readers will imagine. Perhaps Descartes believed that God could have made two plus two equal to five. John Stuart Mill held that, aside from merely verbal trivialities, mathematics consists of the most universal empirical truths, confirmed in all our actual experience. They are not necessarily true, and so, perhaps, they could have been false. Philip Kitcher (1983) suggested an updated version of Mill's idea. And W. V. Quine, in his famous rejection of the distinction between the analytic and the synthetic, can be read as rejecting the very notion that some significant truths, including mathematical ones, are logically necessary whereas other truths are contingent. I am not concerned with any of these issues. Analyticity, necessity, certainty, and the a priori are quite distinct concepts (for my own explanation of this obvious fact, see Hacking 2000b, §5). The Vienna circle, followed by Quine's animadversions, led a couple of generations of philosophers to blur the distinction. But that does not matter here: None of the questions arising from those concepts matters much in what follows. I am wondering about something prior to most philosophizing: *What makes mathematics mathematics?* Is it something intrinsic to the nature of the subject, or is mathematics defined by a series of historical events, which might have been different?

Inevitability—or Not

Reviel Netz caused me to embark on this inquiry. I have learned a great deal, as every reader must, from his pathbreaking book, *The Shaping of Deduction in Greek Mathematics* (1999). I use it below, but his bearing on these notes is more direct. The first time we met he kindly said he liked what I had written about social construction. There was one thing he could not understand. Hacking (1999, ch. 3) describes three "sticking points" between (social) constructionists on the one hand and most scientists on the other. The first sticking point was over contingency. How on earth, he asked, given my own analysis, could I score myself only 2 on a scale of 1 to 5 in favor of the contingency thesis? This would mean that I thought scientific development and in particular mathematical development almost inevitably took the path it did, would it not? It took me too long to realize that he meant something different from what I did.

The last time I met him (before preparing this chapter) was in the spring of 2008. I talked at his seminar in Stanford and explained how I agreed, and yet disagreed, with Bruno Latour's (2008) review of his book. I had introduced Latour to Netz's work, and he was bowled over by it, as I was, but he drew philosophical conclusions quite different from mine (cf. Hacking 2009b, 104–9). After the seminar, I came to understand that both he and Latour, for quite different reasons, hold that what we call mathematics is not determined by the very nature of the science but by the result of a contingent history. Once a problem is determined—for example, which polyhedra are regular?—it is pretty inevitable, I think, that we shall find first that there are exactly five (the Platonic solids) and later realize we had only convex polyhedra in mind, and so be led to the star polyhedra. (This is the basis of the famous example that runs through Lakatos 1976.)

But on the more general question, I think that Netz and Latour are right. Lots of local inevitability (I say), but overall contingency. Why should this matter? It sounds like a mere matter of a word, the noun "mathematics." Or a question of how to draw disciplinary boundaries. Who cares, other than bureaucrats? My philosophical self. For I was, in my youth, a philosopher of mathematics, a topic I find so difficult that I have published almost nothing about it.

And what are these problems (for me)? In Bertrand Russell's words of 1912: "The question which Kant put at the beginning of his philosophy, namely 'How is pure mathematics possible' is an interesting and difficult one, to which every philosophy which is not purely sceptical must find some answer" (Russell [1912] 1946, 84). We know what troubled him. "We do not know who will be the inhabitants of London a hundred years hence; but we know that any two of them and any other two of them will make four of them. *This apparent power of anticipating facts about things of which we have no experience is certainly surprising*" (85; emphasis added).

We can single out two other problems that have fascinated some but only some philosophers from Plato to the present, namely, mathematics' richness of content and its necessity. How can we get so much out of so little, and why is what we get necessarily true? And then there is the more recent problem, Eugene Wigner's (1960) "unreasonable effectiveness" of mathematics in the natural sciences.[1] These are among the mathematical phenomena that have made mathematics loom so large in the work of some, but only some, figures in the canon of Western philosophy. In my opinion, these problems would simply not have arisen if Western mathematics had not taken the course it did. Had we taken the Chinese path—which emphasized approximations rather than deductive proof—we would have had no Plato and no Kant as we know them. And it hardly needs saying that ancient Chinese philosophy is rich and deep but it lacks a Kant and a Plato. Lucky Chinese, some may say.

Philosophy of Mathematics

Philosophers have thought a good deal about the nature of mathematics—in my case, for example, about the logicist claim that mathematics is logic (Hacking 1979). But we have usually taken "mathematics" for granted and seldom reflected on why we so readily recognize a conjecture, a fact, a proof idea, a piece of reasoning, or a subdiscipline, as mathematical. We ask sophisticated questions about which parts of mathematics are constructive, or about set theory. But we shy away from the naive question, of why so many diverse topics addressed by real-life mathematicians are immediately recognized as "mathematics."

The dictionary definitions with which I began already suggest a worry. The French seem to regard mathematics as a collection, an *ensemble* of sciences or disciplines, whereas in the English language, mathematics seems to be regarded as a single science, divided according to if it is pure or applied. Here, then, there is a hint of contingency: the French look at mathematics in one way, and the English in another. And German Wikipedia throws up its hands!

But aside from such niceties, the definitions seem plain enough. Why not stop right here, and answer our question by quoting one dictionary or several? Because the kinds of things we call mathematics are, in a word, so curiously miscellaneous. The *Oxford English Dictionary* (OED) already implies as much, by its implication that the concept of mathematics itself has a history, with the name applying in different ways to different categories over the course of time.

A Mathematician's Miscellany[2]

The arithmetic that all of us learned when we were children is very different from the proof of Pythagoras's theorem that many of us learned as adolescents. When we began to read Plato, we saw in the *Meno* how to construct a square double the size of a given square and realized that the argument is connected to our folklore knowledge of "Pythagoras" (about whom in historical fact we know almost nothing). But this is totally unlike the rote skill of doubling a small integer at sight, or a large one by pencil. Both types of example are unlike the idea that Fermat had when he wrote down what came to be called his last theorem. We nevertheless seem immediately to understand his question about the integers. Surely, we think, it is the sort of thing that is either true or false, a theorem, or not, of elementary arithmetic. The situation is very different from the proof ideas that lie behind Andrew Wiles's discovery of a way to prove the theorem. One key notion there is the connection between several old but apparently unrelated ideas, elliptic curves, rational numbers, and modularity (the Taniyama–Shimura–Weil conjecture). Few of us have mastered even a sketch of that argument; for most, even the concepts are unfamiliar. Is the proof "the same sort of thing" as the well-known Euclidean proof that there is no greatest prime? I am not at all sure.

The mathematics of theoretical physics will seem a different type of thing again, but we should not restrict ourselves to theory. Papers in experimental physics are rich in mathematical reasoning. The mathematics is mostly old-fashioned, taken from what has been compared to a physicist's toolkit (Krieger 1987). Lagrangians, Hamiltonians, all the right stuff, which has been around for well over a century, and which is still the basic kit of string theory. The mathematics in the toolbox—and the way it is used—is very different from that of the geometer or number theorist.

The mathematical part of the physicist's toolkit is mostly old, but something entirely new has been added. We have powerful computational techniques to make approximate solutions to complex equations that cannot be solved exactly. They enable practitioners to construct simulations that establish intimate relations between theory and experiment. Today, most experimental work in physics is run alongside simulations. Is the simulation of nature by powerful computers (applied) mathematics, in the same way that modeling nature using Lagrangians or Hamiltonians is called applied mathematics?

Economists also construct complicated models. They run computer simulations of gigantic structures they call "the economy" to try to guess what will happen next. The economists are as incapable of understanding the reasoning of the physicist as most physicists are in making sense of modern econometrics. Are they both using mathematics?

We are not really sure whether to say that programmers writing hundreds of meters of code are doing mathematics or not. We need the programmers to design the programs on which we solve, by simulation and approximation, the problems in physics or economics. What part is mathematics and what part not? We do not dignify as mathematics the solving of chess problems, white to mate in three. Few people will call programming a computer to play chess an instance of mathematics. Arithmetic for carpentry or commerce seems very different from the theory of numbers. What, then, makes mathematics mathematics?

Only Wittgenstein Seems to Have Been Troubled

It is curious that Wittgenstein seems to have been the first notable philosopher ever to emphasize the differences between the miscellaneous activities that we file away as mathematics. "I should like to say," he wrote in his *Remarks on the Foundations of Mathematics*, that "mathematics is a MOTLEY of techniques of proof" (Wittgenstein 1978, III-§46, 176). I would happily dedicate a paper to examining this insight, but I shall content myself here by saying that, in the whole history of Western philosophy,

Wittgenstein is the only memorable figure whose concerns about what makes mathematics mathematics have been well preserved.

This is not be confused with a quite different demarcation problem, of whether computer-assisted proofs should count as proofs. That is an interesting topic that is usually misunderstood in folk philosophy of mathematics, that is, in the philosophy of mathematics of philosophers. If we were to argue that what makes mathematics mathematics is the possibility of perspicuous proof, then we would conclude that reasoning augmented by computer is not mathematics, but such a conclusion would, in my opinion, be too hasty.

Three Kinds of Answer

What makes mathematics mathematics? There are three inviting answers. They represent different attitudes, perhaps three different casts of mind:

> 1. Mathematics has a peculiar subject matter, which people versed in the discipline simply recognize.
> 2. Mathematics is a cognitive field ultimately determined by a domain-specific faculty or faculties of the human mind. It is a task of cognitive science and of neurology to investigate the faculty(ies) or "module(s)" in question.
> 3. Mathematics is constituted less by its content than by disciplinary boundaries that have emerged in the course of contingent historical practices.

These three are compatible. It will occur to any unsophisticated person that mathematics (1) obviously has a peculiar subject matter, which is (2) investigated by means of one or more mental faculties. Perhaps, as Kant thought, there is a distinct faculty for arithmetical reasoning, and another for geometrical reasoning. The disciplinary boundaries in our teaching and our professions mark our present grasp of that subject matter, and so (3) need not disagree with (1) and (2).

There seems to remain, in this terminus of temporary goodwill, a chicken-and-egg question about (1) and (2) that tacitly ignores (3). Is the peculiar subject matter of mathematics a consequence of our mental faculties, so that in some curious sense there is no mathematics without the human brain to process it? Or are the mental faculties simply honed to accord with a human-independent body of fact? Here we get two fundamentally opposed attitudes to mathematics, both of which take present disciplinary boundaries as irrelevant. They are very nicely on display in a famous debate between a mathematician and a neurobiologist, both among the most eminent living contributors to their respective fields (Changeux and Connes 1989/1995). Alain Connes insists that mathematics is just *there*, a strong version of mathematical realism. Jean-Pierre Changeux insists with equal vigor that mathematics is a by-product of human mental faculties: no mathematics without the human brain.

A mild form of (3) is no trouble to either of these two protagonists. But both men would dismiss the idea that mathematics has no essence and is the utterly contingent result of fairly recent historical events (a couple of millennia at most). I do not support that view, but believe it should be taken very seriously by Connes, Changeux, or most present philosophers of mathematics.

Kant

An unexpected paragraph comes right at the start of the survey article, "Mathematics, Foundations of," in the *Routledge Encyclopedia of Philosophy*. It follows the assertion that Greek and medieval thinkers "continue to influence foundational thinking to the present day": "During the nineteenth and twentieth centuries, however, the most influential ideas [in the philosophy of mathematics] have been those of Kant. In one way or another and to a greater or lesser extent, the main currents of foundational thinking during this period—the most active and fertile period in the entire history of the subject—are nearly all attempts to reconcile Kant's foundational ideas with various later developments in mathematics and logic" (Detlefsen 1998, 181). Russell's 1912 model began, as we have seen, with Kant, but Kant does not loom quite so large in most other introductions to the subject. He is not even mentioned in Leon Horsten's article "Philosophy of Mathematics" (2007) in the online *Stanford Encyclopedia of Philosophy*.[3]

I accept that there is something absolutely right in Detlefsen's stage-setting. For among Kant's innumerable legacies was the conviction that there is a specific body of knowledge, mathematics, of striking importance to any metaphysics and epistemology. On this occasion I shall hardly touch on those three topics of richness, necessity, and the a priori character of mathematics that have so impressed the minds of some philosophers. They are not why I pick up on Kant. His question, you will recall, was "How is Pure

Mathematics possible?" I focus on the implication to which we seldom pay attention, the conviction that there is a specific body of knowledge, of striking importance, namely, PURE MATHEMATICS. Perhaps Kant helped to lodge that proposition in our heads, so that mathematics is just a given, a domain that makes some philosophers curious. Of course mathematics has mattered to some philosophers all the way back at least to Plato, but, as we shall see, Plato's own demarcation of mathematics is different from our own.

Kant's Vision of the Ur-History of Mathematics

Kant had already published the *Critique of Pure Reason* when he sat back and reflected that *even reason has a history.* That pivotal moment, between the first and second editions of the *Critique*, took place when Europe turned from the timeless reason of the Enlightenment to the historicist world that we still to some extent inhabit. In his new Introduction for the second edition, Kant betrayed a wonderful enthusiasm for a defining moment in the history of human reason (as he saw it). Kant, of all people, had become historicist. He used such purple prose that I quote it in full:

> In the earliest times to which the history of human reason extends, *mathematics*, among that wonderful people, the Greeks, had already entered upon the sure path of science. But it must not be supposed that it was as easy for mathematics as it was for logic—in which reason has to deal with itself alone—to light upon, or rather to construct for itself, that royal road. On the contrary, I believe that it long remained, especially among the Egyptians, in the groping stage, and that the transformation must have been due to a *revolution* brought about by the happy thought of a single man, the experiment which he devised marking out the path upon which the science must enter, and by following which, secure progress throughout all time and in endless expansion is infallibly secured. A new light flashed upon the mind of the first man (be he Thales or some other) who demonstrated the properties of the isosceles triangle. The true method, so he found, was not to inspect what he discerned either in the figure, or in the bare concept of it, and from this, as it were, to read off its properties; but to bring out what was necessarily implied in the concepts that he had himself formed *a priori*, and had put into the figure in the construction by

> which he presented it to himself. If he is to know anything with *a priori* certainty he must not ascribe to the figure anything save what necessarily follows from what he has himself set into it in accordance with his concept. (Kant 1929, 19)

We no longer countenance the hero in history, "be he Thales or some other." We can now turn Kant's prose into something closer to the historical facts, thanks to Reviel Netz (1999). He would prefer Eudoxus to Kant's Thales, but the important point is that there was a moment of radical change in the human mastery of mathematics. Kant got that right, by present lights. Using the metaphor recently favored in paleontology, Netz suggests "that the early history of Greek mathematics was catastrophic"—a sudden change in the very "feel" of mathematical thinking. In a lower key: "A relatively large number of interesting results would have been discovered practically simultaneously" (Netz 1999, 273). Netz suggests a period of at most eighty years. We have no need to dismiss the Babylonians, Egyptians, and others who taught mathematics to the Greeks, in order to see that at the time of "Thales or some other" a revolution in reason was wrought.

I emphasize that what Eudoxus and company did was not only to establish some new mathematical facts, techniques, and proof ideas. They also discovered a new way to find things out, namely, by reasoning and proof. This was not a mathematical discovery but the discovery of human capacities of which our species had, in earlier times, only glimmerings here and there. It was the discovery and then exploitation of what we were later to conceive of as a mental faculty or faculties. The original capacities are now a hot topic in cognitive science and neurobiology.

Research showed over a decade ago, in ways so well described by Dehaene (1997) and Butterworth (1999), that we have an innate "number sense"—perhaps a number module as in Butterworth, or a general processor as in Dehaene. There is also ample evidence of some dim awareness of symmetry. But only in recorded history (albeit sketchily recorded history) did humans discover how to use these sensibilities in ways that foreshadowed mathematics.

Netz's book is widely admired for its reconstruction of the diagrams that are notoriously missing from surviving ancient texts. Few readers attend to his subtitle, *A Study in Cognitive History*, yet that aspect is exactly what is fundamental and wholly original. It is the first detailed analysis of the history of the discovery of a cognitive capacity. In addition, his reflections on the material conditions of ancient Greek counting (Netz 2002) provide a local understanding of how some people, at a certain time in a certain ecological setting, discovered how to make use of their innate number sense.

Why Should Greeks Have Cared?

Why did some Greek thinkers think that the newly discovered capacity for demonstrative proof was so important? (This is different from the question of why future Europeans such as Kant and Bertrand Russell thought that it was important.) Netz, following Geoffrey Lloyd (1990, 209 ff.), suggests an answer. City-states were organized in many ways, but Athens is of central importance. It was a democracy of citizens, all of whom were male and none of whom were slaves. It was a democracy for the few; but within those few, there was no ruler. Argument ruled. If you could make the weaker argument appear the stronger, you won.

Athenians were the most consistently argumentative bunch of self-governors of whom we have any knowledge. We read Aristotle for his logic and not for his rhetoric. Greeks read him for his rhetoric; his logic was strictly for the Academy. The trouble with arguments about how to administer the city and fight its battles is that no arguments are decisive. Or they are decisive only thanks to the skill of the orator or the cupidity of the audience. But there was one kind of argument to which oratory seemed irrelevant. Any citizen, and indeed any young slave who was encouraged to take the time and to think under critical guidance, could follow an argument in geometry. He could come to see for himself, perhaps with a little instruction, that an argument was sound. He could even create the argument, find it out for himself. In geometry, arguments speak for themselves to the inquisitive mind.

Plato, the Kidnapper

Nobody debating military strategy or the tax on corn in the Agora was able to use geometrical proof. So why should Greeks have cared about proof? An answer may be that hardly anyone did. Netz's book is about an epistolary tradition involving a small band of mathematicians exchanging letters around the Mediterranean Sea. They cared about new discoveries and new proofs, but not about the very idea of proof. Enter Plato, the kidnapper.

I take the label from Bruno Latour's (2008) brilliant critical exposition of Netz's book. Latour rightly takes Netz's analysis as a compelling example of knowledge sustained by a network of creators and distributors of that knowledge. Nowhere is that better illustrated than by Archimedes, who, working out of Syracuse in Italy, created and maintained an unparalleled body of new understanding and yet had only a handful of disciples and correspondents around the Mediterranean. But what specially fascinates Latour is the *isolation* of this network from the rest of the ancient world, be it learned, political, or vernacular. "To the great surprise of those who believe in the Greek Miracle, the striking feature of Greek mathematics, according to Netz, is that it was completely peripheral to the culture, even to the highly literate one. Medicine, law, rhetoric, political sciences, ethics, history, yes; mathematics, no" (Latour 2008, 445). The Greek and Hellenistic mathematicians were a handful of specialists talking with and writing to each other around the Mediterranean basin, and no one else cared: "with one exception: the Plato-Aristotelian tradition. But what did this tradition (itself very small at the time) take from mathematicians? . . . Only one crucial feature: that there might exist one way to convince which is apodictic and not rhetoric or sophistic. The philosophy extracted from mathematicians was not a fully fledged practice. It was only a way to radically differentiate itself through the right manner of achieving persuasion" (445). Latour overstates his case. The philosophical tradition took a good deal from mathematics: what about the golden mean, for example, or the profound role of proportion in ethical theory? That is irrelevant to Latour's case. He proposes that the philosophers focused on proof in order to differentiate themselves from the common herd. Thus their use of proof as above rhetoric was nothing more than a rhetorical trick.

Latour pays little heed to the ways in which the philosophers were profoundly impressed by the human capacity to prove. They were, as I like to put it, bowled over by demonstrative proofs. In consequence they vastly exaggerated the potential of proof. It is easy to argue that the ensuing theory of knowledge impeded the growth of scientific knowledge from the time of Archimedes to the time of Galileo. To defeat the lust for demonstrative proof, we needed what Kant himself called a second "intellectual revolution," one that he associated with Galileo and Torricelli. That was the discovery of other human talents—not purely intellectual ones—and it led to the laboratory style of scientific thinking. Human beings have engaged in experimental exploration forever, but only at the time of Galileo—or following Schaffer and Shapin (1985) I might prefer Boyle—did the laboratory come into being, not just to explore but to create new phenomena using new apparatus. But that is another story (sketched in Hacking 2009b).

Let us here agree with Latour: Plato kidnapped a certain idea of proof and made it a dominant theme in Western philosophical thought. But let us not grant to Latour the idea that proof is unimportant. Let us not allow Latour to kidnap Netz, that is, to allow us to forget Netz's own fundamental concern, cognitive history (which Latour barely mentions).

On the other hand, let us extend Latour's insight. Kant codified, for the modern world, Plato's kidnapping of mathematics. He was not the first to notice the phenomena of the a priori, the apodictic, and the necessary—hallmarks of mathematics. But in one fell sentence in the *Prolegomena* he ran them together to create the philosophy of mathematics. "Here is a great and proved field of knowledge, which is already of admirable compass and for the future promises unbounded extension, which carries with it thoroughly apodictic certainty, i.e. absolute necessity, hence rests on no grounds of experience, is a pure product of reason, and moreover is thoroughly synthetic: how is it possible for human reason to bring into being such knowledge wholly *a priori*?" (Kant [1783] 1953, §6).[4] He made us forget that these phenomena are noticeable only here and there in the motley of mathematical activity! This leads us, for a final observation about ancient times, back to Plato's own demarcation of mathematics.

Plato on the Difference between Philosophical and Practical Mathematics

An important tradition in reading Plato on mathematics derives from Jacob Klein ([1934] 1968). He argued that Plato made a fundamental distinction between the theory of numbers and calculating procedures. Here is a brief summary of the idea, due to one of Klein's students:

> Plato is important in the history of mathematics largely for his role as inspirer and director of others, and perhaps to him is due the sharp distinction in ancient Greece between arithmetic (in the sense of the theory of numbers) and logistic (the technique of computation). Plato regarded logistic as appropriate for the businessman and for the man of war, who "must learn the art of numbers or he will not know how to array his troops." The philosopher, on the other hand,

> must be an arithmetician "because he has to arise out of the sea of
> change and lay hold of true being." (Boyer 1991, 86)[5]

Plato, then, put to one side the daily uses of arithmetic in technologically and commercially advanced societies such as those of Greece or Persia.
They are appropriate for the businessman and for the man of war, who
"must learn the art of numbers or he will not know how to array his troops."
In my opinion we should avoid the notion that computation is for practical
affairs in "the sea of change." That is the philosophical gloss of appearance
and reality all over again, fitting for a freshman course on Plato, perhaps,
but not for thinking about mathematics, even mathematics as read by Plato.
At this juncture I prefer a reading of Plato less redolent of "true being" than
Boyer's. Miles Burnyeat (2000) explains why Plato's *Republic* demanded
that future administrators spend ten years studying mathematics. It was
not for its practical value or even to improve skills at reasoning, in the way
that British higher education once encouraged study of the classics in order
to administer the empire better, but in order to grasp the orders of reality
more clearly.

This highbrow vision suggests a lowbrow, unphilosophical, and sensible
way to distinguish what Boyer calls "the technique of computation" from
perspicuous proofs. It is closer to the experience of doing or using mathematics: computation is algorithmic. It proceeds by set rules. One does not
understand a calculation: one checks that one has not made a slip. There is
no experience of proof as in the theory of numbers or geometry. This is not
intended as a scholarly remark; it is just one line of thought that stems from
this way of thinking about Plato.

These cursory remarks do suggest, however, that Plato (or his heirs) created a disciplinary boundary between mathematics, the science that every
philosopher and every future administrator ought to master, and computation, the technique of commerce and the military. The distinction bears
some relationship to the recent contrast between pure and applied mathematics, but the fundamental difference is that the one involves perspicuous
proof, insight, and understanding, whereas the other involves routine.

Pure and Mixed Mathematics

Francis Bacon was his usual prescient self when he devised that now abandoned term, "mixed mathematics," which appears at the end of the *OED*

definition.[6] Every "branch" of knowledge had to be arranged by division, so as to yield what we now think of as a tree-diagram.[7] He needed a division of mathematics on which optics and mathematical astronomy could flourish, for they were mathematics in the older sense of the term, as noticed by the *OED*. These sciences he called mixed. He also grouped, under this heading, fields of expertise that included music, architecture, and engineering.

They were not mixed, I think, because they mixed deduction and observation. It was rather a matter of the sphere to which they applied. Mixed mathematics was not pure mathematics "applied" to nature but an investigation of the sphere in which the ideal and the mundane were intermingled. Both the mixed and the pure were that part of natural philosophy that fell under metaphysics, namely, the study of fixed and unchanging relations.

Yet "mixed" was truly a bit of a ragbag. Thus Mersenne thought that optics and harmonics did not give knowledge (*connoissance*) partly because they relied on the senses, but also partly because they mixed "physics" in their reasoning (*elles meslent tousjours la Physique dans leurs raisonnemens*).[8]

Probability—Swinging from Branch to Branch

Many a new inquiry had to be forced onto the tree. Where would the Doctrine of Chances, aka the Art of Conjecturing, fit? It was by definition not about the actual world or about an ideal world. It was about action and conjecture; it was the successor to chance as what Hume called "the superstition of the vulgar." There was no branch on a tree of knowledge on which to hang it. In my opinion, probability was uneasily declared a branch of mixed mathematics, less because of its content than because of its practitioners, such as the Bernoullis, who were mathematicians par excellence. The mixed morphed into the applied. Hence the residual place for the "theory of probabilities" as "mixed or applied" mathematics alongside astronomy and physics in the *OED* entry.

Probability was once a paradigm of the mixed, so you would expect it to continue as applied mathematics. That is certainly not what happened in Cambridge, where the Faculty of Mathematics is divided into two primary departments. One is Applied Mathematics and Theoretical Physics, the home of Newton's Lucasian Chair. The other is the Department of Pure Mathematics and Mathematical Statistics. Probability appears to have jumped from branch to branch of the tree of knowledge. In truth, this devious monkey can lodge and prosper anywhere on a tree of knowledge but is not part of its organic structure at all.

Pure and Applied

There is no space, here, to adumbrate the transition of nomenclature from "mixed" mathematics to "applied" mathematics. *Perhaps* the switch was from an idea of mixing mathematics and the study of nature to one of applying mathematics to nature. That picture may well be too anachronistic or at least too simple a vision.[9]

Galileo's own famous image is a compelling alternative. The book of nature is written in the language of mathematics. Galileo did not *apply* abstract structures to nature. He found the structures in nature and articulated their properties, thereby reading the book of nature itself. At the end of his life, in a famous essay published in 1936, Edmund Husserl (1970) rightly seized upon what Galileo was doing as radically new and said that Galileo mathematized nature. Galileo might have retorted that Husserl had things upside down: "I did not mathematize Nature, for she is already mathematical, and waiting to be read."

The situation looks more straightforward half a century after Galileo. Newton distinguished practical from rational mechanics. He took geometry to be a limiting case of practical mechanics, important to builders and architects. Geometry, the very possibility of which so astonished Plato, was placed alongside the practical arts, which Plato did not count as mathematics at all.[10]

There is little reason to think that Newton, the greatest mathematician of his age, cared much about the phenomenon or experience of proof that Plato had made central to his fetishism of mathematics. To continue Latour's metaphor, slightly tongue-in-cheek, we may venture that Galileo and Newton liberated mathematics from the philosophical bonds in which kidnapper Plato had enslaved it.

Pure Kant

"Pure"—*rein*—evidently plays an immense role in Kant's first *Critique*, starting with its title. The primary contrast for both the English and the German adjectives is "mixed."[11] Hence Bacon's branching of mathematics into pure and mixed. The next, moralistic, sense of being free from corruption or defilement, especially of a sexual sort, comes in a close second. At the start of his rewritten Introduction for the second edition of the first *Critique*, Kant emphasized what, for him, was the primary contrast: "The Distinction between Pure and Empirical Knowledge" (Kant [1787] 1929, 41).

Kant's question in the second edition of the first *Critique*, the one that Russell repeated with such enthusiasm, was "How is pure mathematics possible?" What contrasts with "pure" on this occasion? Today we hear "applied." Galileo and Newton did not speak of applied mathematics. Kant's opposite of pure mathematics was empirical. I think that he spoke of applied mathematics only once, namely, in his *Lectures on Metaphysics* (delivered from the 1760s to the 1790s). "Philosophy, like mathematics as well, can be divided into two parts, namely into the *pure* and into the *applied*" (Kant 1997, 307).

In fact Kant asked a pair of questions, one after the other:

> How is pure mathematics possible?
> How is pure science of nature (*Naturwissenschaft*) possible? (Kant [1787] 1929, 56)

Today we are puzzled, and some are baffled, by the idea of a pure science of nature. In his footnote Kant clearly contrasts it to "(empirical) physics."[12] Kant cites, as an example of pure science of nature, primary propositions "relating to the permanence in the quantity of matter, to inertia, to the equality of action and reaction, etc." (Kant [1787] 1929, 56n). On one reading, Kant is talking about Newtonian mechanics.

The great mathematicians of the generation that flourished in the era of the first *Critique*, men such as Adrien Marie Legendre and Pierre Simon Laplace did not see things that way. They were mathematicians. In general the scientists of that era made little difference between pure and applied: I do not mean they did not do what we call pure mathematics, only that the distinction between the pure and the applied was rather immaterial to them.[13] It was the tidy Kant who put a category of pure mathematics up front. If the likes of a Laplace accepted the very notion of applied mathematics and had to choose between it and pure mathematics, of course he would have said applied mathematics is what matters and is the source of all mathematical creativity. But Kant the philosopher made pure mathematics the Queen of the Sciences.

That is the core truth behind Detlefsen's starting point in the *Routledge Encyclopedia*, the statement that, during the nineteenth and twentieth centuries, the most influential ideas in the philosophy of mathematics have been those of Kant. Analytical philosophers interested in the philosophy of mathematics started from an unquestioned assumption that there *is* pure mathematics. We then proceed with Plato's and Kant's visions as something to accept, to modify, to explain, or to reject. The philosophy of mathematics is implicitly about the philosophy of pure mathematics, with a coda, asking how some of it is so applicable to nature.

The separation of the pure from the applied could not happen on Kant's say-so alone. It also called for some highly contingent events in disciplinary organization.

Applied Mathematics

Our idea of applying pure mathematics to nature should not be read back into Kant or Newton. We do have a convenient benchmark for the distinction between pure and applied in exactly those terms. In 1810 Joseph Gergonne (1771–1859) founded what is usually regarded as the first mathematics journal, *Annales de mathématiques pures et appliquées*. In the prospectus for the journal he listed the fields that constitute applied mathematics: *The Art of conjecturing, political Economy, military Arts, general Physics, Optics, Acoustics, Astronomy, Geography, Chronology, Chemistry, Mineralogy, Meteorology, civil Architecture, Fortification, nautical Arts*, and *Mechanical Arts*. That is pretty much d'Alembert's list of mixed mathematics.

Gergonne, it should be said, founded the journal because he could not get a post at the École polytechnique, where he had a feud with Poncelet over projective geometry. As is the wont of French "losers,"[14] he took a provincial post, in Montpellier. A passage in J. S. Mill's *Autobiography* may dispel potential misunderstanding of what Gergonne was likely to be doing there: "at Montpellier [writes seventeen-year-old Mill] I attended the excellent winter courses of lectures at the Faculté des Sciences, . . . [including] those of a very accomplished representative of the eighteenth century metaphysics, M. Gergonne, on logic, under the name of Philosophy of the Sciences" (Mill 1981, 58).[15]

In fact most of the articles in his *Annales* were contributions to geometry, Gergonne's own field of expertise. Germany followed suit in 1826, when A. L. Crelle (1780–1855) founded the *Journal für die reine und an-*

gewandte Mathematik. Here we have one more indication of the highly contingent ways of how the very concept of applied mathematics became institutionalized. Without that history, I doubt if we would have raised Wigner's question (1960) about the "unreasonable effectiveness of mathematics" in the way in which he is usually taken to have raised it. However Wigner actually concluded his famous paper with a question that seems less dependent on the distinction between pure and applied: he spoke of "the miracle of the appropriateness of the language of mathematics for the formulation of the laws of physics" (Wigner 1960, 14).

Pure Mathematics

Now I shall be insular: Back to Russell, which means on to Cambridge, England. I am interested in the presuppositions that framed the analytic philosophy of mathematics. The analytic tradition in the philosophy of mathematics is properly traced back to Frege, but a lot of the stage-setting is thanks to Whitehead and Russell. Their opus is a third benchmark. Compare their book with Newton's. Two great works are titled *Principia Mathematica.* They are entirely different in content and project.

Perhaps nobody really believed in Whitehead and Russell's great book. Possibly only the authors read all three volumes. But even the great German set theorists set themselves up with that work as a monument, although it turned out, to everyone's surprise, to be essentially incomplete. So it is worth the time to consider the mathematical milieu that Whitehead and Russell took for granted—and their conception of pure mathematics for which they hoped to lay the foundations.

At least in their British curricula we can locate the point at which "Pure Mathematics" became a specific institutionalized discipline. In 1701, Lady Sadleir had founded several college lectureships for the teaching of algebra at Cambridge University. In 1863, the endowment was transformed into the Sadleirian Chair of Pure Mathematics, whose first tenant was Arthur Cayley. This coincided with an important shift in the teaching of mathematics. This is said to have been the first job, anywhere, named "Professor of Pure Mathematics."

This did not change all that much. Before and after that chair was established, the young students who excelled in the mathematics tripos and in the Smith's Prize (a competition for the best young Cambridge mathematician) had names such as John Herschel, G. B. Airy, G. G. Stokes,

William Thomson (Lord Kelvin), P. G. Tait, and James Clerk Maxwell. We remember these as physicists, but they were, in the classifications of their youth, *mathematicians*.

Nevertheless, after 1863, what was called mathematics at Cambridge slowly became what we now call pure mathematics rather than natural philosophy. It was within this conception of mathematics that Russell came of age. Likewise it was in this milieu that G. H. Hardy became the preeminent local mathematician, whose text, *Pure Mathematics*, became a sort of official handbook of what mathematics is, or how it should be studied, taught, examined, and professed at Cambridge.[16] Russell's vision of mathematics was not determined by Hardy's, or vice versa, but the two visions are coeval, a product of a disciplinary accident in the conception of mathematics. Much later Russell's former colleague A. N. Whitehead was to say that "The science of pure mathematics, in its modern developments, may claim to be the most original creation of the human spirit" (Whitehead 1925, 19). Its only rival, in his opinion, was music.

On another occasion I have argued that Wittgenstein's philosophy of logic and of mathematics was deeply influenced by his arriving in town where pure mathematics ruled.[17] He stopped being an applied mathematician studying aeronautical engineering at the same time that he encountered Russell and "pure mathematics." That makes it sound as if Wittgenstein's life as a philosopher of mathematics was molded in Cambridge. But he was Viennese! And there he knew Richard von Mises, who was the greatest exponent of applied mathematics. Both men began as aeronautical engineers.

Applied Mathematics: von Mises

There was a great debate about pure and applied mathematics even within the Vienna circle.[18] The chief protagonist was von Mises (not to be confused with his brother Ludwig, the economist). Von Mises is now best known to philosophers for his frequency theory of probability, a thorough work of, among other things, logical positivism. He himself strongly identified as an applied mathematician and regularly insisted, against some other members of the Vienna circle, that mathematics could be properly understood only by its applications. His dissertation was on the determination of flywheel masses in crank drives. He was an aeronautical engineer, giving the first university lecture course ever, anywhere, on powered aircraft in Strasbourg in 1913, and himself becoming a test pilot during the Great War, and designing the Mises-Flugzeug, a 600 horsepower flying machine that was too late for development into a fighter plane (Siegmund-Schultze 2004.)

Immediately after the war von Mises became head of the new Institute of Applied Mathematics in Berlin, and in 1921 he founded the *Zeitschrift für angewandte Mathematik und Mechanik*. Although most of the Vienna circle were loyal to, or even formed by, "Whitehead and Russell," and thus thought of mathematics in terms of pure mathematics, the residual effect of von Mises was strong. It must have had a profound influence on a thinker of Wittgenstein's stripe. In a fanciful mood I would suggest that it may be helpful to look at the man stereoscopically, with one lens focused on Vienna and the other on Cambridge. Through the Vienna eye he sees application. Through the Cambridge eye he sees purity. Maybe his realization that mathematics is a motley originates from living in both worlds and not being blinded by single vision.

Contingency, Necessity, and Neurology

I have sketched only the beginning of an argument that what is counted as mathematics depends in part on a complex and very contingent history. I do not mean to imply that the history could have gone any way whatsoever. It was constrained by its content and by human capacities. They no longer constrain in the same way. The advance of fast computation is changing the entire landscape of human knowledge, including that of mathematics. This is a topic for the future. Here I have been concerned with the past.

Our picture of the philosophy of mathematics is of philosophical reflection on a definite and predetermined subject matter. I suggest that the subject matter itself is much less determinate than we have imagined. This does not undercut the debate between the two attitudes I have mentioned, (1) platonic and (2) neurobiological, or the traditional philosophical debate between "realists" and "antirealists" of mathematics. It will surely go on as before. I urge only that the more difficult but perhaps more answerable question should now become: how have the platonic and neurobiological constraints jointly interacted with the contingent history of mathematics from "Thales" to now? More locally, does the fact that philosophers of mathematics during the twentieth century were obsessed with a priori knowledge and necessity result from the fact that they all supposed that mathematics is pure mathematics, which is, I suggest, a pure historical accident?

PART V

Widening the Scope of Contingentist/Inevitabilist Targets

Scientific Practices and the Methodological, Material, Tacit, and Social Dimensions of Science

CHAPTER 12

The Science of Mind as It Could Have Been

About the Contingency of the (Quasi-)Disappearance of Introspection in Psychology

MICHEL BITBOL AND CLAIRE PETITMENGIN

A ccording to a widespread tale (Lyons 1986; Costall 2006), the science of mind took four well-defined steps during the past century:

1. Across the turn of the nineteenth and twentieth centuries, mind was identified with conscious experience, and introspection was accordingly used as a primary tool to explore mental activities.

2. Around 1913, there occurred a brutal dismissal of introspection as an acceptable method for scientific psychology, and psychologists undertook a systematic study of "overt behavior" instead; the very reference to "internal" events was banished, and "consciousness" was denounced as a prescientific term.

3. From the late 1950s on, the idea that mind involves internal processes became fashionable again, but it first assumed a strictly objectivist form: internal processes were identified with neurological working, information processing, or cognitive functioning. Introspection itself (or some disguised form of it) was given a respectable objectivist status: that of "meta-cognitive access" (Nelson 1996), "higher-order thoughts" (Rosenthal 2005), or neural "reentry" (Edelman and Tononi 2001). Yet, it was still suspected

of pervasive mistakes (Nisbett and Wilson 1977; Johansson et al. 2006).

4. Finally, during the mid-1980s and 1990s, there was an outburst of "consciousness studies," and systematic introspection arose again from the ashes (Petitmengin 2009) in close association with the quest for "neural correlates" of experiential events.

This common account is clearly incomplete, insofar as it does not mention momentous disciplines such as psychoanalysis, phenomenology, or *gestalt-psychologie*. It is also distorted, because it posits sharp boundaries and definitive judgments of history usually pronounced by a few propagandistic authors. We shall then try to correct the former picture, by relying on the work of serious historians of introspective psychology.

But, incomplete and distorted as it is, this picture at least expresses a tidal movement of epistemic values in the history of the science of mind. Introspection was credited with the highest potential as a method for psychology, and then scorned as radically misconceived and unreliable. What explains this about-face? Was the long eclipse of introspection unavoidable? Could the science of mind have directly jumped to the present phase of rebirth of interest for consciousness and first-person experience, without going through the strange episode of self-denial that started with behaviorism and blossomed with eliminativism? Has something important changed between turn-of-the-century introspection and current introspection that can account for its capacity to resist the usual criticisms? This urge for alternative historical scenarios of the science of mind, and even more for explanations of its actual course, provides us with a promising case study in the contingency of scientific programs (Sankey 2008; Hacking 1999). The case we are interested in does not bear on a path that could have been taken by science's history yet has not been taken but, rather, on a path that could have been much shorter, thus avoiding a surprising doctrinal eclipse of what is most immediately present to us. It is not only a case of contingency of thought or procedures. It is also a case of contingency of attitudes toward the very source of our knowledge. It may represent a much deeper challenge to scientific realism (Sankey 2008; Hacking 1999) than the standard contingency arguments, because it bears on the root presuppositions of this thesis rather than on its arguments.

Our roadmap is the following: we criticize some aspects of the official history of the science of mind, and then we list the major obstacles to a proper use of introspection, and briefly suggest how they could have been overcome.

On Some Amendments of History

As we just mentioned, and as several authors have already pointed out (Boring 1929; Danziger 1980, 1994; Vermersch 1999; Kroker 2003; Brock, Louw, and van Hoorn 2004), the official history (or tale) of introspection is flawed in many respects. Here we shall only focus on three aspects about which the fate of introspection diverged from what is currently believed. They concern: (1) the distinction between the inner and outer domains; (2) vigilance about the reliability of introspection during the heyday of this method; (3) the persistent use of introspection during the reign of behaviorism.

An Inner World?

The word "introspection" is derived from the latin *intra-* (within) and *specere* (to look at). Relying on this etymology, it is tempting to infer that the whole paradigm of introspection relies on a doubtful divide between the internal recesses of the mind and the external world. But was the criticism of the internal/external dichotomy really unknown at the time of the introspectionist wave?

To begin with, the philosophy of mind of the turn of the nineteenth and twentieth centuries was already buzzing with criticism of methodological and substantial dualism. Edmund Husserl was one of the most prominent critical thinkers of that time, denouncing (in the appendix to his sixth *Logical Investigation* of 1901) the distinction between inner and outer perception he had inherited from Brentano. The right distinction, he claimed, was instead between *certain* (immediate and complete) and *uncertain* (mediate and incomplete) perception, within an undifferentiated flux of lived experience. To this, the German neo-Kantian philosopher Paul Natorp (2007; see also Bitbol 2008a, 2008b) added a detailed account of how the dual organization of knowledge (object and subject, outer and inner) may arise from this undifferentiated continuum. According to him, this occurs by way of a double-faced process in which objectivation comes first, and subjectivation arises as the by-product of the former. Objectifying means picking out the component of experience that remains invariable

across personal, spatial, or temporal situations; or at least the components
of experience that vary in the same way (i.e., in a law-like way) irrespective
of the personal, spatial, or temporal situations. The "subjective" domain
is then marked off by contrast and difference from the objectified part of
experience. It includes whatever is left in experience after the objective domain has been circumscribed. Accordingly, the subjective domain evolves
with the process of objectification, and it receives as many characterizations as there are delineations of objectivity.

Thus, accessing the domain of subjectivity is not just a gift but a *discipline* symmetrical to the discipline of objectification. One can access this
domain by reflecting on the (subjective) conditions of possibility of objective knowledge. One can also reach it by relaxing the interest of knowledge
initially directed toward restrictive parts of experience, and eventually by
suspending the activity of fragmentation of the field of experience.

Husserl's and Natorp's deconstructions of the inner/outer dichotomy
was part of the unconventional and antidualistic philosophical atmosphere
in which psychological introspectionist inquiries developed. But what
about psychology itself? There are signs that the philosophical unrest of
the turn of the century influenced the development of psychology. William James is a striking example of a synthetic thinker. On the one hand,
although somehow diffident, he was very much involved in introspective
research.[1] On the other hand, he developed what he called "radical empiricism" (James 1976), in which he construed the objective and subjective
sides, the material and mental domains, as two constructs arising from a
single plane of pure experience. He explicitly drew inspiration from Ernst
Mach's ([1897] 1984) *neutral* monism and anticipated Bertrand Russell's
(1921) *Analysis of Mind.*

Even the two most emblematic users of introspection, Wilhelm Wundt
and Edward B. Titchener, looked somehow uncomfortable with the literal
consequences of the dualist vocabulary they used. Thus, whereas the German psychologist Wundt repeatedly characterized introspection as "inner
perception," he found himself compelled to describe the diverging interests
of natural science and psychology as two modes of "arranging" one and
the same continuum of experience, rather than as two mutually exclusive
(inner and outer) spheres of being (Wundt 1901). Just as Natorp did, Wundt
considered the domain of subjectivity as what is left in experience when
an objectifiable material has been extracted from it, not as some enclosure
separated from the objective world.

Even more striking is the case of the American psychologist Titchener, who (unlike Wundt) saw introspection as the *only* legitimate approach to his field. Despite a thoroughly dualist vocabulary, Titchener told a very different story when he had to orient the reader toward the appropriate techniques of disciplined introspection. He then characterized the difference between introspective psychology and natural science in terms of stances and standpoints, not in terms of directions of some (mental) gaze (Titchener 1916). Moreover, the theme of study of introspective psychology is described as dependent on the state of the introspector, and also as "transient, elusive, slippery"; it is contrasted with the theme of study of natural science, which is both independent of the state of the observer and stabilized in invariant structures. Here again, by reading carefully some paragraphs of the work of a leading introspective psychologist, we have the feeling that a remarkable kinship with phenomenology or with Natorp's blend of neo-Kantianism is latent.

This is not surprising in view of one of our recent studies in introspective psychology, which provides fresh arguments against the dualist view (Petitmengin 2007). In such cases, the reflective attitude, which is conveniently (though inappropriately) described as "turning one's attention inward" helps one contact a dimension of experience in which the very distinction (inward/outward) is seen to vanish.

The Quest for Experimental Rigor

The second point on which the standard history of introspection is flawed is that it underrates the methodological care of the psychologists who used it. Those psychologists of the turn of the nineteenth and twentieth centuries were extremely concerned about the possible pitfalls of introspective inquiry, and they designed procedures by means of which they hoped to reach the standards of reliability and reproducibility of experimental science. They hardly fell under the behaviorist reproach of indulging in speculation rather than sound methodology (Watson 1913).

Wundt thus insisted that it is possible, under certain stringent conditions, to found what he called an *"experiential* science" with the help of the tools and laboratories of *experimental* science. These stringent conditions, as he posited them, were aimed at providing the psychologist with truly reproducible data.

First, according to Wundt, only those mental processes that are directly triggered by physical controllable stimuli generated by mechanical apparatuses and timers are to be considered. Indeed, Wundt considered that the experimental method (as opposed to pure observation) is even more indispensable in psychology than in natural science because psychology must always cope with "transient" phenomena. The repeatable action of instruments here counterbalances the instability of experiences. The selection of acceptable mental material by Wundt was in fact so restrictive as to exclude anything except elementary discriminative judgments about sensations (Wundt 1901, 20). Even less accepted as possible fields of study were elaborate mental processes such as thought or emotion. Thus, under the name of "experiential science," Wundt was committed to an expanded version of Gustav Fechner's "psychophysics."

Second, Wundt insisted that the elementary mental processes triggered by physical stimuli must be studied ex post facto (out of short-term memory), in order to avoid direct interference of the introspective inquiry on the processes to be analyzed.

These two rules were meant to avoid some of the most widespread theoretical objections to introspection, which Wundt was very familiar with and which we shall document in In-Principle Objections and Replies.

However, Wundt's almost narrow-minded care and reliance on technological contraptions were soon perceived as an unbearable yoke by many psychologists. Some of them soon discovered how useful unprepared qualitative descriptions of their experience by subjects can be, in order to disentangle the intricacies of mental events that were wrongly construed as "elementary" in Wundt's laboratory (Binet 1903). This finding generated an outburst of "systematic introspective" studies during the first decade of the twentieth century, in which much more extensive reports of subjective experience were allowed (including about thought processes, emotions, motivations, etc.). The two main groups who developed "systematic introspection" were the so-called Würzburg school of psychology in Germany and Tichener's group at Cornell University in the United States. Yet, in freeing themselves from Wundt's rules and physiological methods, these psychologists did not renounce the use of a strict methodological rule altogether. They only shifted the burden of method from the physical tools of the laboratory to the training of both psychologists and subjects of experiential inquiry (Watt 1905; English 1920). They adopted rules that were tantamount to implement *maximal exhaustivity of description, pondering on the "how" rather than the "why," and careful refocusing of attention on the plane of experience.*

Of course, application of these rules requires serious training; either training of the subjects or training of the psychologists (or both).[2] This was one of the reasons John B. Watson (1913) provided for rejecting introspection as a legitimate approach in psychology: how can we rely on introspective data, he asked, if each time a divergence occurs we are told that the subjects or the psychologists were not properly trained? And does training not amount to prejudice?

As we shall see, the former rules are mostly accepted in contemporary versions of this inquiry, yet sometimes given a different significance. Moreover, the issue of training has been revived in a very different cross-cultural context, with new concern for the efficiency of contemplative disciplines (especially Buddhist) where lineages of experts can transmit their skill by calibrated series of exercises (Wallace 2006a; Nauriyal, Drummond, and Lal 2010; Thompson 2014; Gupta 2004; Shafii 1973).[3]

Here again, we shall have to raise questions about what was (apparently) missing in the strategy of the old schools of introspection despite their methodological precision. This will be done in What Was Needed to Avoid Nearly One Century of Schizophrenic Ostracism against Introspection.

The Underground Life of Introspection during the Twentieth Century

The third main point on which traditional historical presentations of introspection turn out to be flawed is nothing less than the truth of its alleged disappearance. That this claim of the extinction of introspective methods is plainly wrong has been repeatedly stressed by psychologists of the mid-twentieth century who noticed that they had never ceased to use introspection (although usually a shy, truncated, almost invisible version of it) (Price and Aydede 2005). They just could not believe that radical behaviorists were serious when they declared a ban on first-person access to experience in the name of a narrow conception of nature and natural science. After all, these psychologists noticed, "a conscious memory or a dream is as much a natural phenomenon as a star or a starfish" (Warren and Carmichael 1930, 58; see also Moore and Gurnee 1933). As a consequence, the myth of the abolition of introspection by behaviorism was soon replaced by a question about what its "scientifically correct" name was: "Introspection is still with us, doing its business under various aliases, of which *verbal report* is one" (Boring 1953, 169). With the exception of some studies in applied psychology and philosophy where concern for first-person experience had never faded away

(Gendlin 1962; Stern 1985), "verbal report" became the only licit way of evoking introspective access in such a way that no explicit reference to its expressing "inner life" had to be made. The name had changed, but the real practice of psychology (especially in education and psychotherapy) did not exclude something surprisingly similar to introspection. What subjects could report about the various aspects of their lived experience had never ceased to be taken into account to guide the formulation of hypotheses about mental life (Nigro and Neisser 1983). Even hard-nosed eliminativists of the end of the twentieth century could not deny that the very meaning they ascribed to the neurological categories by which they wished to replace "folk-psychological" categories relied on the application of the latter by living human beings reporting about themselves.

One can spot this tendency of using a tamed version of introspection even in the writings of one of the most emblematic supporter of behaviorism: B. F. Skinner. One of his major works was titled *Verbal Behavior* and contained a remarkably accurate study of the "verbal reports" in what he called (with a touch of paradox) "radical behaviorism" (Holland and Skinner 1961). In this comprehensive version of behaviorism, Skinner went so far as to declare that private processes such as emotions and silent thinking should themselves count as behaviors.

More recently, a careful and systematic study of the methodology of introspection has been published under the heading "verbal reports": the celebrated *Protocol Analysis* of Ericsson and Simon (1984). This book deserves to be considered as the true turning point between the long-term behaviorist dismissal of introspection and its recent rebirth. It belongs to the former period by the tribute it pays to the strongly objectivist tendencies of behaviorism, yet it also paves the way for the new epoch by its fine-grained analysis of the way one can obtain reliable reports about first-person experience. Ericsson and Simon thus explicitly criticize what they call the "schizophrenic" attitude of behaviorism toward introspection, between official rejection and unavoidable use of "yes-no" reports. And they undertake a nuanced defense of a fleshed out version of introspection. One major element of this defense is the remark, adduced against Nisbett and Wilson's (1977) devastating criticism of introspection, that since the subject has no complete conscious access to her own cognitive processes, she should not be asked to provide the reason (the "why") of her choices or feelings. Instead, in good agreement with one major prescription of Titchener's introspective school, one should ask the subject questions about the plain facts of her experience, about the "how." As we shall see, a crucial condition for the current revival of introspection is to promote this sort of lucidity about what can and what cannot be expected from a subject's description.

It then appears that there is a component of inevitability in the use of (some variety of) introspection in the science of mind. This component of inevitability is not strictly connected to the object of this science (the mind), since, as we witness in modern cognitive theory, the mind can be redefined so as to exclude any overt reference to consciousness. The inevitability of the use of introspection, rather, arises from the elementary fact that scientists partake of the same human condition as their subjects endowed with mind. If they want to endow their studies of objective processes such as behavior and neural dynamics with meanings that matter for them qua conscious human beings, they are bound to connect these processes with reports of lived experience.

Attitudes in psychology range from (ideally) complete dissociation between the investigator and the domain of her study to full identification of the investigator with the subject who is reporting her lived experience. An ongoing debate about the status of folk-psychology illustrates this difference of attitudes by distinguishing three steps on the scale that goes from distance to coincidence. To begin with, according to eliminativists (Churchland 1986), introspective folk-psychology is a sort of primitive scientific *theory*. Reacting to that, various authors have suggested a second, very different conception (Goldman 1992; Warren 1999). Here, folk-psychology is no theory; it is a system of categories helping us to *simulate* others' mental states. But one can also think of a third conception of folk-psychology, which is even closer to the empathic pole of the range of attitudes toward introspection than the simulation theory because it does not even take for granted that I and the other are to be construed as mutually exclusive beings and standpoints. This conception is inspired by Jean-Paul Sartre's analysis of "other minds." According to Sartre ([1943] 1973), it is pointless to ask for a proof of the *existence* of other minds; this demand can only yield skepticism. Instead, one should realize that the existence of the other is just as immediately certain as my own, because I am, so to speak, woven of the other(s). Along with this perspective, the vocabulary and categories of introspective folk-psychology are *more* than a guide toward simulation of what it is like to be the other: rather they offer material for elaborating a common field of shared meaning and shared experience. Moreover, in this role, the categories of introspection are by no means static; they can be enriched at any moment by the use of disciplined metaphors that are validated by the mere fact that they are spontaneously *recognized* by (at least some) other subjects as a faithful description of a shareable experience (Findlay 1948).

The true question we shall have to address then bears on the contingency or inevitability of the psychologists' diffidence toward the *participative* and *empathic* aspects of introspection, rather than on the contingency or inevitability of the use of introspection in psychology.

What Was Needed to Avoid Nearly One Century of Schizophrenic Ostracism against Introspection

In this section we shall list and discuss some of the main epistemological features that were missing (or poorly understood) in early introspective psychology, and that may explain its vulnerability to criticism as well as its apparent eclipse. We also wish to show, whenever possible, the ability of the new wave of introspective studies to cope with most of these deficiencies. The deficiencies of the first wave of introspective psychology are:

>1. Lack of a universally accepted understanding of what exactly the act of introspection *is*;
>2. Correlative lack of a convincing set of answers to a traditional list of in-principle objections against the very possibility of introspection;
>3. Uncertain conception of (and uncertain criteria about) what makes introspective reports reliable, and possibly *true*;
>4. Difficulties about where the regulative ideal of objectivity should be applied; and
>5. Absence of a proper third-person correlate of detailed first-person experiences and of the possibility of elaborating a triangulated approach by combining them.

The following five subsections will address the above-mentioned deficiencies in turn.

Introspection without "Intro" and without "Spection"

So, what *is* introspection, and what should it be, exactly? As we have seen in On Some Amendments of History, the introspectionist psychologists of the turn of the nineteenth and twentieth centuries had doubts about the dualist picture of inner and outer realms that would fully justify using the term "intro-spection" about a certain mental act of meta-awareness. Yet, despite this widespread doubt, most of their overt characterizations of introspection remained in line with dualism. The two-realms and two-directions-of-gaze model was still dominant. Wundt thus wondered "how can our own mental life be made the subject of investigation like the objects of this external world?" (Wundt 1901, 11). Similarly, Titchener approved the idea that "introspection is simply the common scientific method of observation" (Titchener 1912, 493). Titchener accordingly stated the different directions of gaze by which one should characterize the two kinds of "observation": "the method of psychology is observation. To distinguish it from the observation of physical science, which is inspection, or a *looking-at*, psychological observation has been termed introspection, or a *looking-within*" (Ticthener 1910, 20). Later textbooks of psychology usually retained this standard definition of introspection as *observation of some internal occurrence* (Moore and Gurnee 1933). Such a definition is especially significant in view of etymology. The Latin prefix "ob-" means "facing," and the Latin verb "servare" means "to watch over." Observing then means "watching over what is facing us (and is therefore different from us)." The paradigm of detachment here clearly dominates any other view of introspection.

It is on this unsophisticated epistemological ground that nuances, inflections, and even skepticism grew up. Wundt resisted from the outset the rough definition of introspection as "inner *observation*," and instead referred to "inner *perception*," thus accepting a distinction previously introduced by Brentano (1995). According to Brentano, inner observation cannot be the "true source of psychology," for observing a mental event by fully focusing one's attention toward it would just lead to its disappearance. The true source of psychological inquiry is then inner *perception*, which does not require attention to be focused on some mental object but only that, when attention is focused on some (usually external) object, it remains broad enough to notice other events such as the mental processes that underly the act of attending. One can thus *perceive* a vibration of the telescope while *observing* a planet. As for Titchener, he was also aware of the paradoxical nature of inner observation, which, together with its description, disturbs the process to be observed: "If you try to report the changes in consciousness, while these changes are in progress, you interfere with consciousness" (Titchener 1910, 22). He then suggested two solutions. The first one, which he did not like very much, consisted in relying on retrospective observations of past experiences. This was a widespread strategy at the time, already advocated by William James and John Stuart Mill. The second solution was tantamount to relying on the "introspective habit" of trained subjects, who were able "to take mental notes while the observation is in progress, without interfering with consciousness" (Titchener 1910, 22). But what is this special ability trained subjects acquire not to interfere with their own consciousness while they are observing it? A reasonable assumption, in line with Brentano and Wundt's characterization of "inner perception," is that it is the ability to detect occurrences that are not in the main focus of interest, by extending attention (so to speak) laterally.

Notwithstanding several momentous differences between introspective psychology and phenomenology, this description fits well with Husserl's characterization of phenomenological reduction, which is the chief method to give access, not of course to the "inner world" but, rather, to the whole field of pure experience before exclusive intentional focusing has narrowed down the region of our full awareness. Phenomenological reduction, writes Husserl (2002, 11), helps to reveal the "sides" (or the margins) of our experience that are overlooked as long as exclusive concern for objects prevails. Husserl insisted on the full openness of the subject to the diversity of lived experience during phenomenological reduction or on the quality of expansion rather than refocusing (toward some "internal object") that is given to experience by reduction (Depraz 2008). Even when Husserl spoke of the metaphoric "splitting" of the subject in reflection (as he often did), he mentioned that in the phenomenological variety of reflection I become "*at the same time* plainly seeing subject, and subject of pure self-knowledge" (Husserl 1972, 156).[4] The so-called splitting therefore tends to be all-encompassing rather than discriminative; it represents a stretching of the natural attitude rather than a restriction and a redirection of it. Accordingly, the "splitting," if any, is symmetric rather than asymmetric. It does not give any (reflective) priority to the "subject of self-knowledge" over the elementary "seeing subject" (or subject of object-knowledge) but, rather, puts both of them on the same footing. It does not impose to define the "subject of self-knowledge" as a seer of the "subject of seeing" but, rather, to codefine both subjects within the broadened experiential field of an "I" who has undergone the phenomenological "reduction." Later on, this momentous move was confirmed by Maurice Merleau-Ponty, according to whom the phenomenological attitude means (in terms borrowed from Bergson) that, "instead of wanting to raise ourselves above our perception of things, we plunge into it to dig it out and *enlarge* it" (Merleau-Ponty 1953, 22; Bergson 1934, 148).

Of course, this is not meant to neglect Husserl's own forceful denial that the phenomenological inquiry relies on some variety of introspection. He gave three major reasons for this denial: (1) Introspection, he wrote in his *Ideen I*, arises from a state of *positional* consciousness (which means that in this case consciousness *posits* an intentional object, be it in the focus or in the margin of attention); by contrast, in the genuine phenomenological stance, consciousness becomes "nonpositional" (Flajoliet 2006). Whereas the positional reflection of introspection aims at describing mental processes qua objects, the nonpositional approach of phenomenology tends to reveal the field of transcendental subjectivity that underpins any object-directedness. (2) Being "positional," and therefore directed toward some sort of transcendent object, introspection remains fallible as any empirical investigation is. By contrast, being nonpositional and therefore immerged in immanence, the phenomenological stance is supposed to reach absolute certainty. (3) Phenomenology is not concerned by single events of mental life, unlike the primary step of introspection; it aims at elucidating the invariants (or "essences") of lived experience.

Despite these differences, some psychologists have argued that Husserl's characterization of the phenomenological stance supports a new understanding of introspection (Vermersch 2011). After all, Husserl himself acknowledged that some criticisms of introspection were indirectly aimed at phenomenology and that they had to be addressed in order to defend the latter discipline (Husserl 1983, §79). According to this new understanding, "intro-spection" appears as (or is replaced by) a mental state in its own right, a state of broadened awareness, rather than being taken as a homuncular act of observation of some other mental act or mental state. A new concept of "reflection" is introduced and defined, instead of being squarely rejected in view of its spurious dualist connotations (which imply a mirror and a gaze). "Reflection" in a phenomenological sense no longer means a sort of specular (transcendent) observation but, rather, a *modification of consciousness*, a *transmutation of lived experience as a whole* (Husserl 1983, §78). To stress the difference without breaking lexical continuity, we can give a slightly different name to this renewed concept of "reflection": "coreflection." The latter neologism may prove useful to convey two semantic shifts. According to the first shift, we are no longer concerned by a mere asymmetric revelation of the "seeing subject" by the "subject of self-knowledge," but by their symmetric codefinition within the experiential field of somebody who has practiced the phenomenological "reduction." According to the second semantic shift, the variety of reflection at stake represents in fact an enlargement of the span of experience, and this can be evoked by the three first letters of the word "coreflection": "cor" for the Greek "khôra," which Plato used in the *Timaeus* to mean space, or interval (Bitbol and Petitmengin 2011).

Full realization of this alternative status of introspection, in line with the long-term phenomenological and neo-Kantian traditions, is commonplace today. Gregg Ten Elshof (2005) thus claims that introspection can still be considered as a kind of perception, provided one recognizes that the essential act of any perception involves redirecting attention or *changing its span*. Similarly, Jerome Sackur (2009), making a cogent synthesis of Brentano's and Wundt's reflections, defines introspection as a process of perception *expanded* to what is usually neglected or to what is usually at the periphery of the attention field. More radically, we are invited to discard any remnant of the metaphor of vision and to accept that introspection, far from being like a gaze on some object (be it focused or expanded), is tantamount to (re)establishing an intimate and close *contact* with what is to be explored (to wit, the field of lived experience) (Petitmengin and Bitbol 2009). The metaphor of the sense of touch (with closed eyes) here replaces the metaphor of the sense of vision.

Two major developments of our *Weltanschauung* and of the cognitive sciences can explain why this alternative, nonobservational, and nonvisual conception of introspection is now much easier to accept than it was at the beginning of the twentieth century. One of them is our growing familiarity with contemplative methods whose aim is to stabilize attention and then to use this stabilization in order to get a precise knowledge *by acquaintance* of the subtlest aspects of mental processes.[5] Along with this perspective, the idea of "nonpositional" consciousness, or of intimate contact with experience as opposed to the old-fashioned observational view of introspection, is no longer problematic. Thus, according to B. A. Wallace, "Unlike objective knowledge, contemplation does not merely move towards its object; it already rests in it" (Wallace 2006b).

The other development that makes the nonobservational conception of introspection easier to accept concerns the cognitive sciences. It is the widespread recognition (Schooler 2002) of a background short-term cognitive unconscious (Hassin, Uleman, and Bargh 2006), in addition to the long-term pulsional unconscious delineated by Freud (1976). This allows one to take at face value the image of focus and margin of conscious awareness that sounded so problematic during the first wave of introspective psychology (Bode 1913). This also opens the possibility of applying some (though not every) feature of the model of perception to introspection. Let us remember that a relevant feature of perception is incompleteness: according to Husserl, the act of perception combines a central profile with a surrounding "horizon" of anticipated or altogether hidden profiles. Such a feature is connected with transcendence (because incompleteness can be interpreted in terms of excess of what appears with respect to any appearance). But nothing prevents one from disconnecting transcendence from object-like separation: one can perfectly figure out that some parts of the field to be explored by introspection elude full awareness at a certain moment, and yet that the introspector remains in close contact with this field throughout. After all, when we are in contact with (or immersed in) the field we wish to explore, we are bound to remain unaware of aspects that would immediately become apparent if we stood back with respect to it. The introspective process is thus likely to require a careful process of unfolding or (to use another metaphor) of crawling across the experiential process to be analyzed.

Recent methods of verbal report and introspection fully take this into account. The elicitation method that we currently practice can be characterized as a strategy for progressively unfolding initially "prereflective" aspects of lived experience, by asking subjects to rehearse and even to *reenact* this experience while broadening their field of attention (Vermersch 1994; Depraz, Varela, and Vermersch 2003; Petitmengin 2006; Petitmengin et al. 2009; see also Hurlburt and Heavey 2006). Here, *retrospection* is systematically used (as opposed to "thinking-aloud" protocols). But this is not to meet the traditional objection according to which observation disturbs the observed process if it occurs simultaneously to it (an objection automatically inactivated by the rejection of the observation conception of introspection). This is to enable patient expansion of awareness in a selected slice of experience.

To sum up, two crucial points on which the current definition of introspection differs from the classical one, and which may offer it a better opportunity of development are: (1) overt cultivation of contact with, immersion in, and mindfulness about an all-pervasive experience, rather than narrowly focused observation directed toward some inner sphere of processes; and (2) techniques for encompassing prereflective (or cognitive-ly unconscious) parts of experience in successive fields of attention. Both moves might motivate rejection of the word "intro-spection," but it is con-venient to keep it with us in order to avoid minimizing a certain amount of historical continuity.

In-Principle Objections and Replies: Can Introspection Be Impossible Yet Real?

Introspectionism of the turn of the nineteenth and twentieth centuries stumbled on an impressive list of in-principle objections. Although such objections could not have been sufficient by themselves to destroy the very project of an introspective psychology (because practical success usually bypasses in-principle criticism in the history of science), they contribut-ed to its disrepute. We shall thus list these objections shortly,[6] and outline some replies that new introspection has in store for them.

> **A.** The most archetypal objection has already been met, by way of a phenomenological-like redefinition of introspection. This objection is that it is impossible to observe one's own experience, because this presupposes a split between subject and object whereas in this case the object is nothing other than the subject itself. A very early form of this objection was formulated by Socrates himself, in the *Char-mides* (167 c–d), in order to challenge a widespread conception of wisdom as self-knowledge: "Suppose that there is a kind of vision . . . which in seeing sees no color, but only itself and other sorts of vision: Do you think that there is such a kind of vision? Certainly not!" (Roustang 2009, 78). According to one of the Platonician dialogues that is most likely to express Socrates's position, then, there is no such thing as self-knowledge because the object must be distinct from the mode of access. But the most well-known version of the objection was stated by Auguste Comte (the creator of positivism): "The think-ing individual cannot split himself in two parts, one who reasons and the other one who looks at the reasoning. The observed organ and the observing organ being in this case identical, how could observa-tion take place?" (Comte 2001, 34). James, after Mill, later echoed this objection, although the way he did so paved the path to his reply in terms of retrospection: "The attempt at introspective analysis in

these cases is in fact like . . . trying to turn up the gas quickly enough to see how the darkness looks" (James 1890, 244).

In view of the remarks in Introspection without "Intro" and without "Spection," we realize that this kind of objection is directed against introspection as prejudice says it *should be,* rather than against introspection as it *is* in fact practiced. The prejudice is that part of the subject engages in second-order observing or monitoring of first-order mental processes. Against this prejudice, many results, including from neurophysiology (Overgaard et al. 2006), are consistent with the idea that introspection merely involves a modified version of those very first-order mental processes.

However, we do not want to discard Comte's objection too quickly. Instead, we shall develop this objection and this prejudice one step further, and then compare it with a similar problem in the history of the interpretation of quantum mechanics.

An important development of the alleged splitting of subject and object in introspection was stated repeatedly in the history of psychology: "suppose a particularly persistent introspectionist should desire to introspect the reporting or secondary series, would he not have to assume a third series, and so on, *ad infinitum* and *ad nauseam?"* (Ten Hoor 1932, 324). This threat of infinite regress of "inner observation" had been identified and discussed much earlier by Harald Høffding (1905), a Danish philosopher who was a major inspiration of Niels Bohr. This is how an unsuspected bridge was established between introspection and quantum mechanics, at the deepest epistemological level. Niels Bohr (1934) indeed tended to make a strong analogy between: (1) the situation of an introspector who wishes to observe herself by splitting into a subject part and an object part, and (2) the situation of an experimenter in quantum mechanics who is (instrumentally and interpretationally) intermingled with microscopic phenomena. In both cases, one witnesses a kind of dialectic between the actual inseparability and the alleged necessity of separation between subject and object.

As soon as some divide between object and subject is conventionally imposed despite actual inseparability, part of the object to be known happens to be cut off (because it has been retained on the side of the subject that is narrowly intermingled with it). Then, full characterization of a micro-object can be obtained only by means of several "complementary" (mutually exclusive and jointly exhaustive) experimental approaches, each one being associated with one given position of the conventional divide. Similarly, according to Bohr, full characterization of oneself can be reached only by means of several "complementary" introspective approaches.

However, this dialectical strategy advocated by Bohr is very disputable. Is it not possible to do without any artificial separation of subject and object, yet to approach microphysical and experiential phenomena in a scientific way? As argued in previous work (Bitbol 1996, 2000, 2002), this can perfectly be done provided one does not attempt to objectify a putative property behind each *token* phenomenon, but only the structure that enables us to anticipate phenomena of each *class* and under each *type* of circumstance.[7] Such an alternative approach will be developed in The Quest for Objectivation, as part of our discussion of the kind of objectivity that can be reached by introspective inquiry.

B. Let us come now to a second set of objections: essentially the objection that introspection alters the mental process to be known. There are at least three varieties and many subvarieties of this objection.

B1. *Observational distortion*: The attitude or operation of introspection *disturbs* the mental flux to be known (Hume 1962, Introduction).

B2. *Temporal distortion*

B2.1. One problem is a discrepancy between the fluent nature of experience and the request for stability of knowledge contents. Kant (2002) thus claimed that there can be no knowledge of the soul because the latter develops in time, whereas one should be able to immobilize it somehow in order to extract some knowable invariant. Similarly, Wittgenstein (1980) insisted that language, whose use is extended in time, can by no means catch experience in its present unstable actuality.

B2.2. Another problem is that what can be captured and mastered in experience is only its *past* unfolding. George Herbert Mead and Sartre (2000) thus pointed out that the "I" itself can only be considered as a reconstruction, or that the "I" is always in the past. But if this is the case, is there not a risk of deformation or oblivion? Can there not be a posteriori falsification of the history of lived experience, by the processes that Daniel Dennett (1992) calls "Orwellian" and "Stalinesque"?

B3. *Interpretative distortion*: The categories that subjects apply when they describe their own experience are theory-laden (Gopnik and Meltzoff 1994; Robbins 2004).

Moreover, the use of words alters the experience to be described, and they are even likely to be unable to capture anything properly in experience (this is the charge of *ineffability*).

This series of objections is not as threatening as it looks. Indeed, observational, temporal, and interpretative distortions can only be called "distortions" with respect to experience *an sich*, previous to any attempt at observing, catching, and interpreting. In other terms, the previous objections rely on some version of the "myth of the given" (Garfield 1989).

Let us take the issue of "disturbance" (Jack and Roepstorff 2002) as a paradigmatic example. Speaking of a process *an sich* that is unfortunately disturbed by the coarse instruments we use in order to have access to it, only makes sense if there is a way of accessing it independently of these coarse instruments. In any other case, this is wild speculation. Such a remark is (or should be) a keystone of the interpretation of quantum mechanics. True, the metaphor of an object disturbed by the experimental contraption was usually accepted by physicists in the first years after quantum mechanics was formulated; and it is still in use in popular science books. But it was soon clear that, if taken seriously, it can only lead to the accusation of the "incompleteness" of quantum mechanics (what is this theory that says nothing about the objects as they are *before* instrumental "disturbances"?). And this accusation in turn feeds the persistent dream of a "hidden variable theory." The metaphor of disturbance was then soon discarded by Bohr and replaced by the claim that a phenomenon is co*defined* by the experimental conditions of its manifestation, rather than *disturbed* by them. Here, the phenomenon is taken as inseparable from its experimental context. A similar move has been suggested for introspection: the new introspection bears immediately on reflective experiences rather than on the experience the reflection is about. "Even if the products of introspection are not the direct reflections of underlying thoughts, they are still manifestations of the workings of the mind. Thus, to the claim that spontaneous, unsolicited thought sequences are not reliable documentations of thoughts proper, we retort that we simply do not care" (Shanon 1984, n.p.). Saying that "we do not care" is certainly provocative, but it has the merit of pointing toward alternative epistemologies and alternative strategies (see An Inquiry into the Meaning and Truth of Introspective Reports).

C. The third set of objections claims that one is systematically mistaken (even apart from the attempt at formulating it in words) about one's own experience.

Part of this objection is grounded on the observation that it is very easy for subjects to go astray about the *stimulus* that was applied to them in order to trigger a certain experience. Titchener himself was extremely diffident about the ability of subjects to identify a stimulus: "introspection adds, subtracts, and distorts" (Titchener 1912, 493). More recently, criticisms have been formulated against the propensity subjects have to say that they have seen more than can be evidenced (Cohen and Dennett 2011), or against their inability to see major parts of what occurs in front of them if their attention is distracted (as shown by experiments of "change blindness") (Silverman and Mack 2006).

This objection is an amplification of the charge of distortion or incompleteness against introspection formulated in objections B. However, as we shall soon see (An Inquiry into the Meaning and Truth of Introspective Reports), this charge might well be excessive or misplaced.

D. The fourth and final group of objections focuses on the purely subjective status of introspective descriptions, and on the fact that the situation it concerns is irreproducible. Thus, according to Wundt's early but harsh criticism, unless it is constrained by a strong experimental environment of control, introspection is doomed to extreme idiosyncrasy: "the reports are irreplicable not only by others but even by the particular introspector himself" (Shanon 1984, n.p.). If this is so, a verbal report of introspection only concerns the person who reports at a certain time; it teaches us nothing about other people and not even about the same person at other times.

This is probably the most serious objection of all, but as we shall see (The Quest for Objectivation), the renewed conception of objectivity that arises from a nonrepresentational view of science similar to the view favored by standard quantum mechanics, also suffices to meet it.

We gather from all these objections and sketchy replies that the most crucial weakness of the introspectionist wave of the turn of the nineteenth and twentieth centuries is likely to be its unconditional acceptance of the classical, representationalist theory of knowledge. Since, despite so many blows (including from the contemporary cognitive science [Varela, Thompson, and Rosch 1991; Thompson 2007]),[8] this theory still remains very present in our philosophy of science, it is worth insisting on its deficiencies.

An Inquiry into the Meaning and Truth of Introspective Reports

In a nonrepresentationalist epistemological framework, the issue of the truth or reliability of introspective descriptions is likely to be given a completely new meaning.

The first criterion of truth that comes to mind under the presupposition of a representationalist theory of knowledge is that introspective descriptions should be faithful to the experimental or environmental input that triggered the experience reported. This (too) simple idea has long been criticized in old introspectionism and replaced with the criterion that an introspective description should only be faithful to a slice of experience. Titchener thus wrote: "The question, . . . so far as the validity of introspection is concerned, is not whether the reports tally with the stimuli, but whether they give accurate descriptions of the observer's experimental consciousness" (Titchener 1912, 487). Here, it looks as if Titchener accepts the correspondence theory of truth that goes along with a representationalist epistemology, although he applies it to "conscious contents" rather than to "stimuli." We shall come back to this point soon, but let us first dig more carefully into what the followers of the American introspectionist school called "the stimulus error" (Boring 1929), namely, the error of asking for faithfulness to an external stimulus.

This prescription *not* to seek correspondence between introspective data and stimuli might well have been directed against the first German school of introspection, namely, Wundt's. But in this case, the criticism is probably excessive. Indeed, with the help of the many instruments of his laboratory, Wundt focused his inquiry on very limited introspective reports having the form of judgments of time characteristics (duration or simultaneity), number, and intensity of *stimuli*. Moreover, under strict experimental control, his introspecting subjects turned out to be reasonably faithful to the stimuli that were imposed on them (Wundt 1901, 31).

A modified version of Wundt-like introspection has been revived recently with considerable success (under the name "quantified introspection") (Corallo et al. 2008), and it also yields a positive outcome about the accuracy of simple reports. In this case, the reports bear not on the stimuli themselves but on the time spent by subjects to perform a certain task involving simple stimuli. It appears then that there is a very strict correlation between the measured response time and the subjectively assessed response time (although there is a systematic discrepancy between the *absolute* values of these times). However, this good correlation is disrupted when a second tasks interferes with the first, which is interpreted by the authors as the sign of a competition between the two tasks for their access to the global workspace of the brain cortex. The suspicion of inaccuracy about stimuli, being partly misplaced, is then not sufficient to motivate the rejection of introspection.

Another indication that introspective reports may be less inaccurate about their stimuli than is usually thought can be found in disguised introspective work of the allegedly behaviorist era. One such line of research casts doubts on a widespread anti-introspectionist prejudice of cognitive scientists (after Dennett 1992). According to this prejudice, subjects are systematically wrong about their pretending to see a whole scene extended in space, for they are in fact unable to describe most details of this scene when they are asked to do so. But George Sperling (1960) showed that things might be much more intricate than this, and less challenging for first-person access. Sperling briefly confronted subjects with a table of letters and asked them to report the letters they could remember. Subjects usually claimed they had an iconic memory of the whole table, but, irrespective of the size of the table, they could hardly report more than four letters out of twelve or sixteen. Was their claim of being able to *see* the whole table after its presentation completely illusory? Further inquiry ruled out this negative interpretation of the initial reports. Subjects were asked to concentrate on a single line in the table, and to list the letters of *this* line. The outcome is surprising: subjects were able to report about four letters of *any* line chosen at random by the experimenter. Although the debate is still raging (Kouider et al. 2010), there are stronger and stronger reasons (Block 2011) to accept that subjects indeed have a short-term iconic memory of the whole table, with all its details, but that this memory begins to fade away as soon as a few letters are mentally attended and listed by them. To sum up, the initial introspective report of the subjects was much more accurate than usually suspected.

The way this accuracy was brought out is also very instructive: (1) put subjects in a situation of success rather than a situation of failure (i.e., choose the task in which subjects display optimal performance, and thereby substantiate their claim of seeing the whole table); (2) help them by asking focused questions about what they lived, rather than dispersing their attention by questions either too abstract or too broad in scope. Sperling's experiment conveys, by contrast, an important lesson about the way we should interpret Nisbett and Wilson's (1977; Johansson et al. 2005) influential negative result about the accuracy of introspective reports: this negative result was precisely obtained by systematic avoidance of the two former rules. Nisbett and Wilson's subjects were intentionally put in a situation of failure regarding their aesthetic choice of a human face, among the two that are briefly presented to them; and their attention was diverted from contact with lived experience by abstract why-questions ("what is the *reason* for your choice?"). In good agreement with this evaluation, one of us (CP) recently achieved a clear experimental confutation of Nisbett and Wilson's anti-introspectionist claim (Petitmengin et al. 2013). The method consisted in rehearsing their protocol,[9] while complementing it with a careful intermediate explicitation of the "how" of the experience of subjects, thus automatically bringing into play the two rules of maximal accuracy of reports.

Another locus classicus of the criticism of introspection, from which Watson inferred that a true science of mind could only be grounded on the study of behavior, is the famous unresolved quarrel of "imageless thought" (Ogden 1911; Woodworth 1906). This time, the threat to introspectionism looks even more serious than before, insofar as the issue no longer bears on the ability of introspective reports to be faithful to the stimulus that triggered experience but on their faithfulness to experience itself. In the heyday of introspectionism, the researchers of Titchener's school at Cornell University claimed to have brought out the presence of sense elements, kinesthetic feelings, and images associated with *every* thought process (Titchener 1909), whereas the researchers of the Würzburg school, such as Oswald Külpe, August Mayer, Johannes Orth, and so on (Humphrey 1951, 30), declared that imageless and even "nonsensory" thought exists. However, careful examination of the texts in which the debate about imageless thought developed has shown that the nuclear proto-interpreted data could after all be isolated from the school-related theoretical bias, and that in this case, no true divergence persisted (Hurlburt and Heavey 2001; Goldman 2001). Subjects of both schools indeed reported the existence of "vague and elusive processes, which carry as if in a nutshell the entire meaning of a situation" (Titchener 1910, 505–6) and which involve kinesthetic sensations. But they did not interpret these reports the same way (one school assimilated this to some sort of blurred image, whereas the other one rejected that reading); and both schools probably missed a more faithful description of them in terms of "felt meanings" (Gendlin 1962).

More than a failure of introspection, the outcome of this controversy clearly indicates what kind of work should be done in order to reach a possibility of intersubjective agreement: stepping down as much as possible on the scale of rational reconstructions, explanations, or generalizations and, once again, sticking as much as possible to the "how" of experience. Phenomenological-like reduction is a basic requirement, which relies either on the expertise of subjects or on the expertise of experimenters who can induce it in their subjects by means of a series of basic instructions and carefully selected questions.

In any experimental science, identifying "facts" requires a process of descent along the hierarchy of theory-ladenness; not of course in order to reach a realm of "pure uninterpreted content" but only to pick out a level of interpretation that is beyond discussion in a certain state of culture and research. Consensus on facts can be reached either by relying on a level of theorizing that is unanimously accepted because it is "paradigmatic" or (in revolutionary science) by coming as close as possible to the tacit presuppositions of elementary embodied know-how (in the sense of knowing how to *act*). In introspection, the process of descent must be pushed even further because the level of possible consensus to be reached does not concern our knowing how to *do* but, rather, our knowing how to *be* (in order to gain extended access to one's own experience). Just as ordinary know-how-to-act is learned by nonverbal interaction, imitation, and acquisition of a skill rather than by transmission of ideas, knowing-how-to-be can be learned by direct contact with experts and appropriate training (Wallace 2000) rather than by transmission of theories about the status of phenomenological-like reduction.

But how exactly can one ascertain the "faithfulness" of first-person reports, independently of any relation with the stimuli that triggered experience? One may distinguish two levels of faithfulness assessment: (a) signs of reliability and (b) criteria of validity.

a. As we have just seen, there is at least one index whose presence would lead to strong suspicions against faithfulness of first-person data: lack of consensus about general structures of lived experience. Conversely, one may take consensus about structures as an index of faithfulness, although this consensus might well be partly induced by theoretical (or subtheoretical) prejudice as in the Würzburg and Cornell schools. To avoid the latter bias as much as possible, we need *individual* signs of reliability that may help us to increase the degree of confidence of each interview taken separately. Such signs are currently in use, and their significance has been carefully discussed (Vermersch 1994; Petitmengin 2006; Hendricks 2009). They are detected in the form of bodily attitudes and rhythms of speech that indicate actual contact with one's experience during the process of reporting. However, one must keep in mind that such signs are taken as good ground for reliability only because they are connected with first-person access of the interviewers to the experiential correlates of similar signs within their own bodies. This confirms that faithfulness of first-person reports can be ascertained only by intersubjective criteria; there is no external "absolute" evidence.

b. The same can be said when criteria of validity, or even *truth*, of these reports are sought. Indeed, we can say at least one thing for sure: there is no way of comparing directly an experience *an sich* and its alleged report; neither for experimenters nor for interviewers, nor for the subjects themselves. This is obvious for experimenters, but it is also clear for subjects themselves insofar as their own act of "comparison" is a new experience in which the former experience to be reported is merged and recast. So, how can we sort out this difficult epistemological situation? By relying on sound epistemology rather than on the old representationalist and dualist epistemology.

To take a significant step in this direction, we may conveniently come back to Kant, who was lucid enough to see that the dream of direct comparison and one–one correspondence, far from strengthening ordinary representationalism and dualism in the experimental sciences of nature, undermines it. The age-old objection of skeptics, according to whom we have no "absolute" access to things (no access apart from the *relations* we have with them) and that therefore we can say nothing about what they are *in themselves,* was addressed by Kant in a very innovative way. He first acknowledged that we indeed have no apprehension of objects apart from our very procedure of access (Kant 1988, Introduction). Then, instead of trying to prove the correspondence between knowledge contents and some independent object "out there," he *defined* the object as whatever phenomenon is shaped by the class of intellectual operations used in knowing. The appropriate intellectual operations here aim at picking out the aspect of experience that vary in the same way (i.e., in a law-like way) irrespective of the personal, spatial, or temporal situations. This stable component of experience is considered "objective" *by definition,* and not by virtue of its (doubtful) correspondence with some extraexperiential reality. This suggests that skepticism about any region of knowledge cannot be overcome by relying on some external warrant, but only by using *internal criteria.*

Accordingly, when we look for criteria of validity of first-person reports that are able to resist skeptical doubts, we completely bypass the fruitless search for their *correspondence* to putative "private objects" and instead try to establish criteria of *self*-validation. We also exploit the opportunities of *mutual* validation offered by articulating the domain of first-person reports with cognitive science.

This strategy fits remarkably well with the current philosophy of science, which is undergoing a major paradigm shift. Many philosophers of science now realize that describing science as a passive face to face between the purely mental realm of theories and an extramental reality is a highly implausible picture. Experimental gestures, mathematical practices, and social debates, despite many controversies, are no longer seen as mere neutral windows opening on "pure" reality. Instead, they are understood as an interfacial matrix of ongoing agency, out of which strategies of theoretical prediction and conceptions of reality that are able to guide their use coemerge (Pickering 1995a; Gooding et al. 2005; Galison 1987). Here, as in Kant, answering skeptical doubts no longer amounts to showing that a one-to-one correspondence exists between theoretical symbols and real properties. Instead, it rather requires the displaying of successfully tested patterns of technological actions that have stabilized, have been adopted collectively, and have then been connected to one another in coherent networks. In other words, the new kind of answer to skepticism relies on a pragmatic coherentist conception of truth rather than on a correspondence theory of truth.

The same attitude toward skepticism can be adopted when the validity of first-person reports is at stake. This was already suggested in a pioneering paper by Benny Shanon (1984), and in an increasing number of articles since then (Petitmengin and Bitbol 2009). These authors pointed out that standard critiques just show that introspective data cannot be evaluated on the basis of correspondence; and that this is not to be wondered about or re-gretted because, after all, no other data, not even in experimental science, are *really* evaluated this way. The alternative is then evaluation on the basis of performative coherence, where "coherence" can concern several levels of practice: internal coherence in self-assessment and report (Bitbol and Petitmengin 2013); interpersonal coherence in dialogue (see above); and triangulated coherence in a network connecting introspective reports with experimental practice in psychology and neurology (see Neurophenome-nology). Just as, according to the second Wittgenstein, language must take care of itself (without foundational security in logic), introspection must take care of itself as much as possible (without foundational security in "correspondence" of any kind).

The Quest for Objectivation

Objection D in In-Principle Objections and Replies appeared to us as the most serious challenge to introspection. Here, we shall address it in the same spirit as the problem of *validity* of introspective reports.

The challenge is expressed as follows: what do these strange tales told by subjects about their own experience teach us about the objective world? Is their significance not restricted to each one of the subjects who provides them? Can one not understand the reluctance of mid-twentieth-century psychology toward the participative, empathic, or idiosyncratic aspects of introspection that only worsen the wandering of the science of mind in the swamp of subjectivity? In order to persuade ourselves that this objection is not as devastating as it seems, we can use once again a certain simili-tude between introspective psychology and microphysics. The questions just raised indeed remind us of two related questions that a Copenhagen quantum physicist might have asked. According to Bohr's analysis, each quantum phenomenon is a unique and irreversible event arising from the interaction between a micro-object and a macroscopic measuring appara-tus; moreover, there are few stringent circumstances in which the phenom-enon can be reproduced when the measurement is repeated on the same object. What do such isolated microphenomena teach us about the object *as it is in itself,* independently of the measuring apparatus and its interaction with it? Is their significance not restricted to single runs of the micro-experiment? This puzzlement by no means hindered the development of quantum mechanics. We then just have to find out what, in the methods of physics, made this overcoming of the (virtual) objection possible even before it was formulated.

To begin with, one must remember the most important consequence of Kant's redefinition of objectivity, as documented in the previous subsection: objectivity is not something to be found ready-made *out there,* but a project of operational extraction of invariant or covariant structures out of a cluster of appearances. What should be the method, in order to reach objectivity qua invariance?

Extracting invariant or covariant structures relies on a process of ascent in generalization and theoretical abstraction, opposite to the initial process of descent that is necessary to reach a nucleus of discourse taken as "factual" or "data-like" (see An Inquiry into the Meaning and Truth of Introspective Reports). In other terms, objectivity is generated ("constituted," writes Kant) by selecting an appropriate level of generality or coarseness, such that invariant structures may arise at that level. In the domain of validity of quantum physics, this procedure is implemented thus. One first renounces objectivation at the level of individual phenomena occurring in space-time (this is the reason that the ordinary concept of minute localized bodies is in jeopardy). This being done, one then ascends toward the level of statistical variables. Indeed, as one soon realizes, the strict reproducibility and indifference to order of measurement that is usually missing at the level of individual values is recovered at the level of their statistics. Finally, one ascends a step further, toward the upper level of those formal tools (called "state vectors") able to generate as many statistics as measurement types. State vectors therefore play the role of objective entities without bearing the smallest resemblance to our archetypal image of the objects of physics, namely, material bodies.

The procedure should be the same for introspection: descent and ascent.

> 1. Descent toward minimally interpreted descriptions of the subtlest lived events, without any attempt at asking the *subject* to reconstitute her own cognitive processes (which are actually just as little accessible to subjects as to scientists), or to explain her "reasons" *in abstracto,* or to stipulate her intended meaning. In other words, a

very careful process of phenomenological reduction must be asked from, or induced in, the introspecting *subject*. This process can be repeated and its outcome reproduced in many subjects.

2. A posteriori ascent of the *scientists* who are analyzing the introspective reports construed as data, toward structures generic enough to be seen as stable and invariant across subjects and circumstances. One must then look for these structures at a level of number and generality where variations progressively vanish: "The corpus reaches a state in which an increase in the number of tokens ceases to increase the variety of types" (Shanon 1984, n.p.).

This two-step procedure is exactly the one we apply when we practice the method of elicitation of experience by interviews: (1) being very careful in guiding subjects toward exquisite contact with their experience and undoing any rational reconstructions or generalizations that may interfere with their task of description; and (2) retrieving the data extracted from these disciplined descriptions and extracting generic structures out of them.

Neurophenomenology: An Extension of the Basis of Coherence

Some authors have proposed relying on neurophysiological studies either to discard or to corroborate introspective reports (Coghill, McHaffie, and Yen 2003; Klasen et al. 2008). But this is a lopsided approach that does not take into account the opposite procedure: namely, the (explicit or implicit) use of introspective reports to ascribe functional meaning to certain areas or processes in the brain. So, we need a more balanced approach. This was first suggested by Owen Flanagan (1993) under the name "triangulation": according to him, we have to study the mind from various angles (phenomenological, psychological, neurological, and even computational), none of which is likely to be reduced to any of the others. The conception of truth that arises from this approach fits well with the idea of performative coherence, yet it paves the way toward an extension of the basis of mutually coherent practices. Instead of seeking internal performative coherence in restrictive areas of research, one undertakes a general interconnection of these areas and posits constraints of mutual consistency.

This program took off with Francisco Varela's neurophenomenology (1996). In neurophenomenology, one neither tries to reduce the subjective to the objective nor the other way around; one instead sticks to the experiential realm out of which the subjective–objective dichotomy arises, and then posits *within it* a system of *mutual constraints*. Mutual constraints are enforced between first-person statements of phenomenal *structures* and third-person descriptions of those phenomenal *invariants* that are established by the collectively elaborated neurosciences. The neuroscientific and experiential categories indeed have to be mutually adjusted in order to become fully comparable and in the end compatible with one another. This requires both formulation of appropriate neurological concepts (such as long-range cortical correlations or temporal binding of neural activity) and full use of methods of introspective report (Bitbol 2002, 2006; Gallagher and Zahavi 2007).

The fruitfulness of this method was soon brought out. Antoine Lutz (2002) probed into the experience of recognizing a 3-D geometric shape after staring at an autostereogram, which is an organized dot pattern with binocular disparities. Subjects were first asked to press a button when the 3-D shape had completely emerged, while their brain activity (especially the long-range correlations of the cortical electrical activity) was recorded. At this level of minimal introspective reporting, many discrepancies arose in neuroexperiential correlation. The usual solution that consists in averaging out the discrepancies was rejected, and the subjects were asked instead (by means of the interview method of elicitation) to describe in exquisite detail the state of mind in which they found themselves while being given the task. It then turned out that one could define clusters of subjects according to the level of their preliminary attention, and that in this case virtually no neuroexperiential discrepancy subsisted. This was a clear illustration of the fact that one has to *look for* the proper locus of neuroexperiential correlation, or in Kantian terms that one must *constitute* this correlation by way of mutual ajustments of neurological and introspective procedures, rather than just "discovering" it. Conversely, individual introspective reports gained credibility by comparison with their introspectively interpreted neurological correlates. A new network of performative coherence was established, thereby providing us with an extended basis for ascribing some sort of truth to introspective data.

So, was the disappearance, or rather concealment, of introspection during the twentieth century inevitable? The question at this point is not whether we could have formulated other successful theories or not, as in many studies of contingency in science; it is whether or not we could have avoided

going astray as we did. We believe this misapprehension was virtually inescapable. But this inescapability was not due to some insuperable obstacle
that doomed any attempt in this direction, or to an inherent flaw of introspection as such. It was due to contingent historical conditions and epistemological misunderstandings that gave precedence to concurrent methods
and paradigms in the science of mind. These circumstances, as we have
documented them in the course of this chapter, are: lack of interest and
expertise in contemplative disciplines that could have promoted the indispensable phenomenological-like "reduction" (too cryptically) characterized by Husserl; poor understanding of how the regulative ideal of objectivity is pursued (rather than found ready-made out there); dominance of
a representationalist view of cognition and science; correlative dominance
of a correspondence theory of truth; lack of an extended basis, including
functional exploration of neural processes, for establishing the truth-as-
coherence of introspective reports; lack of understanding and mastery of
the way nonbehavioral psychotherapic methods (including psychoanalysis)
work and can be improved; and so forth. Had these historical circumstances been different, the whole development of the science of mind and
the resulting philosophies of mind would have been very different.

Here, we witness the effect of a contingency of methods rather than only
theories, of which a celebrated example has been given by Peter Galison
(1997). This author pointed out that the representation of the microworld
was very much influenced by a methodological bias: according to whether physicists made a predominant use of Geiger-type counters or bubble
chambers, they showed a theoretical inclination toward the "logic tradition" or the "image tradition," respectively. Similarly, the representations
and ontology of the mental domain have been hugely influenced by the
long-term dismissal of introspective approaches; and we can hence easily
figure out what these representations and ontology would have been if introspection had been given enough credentials. Instead of overdeveloping
third-person inquiry into behaviorism and brain physiology, and pushing
aside first-person and second-person research as mere tools for folk-psycho-
therapy, one would have promoted an advance of both third-person and
first-person disciplined research locked to each other in thorough interaction and "mutual constraints" (Bitbol 2012). Moreover, instead of favoring
eliminativist, reductionist, or functionalist views of mind and overcorrect-
ing these views by dualist or mysterianist antidotes, one would have found
a dissolution of the well-known problems of the philosophy of mind in a

balanced approach of the many facets, third-person and first-person, objectifying and participative, of "the world as we found it," which is nothing other than lived experience. Finally, instead of endlessly discussing the capacity of our theories to hook onto the real world, one would have reconsidered the very concept of reality by exploring the ground wherefrom it first arises, namely, the process of generating structures that are invariant across the manifold situations of sentient subjects. This alternative project of development of the science of mind, and of science in general, is exactly what awaits us.

CHAPTER 13

Laws, Scientific Practice, and the Contingency/Inevitability Question

JOSEPH ROUSE

The question posed in this book incorporates a nested counterfactual conditional. *If* a science had taken a fundamentally different course, then *if* that alternative science had addressed the same subject matter with comparable empirical success, would its results inevitably resemble our actual science ontologically and methodologically?[1] Counterfactuals are notoriously context-sensitive. In W. V. Quine's (1960, 222) famous example, different contexts yield different answers to how Julius Caesar might have conducted the Korean War. His ruthlessness and single-mindedness suggest use of atomic weapons; his military experience and technological know-how suggest firing catapults.

In our case, however, the problem is not how to answer a well-defined question in a given context. We need to find a context and define the question so as to tell us something important about science as it has been and will be. There are serious challenges. The embedded antecedent was needed to rule out some trivial answers to whether the sciences might have been different. If some science had lost all interest in its prior topics, accepted divination as evidence, forsworn all empirical accountability, or met economic or political catastrophe, of course science would be different. So what? Yet the very defining conditions that block such possibilities might also trivialize the question. Interpret success or sameness of subject matter sufficiently narrowly, and the inevitabilist conclusion is unenlightening. Loosen these interpretations too much, however, and contingency comes cheaply, without offering useful insight. Moreover, it is not enough to pose a question whose answer would be genuinely illuminating, if the evidence available is then insufficient to justify either answer.

This chapter explores some further specifications of this counterfactual question, rather than answering a single version of it. As in Hacking's (1999, ch. 3; 2000a) first presentations of the issue, I will not hide my inclinations, but my aim is to examine considerations pointing in different directions. This examination has two parts. I first pose some thought experiments that reformulate the question and what is at stake in answering it. Whether or not these alternative formulations make the question both interesting and answerable, they highlight important aspects of the sciences. I conclude this part with a hypothetical historical variant on twentieth-century biology that could give the contingency issue a more concrete, empirically accessible site.

The second part of the chapter takes up how the contingency issue is affected by some recent work on laws and nomological necessity. The contingency literature has so far studiously avoided this topic, using the less-unencumbered notion of "inevitability" to circumvent talk of "necessity." The unsettled philosophical literature on laws and nomological necessity probably contributes to this avoidance. That omission is nevertheless somewhat surprising, since two often-cited scientific defenders of inevitabilism, Steven Weinberg (1996a, 1996b) and Sheldon Glashow (1992), explicitly appeal to the importance of laws in science: although they do not say much about their conception of natural laws, it is clear from the context that they take natural laws to be *necessary* truths in some sense, and the necessity of laws seems to play an important role in their commitment to inevitability. In any case, the prominence of natural laws and nomological necessity in many conceptions of science strongly suggests that any serious consideration of the topic of contingency cannot avoid addressing laws and necessity. Moreover, Marc Lange's (2000, 2007, 2009) recent work now provides a more adequate and relevant conception of laws and nomological necessity.[2] The second part of the chapter thus considers whether and how Lange's approach might reconfigure the contingency issue.

Variations on a Theme from Hacking

In their initial papers that posed the contingency question, Hacking and Soler (2008b) were careful to consider multiple ways one might conceive what is at issue. To these alternatives I now add several further variations that I have not seen in the literature. I begin with a concern central to my own work. How is the contingency question affected by a shift of attention from scientific knowledge to scientific practice? Although reflection on scientific practice has contributed to previous reflections on contingency, the emphasis has remained focused on the possible contingency of knowledge claims abstracted from their interpretation and use in practice. We ask whether an alternative, empirically successful physics must preserve an ontological commitment to quarks or an epistemic commitment to Maxwell's equations. Yet any actual compilation of the current state of scientific knowledge, whether in review articles, textbooks, handbooks, or encyclopedias, is typically governed by how that compilation is to be used and is selective and interpretive accordingly. I am tentatively inclined toward an inevitabilist answer to the abstract question about ontological commitments, to the extent that the question is well defined. Yet I am more strongly inclined to doubt that such high-level ontological convergence matters much. Many of the aspects of science that I regard as more philosophically significant are contingent in fairly straightforward ways.

One respect in which attention to practice changes the question is through a different temporal orientation. The more common philosophical focus is retrospective (which achieved results to accept), rather than prospective (which scientific questions to ask and how to pursue them). The significance of this shift in orientation might become clear from a slightly different thought experiment. The canonical counterfactual hypothesis usually presented to pose the contingency issue considers the possibility of a science that took a very different conceptual or methodological trajectory and yet achieved comparable empirical success. But now imagine instead a science that proceeded along a very different conceptual or methodological trajectory yet *did* converge at the present moment on the same theoretical and ontological commitments as its actual counterpart. Would we expect this alternative science to share not only the retrospective achievements of its actual counterpart but also its orientation to future research? Would it ask the same questions, give the same priority to various research programs, or even assign the same plausibility to various unsettled hypotheses? Presumably not, or at least not inevitably.[3] If the subsequent trajectory of a science is shaped by the questions it asks and the judgments it makes about the importance and plausibility of various research agendas, then perhaps any convergence of alternative counterfactual histories of science would have to be short-lived.[4]

This thought experiment, if you accept its premises, nevertheless need not inexorably support contingentism. It could lead those of an inevitabilist bent to draw the line earlier. Instead of concluding that any successful science must eventually converge on our results, they might conclude instead that the very hypothesis of a successful, radically different historical trajectory is suspect: any science that would be as empirically successful as ours could not actually diverge too far from our actual conceptual and methodological history.[5] Such divergences would encounter dead ends that should eventually direct inquiry back toward its actual path, on pain of failure. Indeed, the inevitabilist might argue, the actual path of science is itself replete with dead ends. Two different histories that diverged in their failed lines of inquiry but converged on their empirical successes would not constitute fundamentally different sciences.

The turn to practice might revise the issue in other ways as well. Even the most dedicated scientific realist admits that science does not aim for truth alone. As Catherine Elgin eloquently expressed this point, "if what we want is to increase the number of justified truths we believe, our cognitive values are badly misplaced; if [instead] we seek quality rather than quantity, . . . [we must ask] what makes insights interesting or important [and] what knowledge is worth seeking and having" (Elgin 1991, 197). Most truths about nature are not scientific truths (in the sense of truths that matter to science). Differing judgments about scientific significance thus matter well beyond whether they lead to differences in accepted truth claims. We need to ask which accepted beliefs matter to science, and how they matter. Two alternative sciences that agree completely about the truth or falsity of particular statements could still differ dramatically about which ones are scientifically important, and whence their importance derives.

A more specific thought experiment might heighten the impact of this point. In this alternative imaginary history, physics since Newton looks remarkably similar to its actual theoretical, mathematical, and experimental history. Yet this alternative practice differs from our physics in its resolute and unswerving Newtonian piety. Every physicist would now recognize that God's creation and Christian scripture must be read together. Just as

Newton's empirical research led him to deny the Trinity and risk his professorial appointment out of concern for his immortal soul, so our modern physicists would tremble at the stakes in scientific inquiry.[6] High Church devotees of string theory and of alternative approaches to quantum gravity would worship differently and argue the scriptural as well as empirical and mathematical virtues of their respective hypotheses and models. Their Low Church counterparts would emphasize God's primary concern for physics at lower energies with messier entanglements and celebrate the glories of his dappled creation against those satanically tempted toward the hubris of high theory: just calculate His praises! No matter how closely such a practice resembles contemporary physics in its empirical achievements and theoretical conjectures, such a counterfactual history would comprise a rather different physics. Yet it is hard to deny that our sense of how and why physics matters is historically contingent as well as contested.

The turn to scientific practice also encourages attention to the material practice of science. Philosophers and other science studies scholars now recognize the importance to scientific understanding of alternative experimental systems, canonical theoretical models, and relatively autonomous traditions of instrumentation. Such concerns figured early on in Hacking's (1999, ch. 6) reflections on the role of weapons research in shaping the form of subsequent science, an early foray into the contingency question. Peter Galison's (1997) account of how surplus electronics from World War II built the material culture of postwar physics amplifies Hacking's point. Yet subsequent discussions of contingency have leaned back toward theoretical representation. Several papers (Hacking 1999, ch. 3; Soler 2008b; Franklin 2008; Trizio 2008) cite Pickering (1984a) on the contingency of high-energy physicists' acceptance of quarks, but none talks about the associated shift in experimental focus from soft scattering of hadrons to lepton-lepton interactions and the more rare hard scattering hadron events. Does the emphasis on ontological commitments suggest that, despite all the talk about scientific practice, we philosophers still believe that the really important changes in science concern theoretical beliefs and ontological commitments? Or is the contingency issue itself a new way to reassert the philosophical primacy of theoretical commitments? That could be so if the contingency issue were open to plausible dispute concerning ontology, but not concerning material culture. Grant that the instruments, experimental systems, and models with which scientific understanding is realized are more obviously contingent. It does not follow from the mere recognition that abstract ontological commitments are the only conceivable locus of scientific inevitability that such commitments are all that important. Imagine a counterfactual history of science that retained the high-level theoretical commitments of our actual science but differed substantially in the instruments, skills, models, and experimental practices through which those commitments were articulated and brought to bear on the world. Why not say that such a science differs fundamentally from our science in most ways that matter to scientific work?

A further thought experiment highlights why the issue of ontological convergence may not be the most interesting or important. Imagine now a history of science that closely resembles the actual history in its theoretical commitments and experimental practices. In this possible world, however, strong social norms and powerful institutions blocked the export of scientific materials, instruments, or practices beyond the laboratory. The scientific study of electricity and magnetism, synthetic organic chemistry and petrochemistry, radioactivity and nuclear structure, and cell biology and genetics were then highly advanced. This imaginary world nevertheless has no chemical industry (except laboratory suppliers), no electrical grid, no steam power outside of physics laboratories, and no influence of microbiology, genetics, or cellular immunology on medicine. Once again, theoretical and ontological convergence between alternative-historical sciences would nevertheless permit dramatic differences in meaning, significance, and engagement with other practices and concerns.

A further emphasis on scientific practice might lead to one more transformation of the issue. In asking whether an alternative empirically successful science would converge with our science, we implicitly presume that our current commitments are mostly correct. Yet in emphasizing science as research practice rather than accumulated knowledge, we can acknowledge that the sciences remain open to further development. An alternative science might fall short where current science does well, yet surpass it where (unbeknownst to us) our success is more limited. Moreover, truth is not the only epistemic concern in play. Sciences may differ in the detail of their conceptual articulation (Wilson [2006] offers telling examples of how concepts in some fields of applied physics are developed in very fine-grained ways at the cost of significant conceptual and mathematical discontinuities). They also make different trade-offs between explanatory power and empirical accuracy (Cartwright 1983). Accuracy and precision guide epistemic assessment in divergent ways. Technological applicability and theoretical coherence are distinct virtues. And so forth. When asking whether an alternative science could be empirically successful, we should not forget the variety and complexity of the possible measures of empirical success.

That completes my reflections on scientific practice. I now consider the complications introduced by different scales of analysis of science. An important impetus for contingentism has been the rise of microanalysis in science studies. The more closely one looks at specific scientific projects, the more plausible it seems that they might have been different. Inevitabilists may see themselves as instead standing back from local variations to discern broader basins of attraction. The microhistory of electromagnetic theory in late nineteenth-century Britain and Germany might have unfolded quite differently, they would urge, and yet relations of partial differential dependence among variables and relativistic symmetries between electric and magnetic fields must eventually emerge. Contingentists address this challenge with strategic microstudies. If even the major turning points in scientific history show irreducible contingency, then there may be no inevitable attractors. Pickering's (1984a) study of the November Revolution of 1973 in high-energy physics (HEP) is exemplary here. Yet the inference from such strategically chosen studies follows only if this contingency is temporary, such that the contingent alternative history then remains stable. A non-quark-based high-energy physics in lieu of the November Revolution is less consequential if it merely left open the possibility of a December or January revolution instead. Contingentism needs not only an alternative historical trajectory but also its longer-term stability.[7] In Laws, Nomological Necessity, and the Contingency Question, I will briefly return to the question of what inferences should be drawn from historical or sociological microstudies of science. Yet for now, this difference of analytical perspective threatens to be empirically irreconcilable. The biggest challenge for counterfactual history has always been to assess the plausibility of extended chains of alternative contingencies, and such iterated counterfactuals are not readily accessible to nonmanipulative empirical study (in the absence of dynamical laws that could be used to construct and run simulations, paralleling the use of global climate models to simulate counterfactual climatic histories). Contingentists may dismiss inevitabilists' demand to show them an actual alternative history of a successful, nonconvergent science, but they are usually not well-positioned to meet it. Hacking's suggestion that the contingency question is an irresolvable "sticking point" thus may seem very much in play.

The history of biology over the past century might nevertheless provide an interesting setting for posing the contingentist question in a more empirically tractable way.[8] Most discussions of the contingency question have focused on physics, but taking physics as a stand-in for science in general may again prove misguided. At the outset of the twentieth century, biological conceptions of heredity did not sharply distinguish development from genetics. That situation was dramatically altered by classical genetics, which notoriously relegated development to the study of "how genes produce their effects." This genocentric perspective was enthusiastically taken up within both the neo-Darwinian synthesis and what became molecular genetics, which together dominate the canonical narrative of twentieth-century biology. Yet a small but determined coterie of dissenters remained, some drawn to the holistic approach to genetics in Germany, others from embryology and its successor discipline of developmental biology. Moreover, the molecular revolution and the rise of genomics offer some retrospective vindication to their challenges to the classical gene concept, and the associated reductionism in ontology and methodology.[9] Geneticist Raphael Falk nicely summarizes this historical shift: "For the molecular reductionist what remains of the material gene is largely an illusion. . . . For many practical experimentalists the term gene is merely a consensus term or at most one of generic value" (Falk 2000, 342). Yet for all of its complexity, fuzziness, and internal tension, the gene concept remains and likely will remain entrenched.

In this context, we can imagine a counterfactual history of biology. In this possible world, Goldschmidt and others prevailed in their adamant opposition to any conception of "genes as the atoms of inheritance" (Falk 2000, 324). Development remained integral to biological understanding of heredity. DNA and molecular biology were incorporated from the outset within a more complex and holistic conception of epigenetic regulation. Could this counterfactual biology conceivably have attained the present level of empirical success and theoretical understanding, while doing without the classical gene concept and reductionist methodologies?

Underlying this counterfactual question is a more specific scientific concern. Regardless of whether ontological reductionism is correct, the history of biology has often been taken to vindicate methodological reductionism. We now know, contra Jacques Monod, that what is true for E coli is far from true for elephants or any other eukaryotes, but some comparably reductionist assumption might still be an "obligatory passage point" for any viable biological research program. Moreover, the long, detailed history of dissent and its vindication in more orthodox terms might allow for the reconstruction of a more sustained counterfactual narrative, with specific empirical considerations bearing on whether such a historical trajectory is conceivable. Some available considerations seem to point in different directions. The inevitabilist will highlight the failure of Richard Goldschmidt's and other holistic research programs. The reemergence of epigenetics and developmental genomic regulation drew on the rich technical and conceptual background of molecular genetics. Yet contingentists can respond that often, what was previously lacking was imagination and a willingness to ask the right questions. One review article notes that "Nüsslein-Volhard's and Wieschaus' [experiments on] mutations that perturb embryonic segmentation could have occurred forty years ago" (Ashburner 1993, 1501). Had others been ready to build on their work, the ingenious experiments of embryologists like Conrad Waddington or Barbara McClintock's work on transposition in maize might have guided experimental probing of the complexities of development and heredity without starting from a reductionist commitment to hereditary "particles." Yet inevitabilists could then respond to such contingentist challenges with some empirical specificity. Marcel Weber (2007) points out that the first efforts at DNA sequencing had to rely on the extensive classical-cytogenetic mapping of mutations in *Drosophila* to identify which DNA segments to clone. This and other historical studies may suggest that recognition of complex regulatory dynamics could only emerge from the achievements and limitations of its reductionist predecessors. Without reductionist oversimplification, biological complexity may be initially intractable. I do not pretend to know how this dialectic should come out. I do believe that the history of counterreductionist research programs and the gradual emergence of epigenetic complexity are rich resources for empirical exploration of whether central ontological and methodological assumptions of a highly successful science really were inevitable.

Laws, Nomological Necessity, and the Contingency Question

I now turn to the issue of how natural laws and nomological necessity bear on the contingency question. Contingentist responses to Steven Weinberg's and Sheldon Glashow's avowed inevitabilism have largely focused on their commitment to realism, but in fact, their most pointed remarks concern the role of laws in science. If we are going to cite them as exemplars of many scientists' commitment to inevitabilism, we should at least take their emphasis on laws seriously. In science studies apart from philosophy, there has been relatively little discussion of laws and nomological necessity. Even within philosophy, the topic confronts skeptical currents. The most familiar conceptions of law have well-known difficulties, and the theological overtones of the concept raise suspicions (Giere 1999, ch. 5; Kitcher 2001, ch. 6). The eclipse of deductive-nomological theories of explanation, the emergence of serious philosophical attention to evolutionary biology and other sciences of historical contingency, and a growing recognition of models as an alternative locus for theoretical understanding have encouraged thinking of "science without laws."[10] I was once among the skeptics. Marc Lange's (2000, 2007, 2009) revisionist account of the role of laws in scientific practice nevertheless should lead us to revisit the issue; Lange's work also enables reformulation of the contingency question in some newly illuminating ways.

I will extract three primary considerations from Lange's rich and informative discussions of laws: the role of laws in scientific reasoning, their pervasiveness in science even beyond physics, and the resulting reconception of "necessity" in inferentialist terms. First, Lange emphasizes the role of laws in scientific practice, especially for inductive and counterfactual reasoning. Laws are not just especially important achievements in retrospect but are also integral to further exploration in research. Indeed, one might argue that the difference between laws and subnomic truths primarily concerns their prospective rather than retrospective import. A long empiricist tradition has claimed that a retrospective assessment of empirical data could justify no claim stronger than the subnomic counterpart to a law; the long history of empiricist skepticism about laws that would reduce them to actual regularities (i.e., to subnomic truths) largely turns on such retrospective considerations of empirical justification.[11] Lange (2000) instead introduced his reflections on the role of laws in scientific practice by asking what (prospective) *commitment* is expressed by taking a hypothesis to be a law, and not merely a subnomic truth. The most crucial difference is then inductive-inferential, and the laws then express norms of reasoning within scientific practice as well as truths about the world. In taking a hypothesis to be a law, scientists thereby implicitly claim that the best inductive *strategies* to pursue in this context will vindicate the reliability of the inference rule corresponding to that hypothesis.[12] A law expresses how unexamined cases would behave "in the same way" as the cases already considered. The familiar difficulty, of course, is that *many* inference rules are consistent with any given body of data. Lange therefore asks which inference rule is *salient* in this context. The salient inference rule would impose neither artificial limitations in scope nor unmotivated bends in subsequent extension. For example, inferring from local electrical experiments that "all copper objects *in France* are electrically conductive" would be an inappropriately narrow scope limitation. Absent further considerations, geographic location is not salient for those experiments. In the other direction, Goodman's (1983) "grue" and Kripke's (1982) "quus" are thought experiments that provide infamous examples of unmotivated bends in inductive strategy.

Even science studies is subject to these considerations. A small number of empirical case studies such as Pickering's history of quark theorizing or Shapin and Schaffer's (1985) treatment of Hobbes and Boyle seem sufficient to generate far-reaching conclusions about science. We need not ask in the wake of Pickering's work whether the November Revolution would have been inevitable after all, if only there had been high-energy physics research centers located in Italy, or all physicists had worn purple shirts. But why not? Lange's point is that any inference from empirical data implicitly invokes laws, that is, patterns of invariance under counterfactual perturbation. Indeed, to pose the nested counterfactual query at issue in this volume is to commit oneself to there being laws of science studies in this sense.[13] What laws of science studies express the inductively salient inferences about scientific contingency to be drawn from empirical case studies? If Pickering's account is correct, it is not merely a peculiar and localized feature of high-energy physics in the early 1970s that its subsequent theoretical course was contingent on a variety of local features of scientific practice; Pickering's case study is intended to provide empirical confirmation for the claim that other sciences were also historically contingent in their methodological and ontological commitments.

The second point I take from Lange, already exemplified by the recognition that there might be laws of science studies, is that this conception of laws is not limited to the physical sciences but instead applies throughout empirical inquiry. There are, for example, laws of functional biology, which allow biologists to infer from anatomical, functional, developmental, or genetic features of an organism that other organisms of the relevant taxon will display the same features, absent evidence to the contrary. How else could scientists reasonably draw inferences about biological function from a limited data set, let alone intelligibly claim to have sequenced *the* genome of *Drosophila* or *C. elegans*? Lange also extends these considerations to more specialized sciences: there may be laws of cardiology (e.g., concerning the effects of epinephrine injections on heart function), island biogeography (the area law for species diversity), or clinical psychopathology (e.g., concerning the effects of a drug on patients who exhibit characteristic symptoms of autism). These laws, as always, are circumscribed by the aims of inquiry. Laws of functional biology admit of exceptions (recognizing the variation present in evolving populations), so long as the exceptions are tolerable background noise. That same variation, of course, would no longer be noise within evolutionary biology, which consequently invokes a different set of laws. In cases outside the scope of a discipline's concerns, *any* inference rule is reliable enough for that discipline. Thus, Lange points out, it does not matter to cardiologists what effect epinephrine would have induced had human hearts evolved differently: *any* inference rule about the effect of epinephrine on differently evolved human hearts would be reliable enough for the practice of cardiology (but not for the purposes of evolutionary biology).

Laws thus have quite different ranges of counterfactual invariance, corresponding to what is at stake in various scientific practices.[14] Moreover, these ranges are not hierarchically structured along familiar reductionist lines, since the laws of the "special sciences" would still have held under some counterfactual suppositions that violate the laws of physics. The laws of island biogeography or of functional biology would still have held even if the laws of physics had been different in some respects or had been violated. Evolution bars creationism, but cardiology need not: had some god created the universe six thousand years ago, epinephrine would still have affected human hearts in the same way. The laws of a scientific practice express norms of inductive inference and counterfactual invariance that govern how one ought to reason in that practice, whether in designing experiments, extrapolating from evidence, or acknowledging limits of scope, accuracy, and relevance. To a significant extent, those norms are independent of other disciplinary concerns.

The third aspect of Lange's view highlights its special relevance to our topic. The laws of various scientific practices carry their own form of nomological necessity. Nomological necessity is a distinctive level of subjunctive and counterfactual invariance. It accrues to the entire set of laws in a scientific domain, not to individual laws directly. To see how this conception works, start with the distinction between laws and subnomic statements: "A claim is 'sub-nomic' exactly when in any possible world, what makes it true (or false) there does not include which facts are laws there and which are not" (Lange 2009, 17–18). What then distinguishes the laws from contingent truths is their "sub-nomic stability." That is Lange's term for the collective inferential stability of the laws' subnomic correlates under counterfactual perturbation. The set of those subnomic truths m for which m is also a law is inferentially stable under any supposition logically consistent with that set.[15] Lange's point in using subnomic stability as the distinguishing mark of laws becomes clearer by contrast. A set of statements that includes even one accident will not be subnomically stable. There will always be some circumstances, logically consistent with the set of laws and one or more accidents, in which an accidental truth would not have held. For example, if the laws of physics were augmented by the contingent truth that I attended the 2009 workshop on contingency at Les Treilles, that augmented set is not subnomically stable. Had my flight from New York to Zurich crashed, then I would not have attended the workshop, yet that counterfactual supposition is *logically* consistent with the set consisting of the laws of physics plus the accidental truth that I attended the workshop. For any accidental truth, there will be some nomologically possible counterfactual hypothesis under which that truth would not have held (that is, what it is for an accidental truth to be "contingent"). By contrast, the laws of physics would still have held under *any* counterfactual hypothesis that did not itself violate one of those laws. To be a law *is* to be a member of such a counterfactually stable set, in a way that confers no privilege on physical laws: if there are laws of functional biology, then they would still have held under any counterfactual hypothesis consistent with the laws of functional biology (even though *some* of those hypotheses would violate physical or chemical laws). That pattern of reliable counterfactual stability that is constitutive of sets of laws is crucial background that enables counterfactual reasoning in the sciences: scientists rely on the laws to determine what would have happened under various counterfactual scenarios.

I have only quickly sketched some key features of Lange's account of laws and nomological necessity, but these should be sufficient to indicate three ways in which accepting that account would affect the contingency issue: it suggests a new thought experiment for the inevitabilist, but also two forms of pluralism that give contingentism another dimension. First,

Lange gives us another revealing way to construe the issue. We can ask whether an alternative science could duplicate the empirical success of an actual science (literally duplicate it), by agreeing about the truth or falsity of every subnomic statement in its domain, and of all counterfactual conditionals whose antecedents or consequents include those subnomic truths, but disagreeing about which of those truths are also laws. Here, the answer is no. Lange (2009, 37–38) proves that for any alternative inferentially stable sets of subnomic truths in a given domain, one must be a subset of the other. The differences in levels of this hierarchy reflect different species of necessity (logical, metaphysical, nomological, etc.), rather than alternative ontological commitments.[16] If sameness of subject matter is identified jointly by the set of subnomic claims relevant to a science and what is at stake in its inquiries (which determines the counterfactual suppositions that are of interest, and what is sufficient reliability), then the inevitabilist is surely correct in this case. Is this conclusion just another form of the question that trivializes the answer? No, because we learn the importance of counterfactual reasoning in scientific practice, and how it makes a difference to the contingency issue. Sciences do not merely propose and examine hypotheses about what actually happens, they also enable reasoning about what might or would happen under counterfactual circumstances. Recognition of this role for counterfactual reasoning strongly constrains how scientific practice proceeds.

Yet Lange's account also offers resources for contingentists. Understanding laws as norms of inductive and counterfactual reasoning encourages a wide-ranging scientific pluralism. Not only are the laws of so-called special sciences independent of the laws of physics, but their range of invariance extends to some counterfactual suppositions that violate physical laws. Moreover, some of these sciences cross-classify. I already mentioned that the same pattern of variance within a population has very different significance within evolutionary and functional biology. Medical science then enters this terrain with different stakes, different sources of evidence, such as patients' first-person reports, and different conceptions of those stakes and which facts matter to the science.

A different kind of cross-classification emerges with patterns of inductive salience from different theoretical presumptions or different demands for precision or accuracy. Consider, for example, the different contexts in which Boyle's law and the van der Waals law are inductively salient and empirically adequate, and those in which there is no general gas law at all,

but only different laws for the chemistry of each gas.[17] Such cross-classifi-
cations also include the differences among quantum mechanics, classical
mechanics, and the various semiclassical models or laws. Such differences
are sometimes pervasive even in a single domain. Thus, Mark Wilson calls
attention to the telling case of "*billiard ball mechanics*": "It is quite unlikely
that any treatment of the *fully generic* billiard ball collision can be found
anywhere in the physical literature. Instead, one is usually provided with
accounts that work approximately well in a limited range of cases, coupled
with a footnote of the 'for more details, see . . . ' type. . . . [These] specialist
texts do not simply 'add more details' to Newton, but commonly overturn
the underpinnings of the older treatments altogether" (Wilson 2006,
180–81). After detailing the many permutations of "classical" treatments of
billiard ball collisions, sometimes overlapping and sometimes leaving gaps,
Wilson concludes that, "to the best I know, this lengthy chain of billiard
ball declination never reaches bottom" (181).

Lange's work thus advances beyond the now widespread recognition
of the multiplicity and mutual inconsistency of theoretical modeling. He
provides a principled basis for asking when such multiplicity actually constitutes a distinct scientific domain. Absent Lange's considerations, such
models would merely illustrate Wilson's suggestion that scientific theory
resembles overlapping facades in an atlas rather than a single, unified map.
The crucial point Lange contributes is the holistic inferential relations
among laws. Lange's account suggests that we enter a different scientific
domain when its theoretical articulations function together as a systematically interconnected, counterfactually stable set of laws rather than just a
collection of models of varied utility for various purposes.[18] This difference
might then radically undermine the inevitabilist's challenge for the contingentist to display an actual alternative successful science, since it shows
that this challenge may have already been met. Lange's account suggests
that physics as we know it is already composed of multiple distinct bodies
of knowledge, including quantum mechanics, neoclassical physics, and
perhaps even various domains of materials science and applied physics.
These domains are conceptually and nomologically distinct, and yet each
has impeccable empirical adequacy within its domain. Attention to the
importance of nomological necessity and counterfactual reasoning in scientific practice might then undermine the inevitabilist thesis on its home
turf, the extraordinary empirical success of science as it actually has been.

PART VI

Contingency and Scientific Pluralism

CHAPTER 14

On the Plurality of (Theoretical) Worlds

JEAN-MARC LÉVY-LEBLOND

The Mathematical Polymorphism of Physics

I will start with a rather uncontroversial statement, at the very heart of standard physics, namely, that many physical theories show a mathematical polymorphism of great import (Lévy-Leblond 1992). By this, I mean that many of the laws and concepts in a given domain may be endowed with several quite different mathematizations. An elementary example is that of uniform rectilinear motion, which may be thought of in chrono-geometrical terms: equal spatial distances covered in equal time intervals (Galileo); in functional terms: linear dependence of the distance on time; in analytical (differential) terms: constant velocity or zero acceleration (Newton). A less trivial example is given by the classical dynamics of point masses, which may be formulated in terms of simple differential equations (Newtonian viewpoint), systems of partial differential equations (Hamiltonian viewpoint), variational equations (Lagrangian viewpoint). Of course, such different formulations are equivalent, mathematically and logically speaking. But they are not so, by far, physically speaking: there are clear conceptual and practical distinctions between them, since the advent of a new formulation is not a mere game of diversification but usually corresponds to the necessity of solving new problems or to the influence of novel ideas (originating either in other branches of physics or from developments within mathematics).

The mathematical polymorphism of physics thus expresses the diversity of the situations in which a given law must be implemented and its more or less efficient working, dependent on which formulation is used to deal with the specifics of the problem. Referring to the example already mentioned, classical dynamics will be used mainly in its Newtonian form if one is interested in computing trajectories, but its Hamiltonian form will be preferred instead if one is looking for conservation laws. Similarly, the choice between local formulations of electromagnetism (Maxwell differential field equations) and global expressions (using integral quantities, such as flux or circulation, in Gauss's and Ampère's theorems) is dictated by the circumstances at hand. This is the reason that physics, in contrast to mathematics—at least in their modern form—is not readily amenable to axiomatization. Given a set of different statements, all of which can be deduced from one another, the physicist is reluctant to impose a hierarchical ordering. "Principles," "laws," and "applications" in physics possess a relative mobility, an interchangeability, to a much higher degree than axioms, theorems, and corollaries in mathematics (Feynman 1965).

The plurality of theoretical physics exhibited by its mathematical polymorphism is even more pertinent with respect to the future of physics than to its past. Among the various formally equivalent formulations of a given physical law available at a certain time, only one or a few of them may in general extend naturally to new phenomena. To take the same example, Einsteinian relativistic mechanics hardly tolerates a Newtonian-like formulation (originally requiring instantaneous action at a distance), whereas the Lagrangian formalism applies quite naturally.

Now, the importance of a particular formulation cannot be attributed solely to its intrinsic, mathematical, merits. Each one of a set of logically equivalent formulations carries with it a complex set of mental representations, even if only through its terminology—a point to which we will return at the end of this chapter. It carries specific worldviews beyond the realm of physics proper and may thus be preferred or dismissed by certain schools of thought. This cultural and ideological matrix can play an essential role in the physicists' adoption or rejection of the considered formulation, at the individual or collective level. For instance, it is most interesting to study in detail the teleological and theological implications of variational principles, from Maupertuis to Feynman (Yourgrau and Mandelstam 1968).

Conversely, a physicist's philosophical or metaphysical preferences may well condition his or her privileged attachment to a certain type of formulation and, according to the case, may propel or impede his work; it may be held, for instance that Einstein's predilection for geometrized physics helped him discover general relativity, while obstructing his acceptance of quantum theory. In order to emphasize the significance of the mathematical polymorphism of physics as a real plurality, let us now investigate in detail the case of the so-called special relativity.

What if Einstein Had Not Existed?

Physicists have become used to the practice of *gedankenexperimenten*, which consists in asking counterfactual questions—"What if . . . ?" It all started when Galileo wondered "What if air resistance did not exist?" and discovered the law of free fall (Lévy-Leblond 2001a). This strategy was later fruitfully extended and popularized in physics, in particular by Einstein. I propose to use it here, not *in* physics, but *on* physics. Understanding the unfolding of any specific stretch of history greatly benefits from imagining how it could have developed otherwise, if a certain crucial circumstance had been different. The investigation of historical contingency through the construction of alternate fictitious narratives has lately become a well-established endeavor (Milo and Bourreau 1991). Of course, this has long constituted a well-known domain of literature.[1] The use of similar strategies in the history of science is less developed,[2] and I wish to contribute here to its extension.

Let us consider the development of space-time physics (usually but unfortunately referred to as "relativity theory") at the beginning of the twentieth century, to which the contribution of Einstein was central. I will show that another, quite plausible, scenario could have taken place had Einstein not been there to make his justly celebrated 1905 breakthrough.[3]

Let us start by recalling the real course of events. Classical physics, as epitomized by Newtonian mechanics, further developed by d'Alembert, Lagrange, Hamilton, Jacobi, and others, implied a view of space and time characterized by an absolute time and contained an implicit theory of relativity, today called in retrospect Galilean relativity (as a deserved recognition of Galileo's first understanding of its essence). This was completed by the emergence of physical concepts such as energy and momentum, expressing the dynamic properties of the Newtonian notion of mass (figure 14.1).

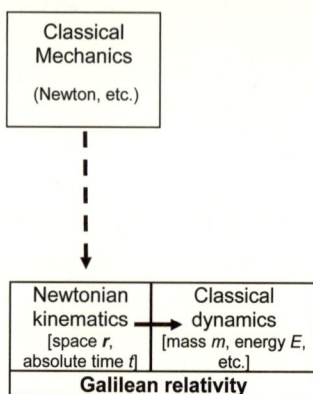

FIGURE 14.1. The development of the classical theory of space-time

However, by the end of the nineteenth century, the advent of a full-fledged theory of electromagnetism, following the seminal work of Maxwell, gave rise to a serious conflict, insofar as one of its main aspects was the appearance of the velocity of light as an invariant magnitude. In other terms, the law of addition of velocities, characteristic of Galilean relativity, was invalidated. After years of various but equally unsuccessful attempts to do away with this problem, it was the great merit of the young Einstein to cut the Gordian knot by sacrificing the classical view of space and time to propose a novel one, consistent with electromagnetic theory. His revolutionary gesture in fact was a conservative one, in that he got rid of the Galilean theory of relativity only to replace it with a new one, thus safeguarding and consolidating the principle of relativity itself (figure 14.2). Einstein's views were fast accepted and developed (most important, concerning the mathematical structure of the new space-time, by Minkowsky). He and others (notably Langevin) then demonstrated how to recast the dynamic notions of mass, energy, and momentum, leading in particular to that most famous of equations in physics, $E = mc^2$, and to the understanding that the inertia of a body increases with its velocity.

Let us now start our *gedankenexperiment*, by asking what could have happened if Einstein had not existed. In fact, by the beginning of the twentieth century, a large amount of work had already been devoted to the question of the relationship between mass and energy within electromagnetic theory. It had been understood by many authors that, due to the electric field surrounding a charged body, its inertia should be a function of its velocity, and various formulas had been proposed (Wien, Abraham, Langevin, and Bucherer, etc.). Max Planck was soon to propose that this relation should be a universal one, valid for all bodies, whether electrically charged or not. It can safely be asserted that if Einstein's paper had not been published in 1905, a new dynamic approach establishing the correct relationships between mass, energy, and inertia would have appeared within a very short time. The structure of space-time would then have been reconstructed in accordance with this new dynamics, inverting the historical order that, to this day, is still considered the logical (and in any case didactic) one (figure 14.2).

But the experiment can be pursued further, so as to reach a deeper result. Indeed, Einstein's derivation of the new relativity theory suffers from a serious weakness. It relies on the analysis of the comparison of space and time measurements performed in different frames, and related by the exchange of luminous signals. The invariance of the velocity of light thus has to be taken as a basis of the theory—the so-called second postulate. The very success of this operational approach, to most physicists (and science philosophers as well), hides its ad hoc nature and limited validity. For how could such a fundamental structure as that of space-time depend on a specific property of a particular physical agent, even if it were light itself? What guarantee do we have, then, that other and independent phys-

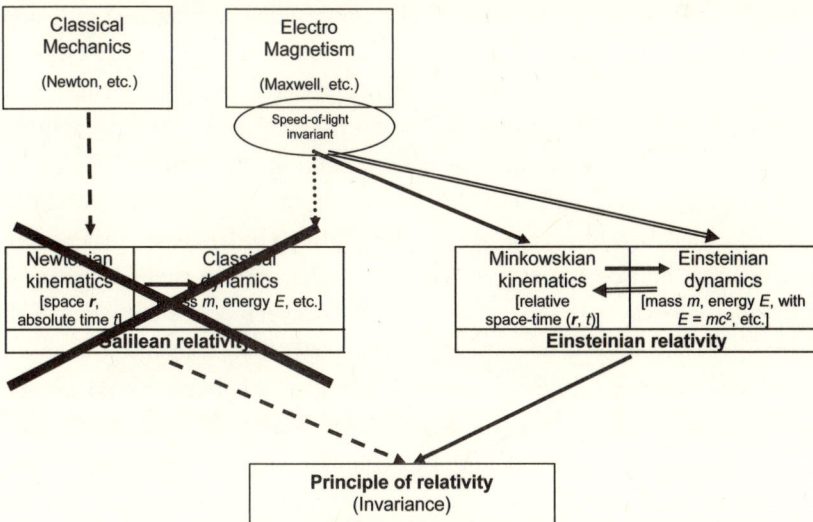

FIGURE 14.2. The development of the modern theory of space-time

ical phenomena should be ruled by the same structure? The question, although somewhat speculative at the time of Einstein's publication, would soon have become quite relevant with the discovery in the following decades of nuclear interactions, which indeed do obey Einsteinian relativity even though they have nothing whatsoever to do with electromagnetism. A second flaw appeared (or should have . . .) when the dynamic notions of mass, energy, and momentum were recast in accordance with the new spatiotemporal order. It was then understood that the invariance of the velocity of light was but a consequence of the massless nature of its elementary constituent, the photon. But, in due rigor, no physical magnitude can ever be known with total precision, so that all we can measure is an upper limit for the mass of the photon. No matter how small this quantity is, the possibility of a nonzero mass for the photon cannot be excluded. In such a case, light would ultimately not travel with the supposedly invariant velocity of light, thereby undermining the very historical foundation of relativity theory. The foundational weakness of relativity theory due to the use of the second postulate was soon noted, and various papers appeared during the second decade of the twentieth century showing how to deal with this hypothesis. The idea is a very deep one, going back to the Erlangen program of Felix Klein, which established the existence of an intrinsic link between a given geometry, that is, a spatial structure, and an invariance group. This idea may be extended via the general and abstract relativity principle to the spatiotemporal structure, basing it on an appropriate invariance group and giving rise to what could (and should) be called, rather than the customary "relativity theory," a genuine "chronogeometry." Such a construction, relying on a few very general assumptions (such as the homogeneity of space-time, the existence of some causal ordering, etc.), irrespective of the nature of the physical phenomena themselves, then leads quite naturally to the so-called Lorentz group, which entails the Minkowskian structure of space-time (figure 14.3). It is my contention that, had Einstein not existed, this alternate way to the theory could well have been consistently followed, leading to success and becoming the accepted view. Instead, the relevant works all remained largely ignored, as the idea was rediscovered decade after decade but soon equally fell into oblivion (Lévy-Leblond 2001b). The situation has only recently started to change, and this more general and more solid presentation of the theory is slowly finding its way into modern textbooks.

Now, of course, the first question is, can these alternate paths be taken

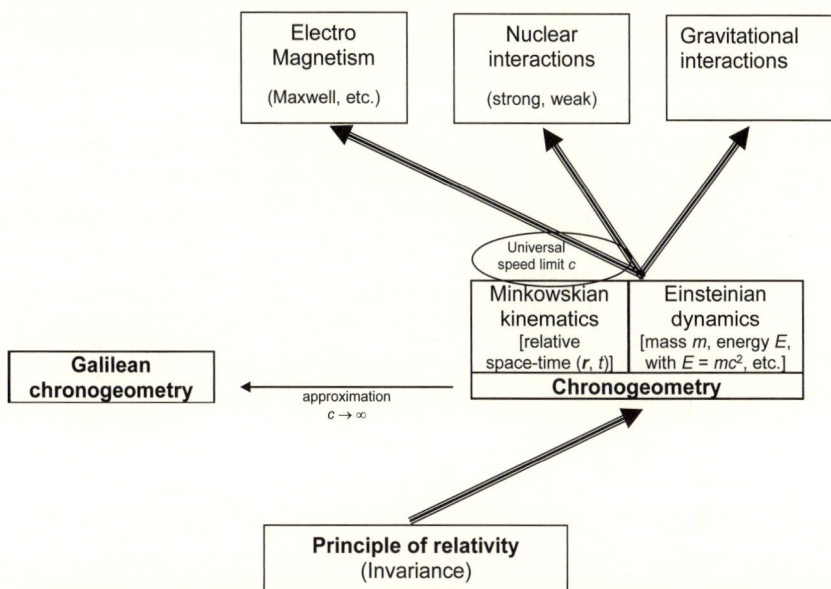

FIGURE 14.3. A completely different fictitious development of the modern theory of space-time

as arguments for the contingency of a most important aspect of modern physics? Or are there good reasons that views akin to Einstein's would have triumphed even through the work of someone else? Based on a detailed study of the state of the art, and taking into account the vast body of now forgotten papers that pointed in alternate directions, I do not hesitate to assert that, indeed, the science of space-time in the twentieth century could have developed differently.

But there is another, deeper question that might now be asked by the inevitabilist: "All right, you tell a different story and make it very plausible, but isn't it the case that it finally ends up with the very same theory, so that my position is vindicated?" To which, the contingentist, that is, myself, may answer:

It is *not* the same theory, unless you identify a physical theory with its formal structure and neglect its intellectual contents. Had the alternate approach been realized, there would be a number of significant differences:

• The set of fundamental notions would be (partially) different and their conceptual status clearer:

—there would be a clear distinction between mass (constant) and inertia (variable) instead of the present confusion around the notion of "restmass" versus mass variable with the velocity;

—the limiting velocity (universal, to be called perhaps "Einstein's constant") would be distinguished from the velocity of light (even though the latter is quite close and perhaps equal to the former);

—the epistemological accent would be placed on invariance (absolute notions) instead of variability (relative),[4] in better agreement with the foundations as well as with the practical use of the theory.

• The foundational solidity of the theory would be much higher; the possible measurement of a nonzero mass for the photon (that is, of a noninvariant velocity of light) would not at all put into question the consistency of the theory, and its validity for ruling possible new phenomena would in principle be guaranteed.

• The generality of the theory would appear much stronger, as its group-theoretical derivation precludes any other possibility, thereby severely restricting the range of alternatives; if a replacement of the theory appeared to become necessary in the future, it would require the replacement of most deep postulates.

• Pedagogical approaches to the theory would be different and certainly more convincing, as many of the weaknesses or loopholes of conventional presentations very often give rise to misunderstandings or nonunderstandings by students.

• Last but not least, the large cultural impact of relativity theory would have been quite different, from its many and often problematic philosophical exegeses (e.g., Bergson 1922),[5] to its vernacular manifestations, for instance in popular iconography (Einstein's tongue!).

Similar counterfactual but consistent narratives could easily be built about almost any specific domain. Admittedly, we are dealing here with cases of contingency in a somewhat limited sense. But this is precisely why I choose to present this argument: although it does not constitute a proof in favor of the possibility of a radically different science, it has the merit of relying on established facts and offers what I hope to be a convincing argument within its restricted area.

Minority Reports: Actual Pluralism

Most arguments in favor of the strong pluralist-contingentist thesis aim at showing that there could be, depending on some circumstances, various inequivalent forms of scientific knowledge enabling us to understand a given body of experience. The accent usually is on assessing the comparable validity of the alternate potential theorizations. I suggest here that it might be interesting to consider a prior question, namely, that of the actual existence of diverse conceptual frameworks. In that respect, some episodes in the history of science already offer good examples of effective pluralism in science not "as it could have been" but as it actually has been. I take issue here with the Kuhnian *doxa*, according to which, between short revolutionary episodes, science proceeds along a "normal" path, following a standard paradigm. In many cases, one can show that, alongside the orthodox majority view, heterodox minority views remain durably alive. Let me sketch two such examples, taken from the history of physics.

Le Sage's Theory of Gravitation

It is well-known that Newton's very successful theory of gravitation met with a serious difficulty, namely, that it required instantaneous action at a distance. The inverse-square law of gravitational attraction between two celestial bodies dealt with the distance of the two bodies. But how could the Earth act on the Moon through the void space separating them? This was the root of the long persisting objection of most continental physicists. Newton himself admitted the existence of the problem, to which he famously answered "Hypotheses non fingo," that is, "it works, but I do not know how." A clever solution to the question was offered by the middle of the eighteenth century by Georges-Louis Le Sage (1782), a physicist from Geneva, following an earlier attempt by Fatio de Duillier, a young close friend of Newton for a time.

Le Sage postulated the existence of a universal isotropic and uniform flux of particles filling the cosmos, of a nature different from those of our ordinary "mundane" matter, which he thus called "ultramundane corpuscles." The impact of these corpuscles on an isolated celestial body would have no effect because any push by a corpuscle in one direction would be compensated by an equal and opposite push by a corpuscle coming in the inverse direction. However, when two such bodies at a finite mutual distance are considered, it is easily understood that each one prevents a part of the flux from reaching the other one. This partial screening effect then gives rise to a noncompensated pressure on the outward-facing sides of the bodies. There is no "attraction" between the bodies but rather an inward push. The great epistemological appeal of the theory is thus that action at a distance is but an appearance, the true forces at work being contact actions, which, at the time, seemed much more natural and required no deeper understanding. Besides, when Le Sage's theory was mathematically formulated, it offered an immediate explanation of the inverse-square law, which, in Newton's theory, was a postulate without physical basis. This is not the place to trace the long history of the detailed and subtle discussions around Le Sage's theory from the end of the eighteenth century to the very beginning of the twentieth, which drew in most great physicists of the period, such as Euler, Bernoulli, Boscovich, Lichtenberg, Laplace, Kelvin, Tait, Maxwell, Lorentz, George Darwin, and finally Poincaré, who dealt, or so it seems, the death blow to the theory.

Although almost no trace of it can be found in the vast majority of textbooks or popular works, Le Sage's theory was discussed throughout the period as an underlying current of physical thought that was foreign and rival to the mainstream. The nineteenth century saw the slow emergence of the field concept, from Faraday to Maxwell, which eliminated altogether the very idea of action at a distance, replacing it with the entirely local conception of field being generated by an agent body, then propagating in space and finally acting on the patient body. Although developed first in the context of electromagnetism, the notion of field as mediating interactions was soon implicitly applied to gravitation, thereby eliminating altogether the need for a purely mechanical process such as Le Sage's. Nevertheless, as we have said, for more than a century no serious argument was leveled against the theory.[6] The long survival of this alternate viewpoint can be seen as an actual case of pluralism, at a very deep level, since the question here is less the choice between two types of theories than the choice of what counts as an acceptable scientific theory.

Indeed, the reasons for the neglect or dismissal of Le Sage's theory well before any convincing criticism could be directed against it appear quite contingent, in the strict sense, as they seem to be based mainly on personal tastes or collective prejudices. Here is what Euler answered to Le Sage at the inception of the theory: "You must excuse me Sir, if I have a great repugnance for your ultramundane corpuscles, and I shall always prefer to confess my ignorance of the cause of gravity than to have recourse to such strange hypotheses" (Wolf 1862, 184). One century later, George Darwin wrote, after a thorough but unconclusive review of the theory: "I will not refer further to this conception, save to say that I believe that no man of science is disposed to accept it as affording the true road" (Darwin 1916, 10). It might even be argued that the corpuscular theme of Le Sage's theory has in some sense survived and flourished, although not as an explanation of gravitation, by underlying the modern discoveries of the many forms of cosmic particles, such as the neutrinos—"ultramundane corpuscles" for sure.

Stationary State versus Big Bang Cosmology

The understanding that the universe was expanding sprang from the observations of distant galaxies by Hubble in the 1920s. The cosmological models proposed first by Friedmann and Lemaître in the same period led to the idea of a primordial state of the universe with a very high temperature and density. Traced backward in time, the history of the cosmos seemed to go back to an initial instant some billion years ago. The expression "Big Bang" was later coined and widely accepted to describe this scenario. Obviously, the notion of a finite age of the universe could not fail to raise some deep epistemological and metaphysical problems, threatening to support a scientifically grounded version of creationism. This is why in the late 1940s a few astrophysicists, notably Fred Hoyle (who, ironically, was the first one to use the expression "Big Bang" as a derogatory term!) and Hermann Bondi, proposed an alternate theory of a universe that both expanded and remained in a stationary state, with constant density. Their views required a deep change in the fundamental laws of physics, namely, the existence of a permanent and spontaneous creation of matter to maintain its density despite the spatial expansion. However, this violation was so small that it had no possible effect on the ordinary scale and could not be directly observed.

Despite its compatibility with the existing astronomical data for some twenty years after its inception, the theory of the stationary state universe remained a minority opinion. The consensus today is that it was refuted by two advances in astrophysics in the 1960s: the observational discovery by Arno Penzias and Robert Wilson in 1964 of the cosmic radiation background as a blackbody radiation at a temperature of a few kelvins, and the explanation of the relative abundance of the elements helium and hydrogen. Indeed, both these facts were readily explained in the "Big Bang" framework as consequences of the universe's very high temperature primordial era. Looking closely at the primary literature of the time, one can show that the story is not that simple (Lepeltier 2010). Indeed, Hoyle and coworkers strove for many years to find alternate explanations in their own theory. They slowly lost ground as the "Big Bang" theory proved able to encompass more and more facts and to allow great flexibility in its detailed scenarios, while remaining consistent with the accepted fundamental laws. The contingency of the actual course of events becomes quite striking when account is taken of two specific facts. First, Thomas Gold, a collaborator of Hoyle and Bondi, had in 1955 already considered within the stationary state theory a mechanism leading to a background blackbody spectrum with the correct temperature; but Hoyle and Bondi were not fully convinced and rejected the idea—a stand that they would bitterly regret later. Second, the "Big Bang" prediction of the blackbody cosmic radiation was advanced by Dicke and Peebles early in 1964 (after earlier but forgotten indications by Alpher and Hermann in 1948), a few months before its observation. Now, consider the quite plausible situation that Hoyle and Bondi had endorsed Gold's hypothesis and predicted the existence of the cosmic background, and that the observation by Penzias and Wilson had preceded the paper by Dicke and Peebles. Would it not have been considered then as a victory for the stationary state theory, thus strengthening it for a long time to come?

Let us add that the present situation of theoretical cosmology, strongly put into question by many observations, today explained away by notions of "dark matter" and "dark energy," which might be quite provisory, cannot be considered a final triumph of the "Big Bang" theory. More to the point perhaps in the present framework is the fact that the usual way of understanding the "Big Bang" theory as endowing the universe with a finite age and thus having had an origin can be shown to be rather preposterous. Indeed, by exercising some reflection on the very notion of time and its "initial" singularity, one reaches in a rather simple and convincing way the conclusion that the standard cosmological theory does not include a proper zero instant, and that the admittedly consistent and objective measure of a finite "age" can (and should) be reinterpreted so as to eliminate any question about the origin of the universe. The crucial point, which we cannot develop here, is that a finite magnitude of some physical quantity does not preclude its correspondence to an infinite concept. A well-known case is that of the "absolute zero" of temperatures, which is both finite on the usual scale (−273.15 degrees Celsius) and infinite as not being attainable and representing an ever receding horizon. The same is rigorously true of the "initial instant" in the customary Big Bang cosmology (Lévy-Leblond 1990).

Wiggleworm Physics

The question of contingency versus inevitability clearly has some connection with the question of incommensurability. It would seem at first sight that admitting the thesis of epistemological incommensurability between various theoretical frameworks would be a strong and perhaps irrefutable argument for contingentism. But "it ain't necessarily so,"[7] as we will see by considering a nontrivial inevitabilist argument that is worth discussing.

More than fifty years ago, Jerome Rothstein published a clever and provocative paper on "wiggleworm physics," which certainly is still one of the best *gedankenexperimenten* on the question of contingency (Rothstein 1962). It is worth quoting the introduction to his article at length:

> We know that our picture of the universe is very strongly influenced by our means of perception. Basically, we are eye-hand-coordinated animals. Perhaps our sensory and psychological apparatus prejudices us in favor of some particular picture of the universe, when in reality the picture is warped by being filtered through our particular sensory and nervous equipment. . . .
>
> In an effort to see how reasonable the possibility that we are all being fooled really is, we shall try to sketch how science might develop for a race of blind, deaf, highly intelligent worms living in black, cold, seabottom muck, and possessing only senses of touch, temperature, and a kind of taste (i.e., a chemical sense). These hypothetical creatures are chosen because they seem to be about as different from human beings as they possibly can be and still be able to learn things about their environment. . . . As a rule of the game, we impose the condition that the only concepts they can use are those that develop by intelligent and imaginative thinking about the experience available to them. (Rothstein 1962, 28)

Rothstein then proceeds to a detailed reconstruction of how these intelligent wiggleworms would build a conceptual worldview, starting with an elaboration of space and time concepts. Deprived by their environment and their very bodily structure of any direct access to rigid bodies, it seems clear that they would evolve topological notions far earlier than metric notions. In mathematics, continuous and analytical considerations would precede discrete and arithmetical ones. Chemistry would mature well in advance of physics. Fluid mechanics would antedate mechanics of rigid bodies. And of course, astronomy, one of the earliest sources of scientific endeavor in our own species, would wait a long time until a very sophisticated technology would allow the wiggleworm scientists to reach the surface of their oceans and detect celestial radiations. The discovery and exploration of land would be as difficult and hazardous for them as the exploration of space (or, for that matter, the exploration of oceanic depths!) is for us. The rational imagination with which Rothstein depicts the evolution of wiggleworm science and, further, of wiggleworm technologies makes for very pleasant reading. Far from being a mere exercise in science fiction, the whole purpose of the paper is to try to answer the very question of contingency in science, or rather, and more modestly, to endow it with a concrete meaning. In Rothstein's words: "At any one time, wiggleworm science would probably not bear a close resemblance to our own. . . . What meaning, then, can one ascribe to the question as to whether Wiggy and we can come up with equivalent pictures of the universe?" (38). The problem of course is that both our and Wiggy's pictures of the universe are constantly changing, so that it is extremely improbable that both these pictures could ever be "the same." The quotation marks around "the same" of course refer to the fact that we have to forget about a complete identity of concepts and/or formalism, even acknowledging the traditional problems of translation between (very) different languages. Rothstein proposes to understand "sameness" as theoretical isomorphism, in the sense that two theories T_1 and T_2 would be considered isomorphic if any consequence of T_1 can be deduced from T_2 and conversely. Since this isomorphism is most improbable, Rothstein proposes the stimulating idea of replacing it with what he calls "dynamic homomorphism," which he explains as follows:

> To introduce this idea, consider a sequence of pictures of the world which we have created: $H_1, H_2, \ldots H_n \ldots$ Let the subscripts number the pictures such that the higher subscripts correspond to later, more inclusive theories. If $i < j$, we write $H_i < H_j$, where the inequality signs means that H_i covers a smaller body of experience than H_j and

that in that portion of experience where H_i holds with reasonable accuracy, it is an approximation to H_j with the description afforded by H_j at least as good as that afforded by H_i. In a similar fashion, we consider a sequence of pictures that Wiggy establishes, designated by W_1, W_2, . . . W_m . . . The same conventions with regard to the subscripts and the relations between the W's are assumed to hold as discussed above for the H's. . . .

We can now explain dynamic homomorphism. By this we mean that given any W_p, there will exist an H_q, where q depends on p, and such that $W_p < H_q$. Also, given an H_r, there will exist a W_s where s depends on r, such that $H_r < W_s$. . . . Expressed in this form, it would appear that dynamic homomorphism applies to independent developments in human science, and basically no more or less than it applies to Wiggy's and our developments. (Rothstein 1962)

In other terms, Rothstein first acknowledges the inescapable inequivalence of two scientific worldviews, that is, the fact that at any given time their intersection is but a small subset of each one. He then astutely proposes a time-evolving scheme in which a mutual nesting of the successive domains encompassed by the two worldviews assures that at any historical time humans know more than wiggleworms at some time in their own history and less than at a later time—and reciprocally of course. Rothstein optimistically concludes:

It seems reasonable to conclude that though we shall always make errors or be mistaken to some extent in our ideas of the universe, we shall improve or correct our pictures sooner or later, and the same applies to Wiggy. Ultimately, when both he and we have covered the same accessible portions of the universe, we should find that the differences between our pictures would be only those of isomorphism. There seems to be no reason to believe, if an objectively "real" world exists, that Wiggy and we would end up with nonisomorphic pictures of the same body of experience. (Rothstein 1962, 38)

I have dwelled at length on Rothstein's proposal because it certainly is one of the cleverest attempts to bypass the contingency question—and one of the funniest as well. As such, the criticisms one may level against it are of special interest because quite a number of considerations show that the underlying conception of "pictures of the universe" is somewhat overly primitive.

Indeed, Rothstein's idea of dynamic homomorphism relies on a rather simple-minded analogy between scientific domains of knowledge and domains in geometric spaces, enabling one to think in terms of inclusions, intersections, and so on. But even if one sticks to such naive topological representations, a few elementary questions immediately come to mind:

> —Can the different "bodies of experience" be mapped in spaces with the same dimensions? A more sophisticated metaphor could well require the immersion of the two pictures to be compared in a space with a higher dimensionality than each one, in which case, the set-theoretic ideas of inclusion, intersection, and so on would lose their relevance.
>
> —Are the contents of domains of intersection necessarily equivalent? There are very many ways to map a given region: compare a purely physical map (orographic and hydrographic) with a social one (cities, roads, etc.). A "picture" (Rothstein's term) is necessarily partial and relies on the choice of some elements considered as relevant. A complete mapping of any complex reality obviously is impossible—and would be of no use whatsoever, since it would be equivalent to the territory itself.[8]

On a more general level, the view of science history as encompassing a monotonously increasing and boundless domain of knowledge seem rather outdated today. We know that episodes of scientific development have a beginning and an end, and that absolute cumulativity is never the case. Science advances by forgetting as much as it does by learning.[9] Furthermore, Rothstein's argument is a purely epistemic (internal) one, which neglects the social and cultural context. Can two theories be considered "equivalent" if they have the same formal contents but widely different meanings as regards their practical applications and symbolic interpretations? This of course applies already within human science: is the so-called Pythagoras theorem really the "same" in Vedic India, in ancient Greece, and in modern mathematics? Considering the final demise of Greek science or Arabic science, to name but two crucial episodes, are we so sure that ours will not meet a similar fate (not to question the very lifetime of humanity

itself)? Then, of course, the vision of wiggleworms and humans "ultimately" reaching isomorphic pictures becomes a sheer fantasy.

Contingency and Relevance: The Hell of a Science

Contingency is usually advocated by trying to prove that a theory different
from the one that came to be admitted could have been (or has been)
equally convincing in explaining the "facts." But should one not point out
first that there is a strong contingency in the very set of ideas that are accepted at given times and places as satisfying our desire for explanation?
Or, to put it still differently, contingency should not be considered to be
concerned only with the validity of the answers and results but with the
relevance of the questions and problems being investigated. A very interesting case at hand is seen in the role that the notion of hell played in physics
and astronomy during the seventeenth and eighteenth centuries. This is a
period when English science was closely linked with natural theology. The
necessity was deeply felt to establish the compatibility and even mutual
support of religion and science and, why not, to consolidate the first by
the second. Newton's work presents a major example of the trend (see his
notion of space as *sensorium Dei* [Snobelen 2004]). In this perspective, the
question of the place and nature of hell came to be considered a perfectly
reasonable subject of scientific inquiry, and several very serious investigations were devoted to it (Almond 1994).

Tobia, a learned clergyman with astronomical interests, was one of the
first to take issue with the old notion of a subterranean location for hell. In
a book published in 1714, he had "two or three Things to object against
the Opinion that it is in the Bowels of Earth" (Swinden 1727, 354–55).
In the first place, he argued that "the Fire of Hell is not metaphorical but
real," which led him to a chemical objection: "Since corporeal Fire doth
require both Fewel to feed upon, and likewise Air to sustain and preserve
it; and since a sufficient Quantity of either of these cannot be reasonably
supposed to be about the Center of the Earth, therefore I conclude, that
Hell cannot reasonably be thought to be placed there." He followed with
a second objection, of a geological nature, to the effect that water and not
fire should be found at the center of the Earth, and he finally advanced
a demographic consideration: "A third Argument against Hell's being at
or about the Center of the Earth, is, that such a supposed Place must be
too small to contain the lapsed Angels, and the infinite Number of the

Damned." Swinden then ended up with "a Conjecture that the Body of the Sun is the Local Hell," which he accompanies with "an Apology for the Novelty of it," before giving quite a number of scientifically, philosophically, and theologically founded arguments for this conjecture, starting with the fact that "it is obvious to each Man's Sense and Reason, that the Body of the Sun is real, corporeal Fire," and strongly relying on the Copernican view of the cosmos, showing that Swinden was not an archaic thinker but, quite the contrary, a modern scientist of his time: "Another Reason for the Sun's being the Tartarus is that it is placed in the Center of the Universe, from which it is supposed not to have moved in the least at any Time. Of all the Affections of Place, Immobility suiteth best with Eternity." His book was well received, several times reprinted, and even translated into French in 1727.

At roughly the same time, William Whiston published in 1717 one of the most important treatises of natural theology, in which he advocated a rival theory, namely, that the comets were the most natural location for hell:

> I observe, that the Sacred Accounts of Hell, or of the Place and State of Punishment for wicked Men after the general Resurrection, is agreeable not only to the Remains of ancient profane Tradition, but to the true System of the World also. This sad State is in Scripture describ'd as a State of Darkness, . . . of Torment and Punishment for Ages . . . by Flame, or by Fire, or by Fire and Brimstone, with Weeping and Gnashing of Teeth; where the Smoak of the Ungodly's Torment ascends up for ever and ever. . . . Now this Description does in every Circumstance, so exactly agree with the Nature of a Comet, ascending from the Hot Regions near the Sun, and going into the Cold Regions beyond Saturn, with its long smoaking Tail arising up from it, through its several Ages or Periods of revolving, and this in the Sight of all the Inhabitants of our Air, . . . that I cannot but think the Surface or Atmosphere of such a Comet to be that Place of Torment so terribly described in Scripture, . . . which will be indeed a terrible but a most useful Spectacle to the rest of God's rational Creatures. (Whiston 1717, 156)

Let us emphasize that William Whiston was no marginal or secondary character. He made quite a number of contributions to the theory of comets, being one of the first, with Edmond Halley, to argue for their periodic returns. He became assistant to Newton and was chosen as his successor in the Lucasian Chair of Cambridge. This is no surprise in view of the fact that Newton's own work on comets was closely associated with his providentialist and apocalyptic views (Snobelen 2004).

One could fruitfully pursue the exploration of this body of literature, but we will be content here with these two authors.[10] The case of the science of hell shows quite clearly the existence of deep cultural elements in the criteria that enable a given question to be considered a relevant scientific one. Now, the wide variety of cultural contexts in turn confer a strong element of contingency to these criteria. It is telling in that respect to compare the state of mind of the English scientists in the seventeenth and eighteenth centuries, dominated by the perspective of natural theology, with that of French scientists, operating in a much more skeptical philosophical environment. To keep up with our example, it is worthwhile to read the article "Enfer" in the *Encyclopédie,* written by de Jaucourt, who, after a thorough and very informative review of the recent research at the time on the location of hell, which we just presented, concludes in a characteristically witty tongue-in-cheek way: "Nous laissons au lecteur à apprécier tous ces systèmes; et nous nous contenterons de dire qu'il est bien singulier de vouloir fixer le lieu de l'Enfer, quand l'Écriture, par son silence, nous indique assez celui que nous devrions garder sur cette matière" (We leave it to the reader to appreciate all these systems, and we will be content to state that it is most strange to pretend to determining the location of Hell, when the Scriptures, through their silence, point clearly enough to the one we should keep in this matter).

One cannot help thinking that these past scientific studies of hell have left some devilish imprint on the folk-culture of physicists today; to wit, a well-known jocular discussion on the temperature of hell (Anon 1972; Nassau 1972; Healey 1979; Mira-Perez and Viña 1998).[11] But the reference also invaded the popular representation of Venus as an infernal place, starting with the accounts of astrophysicists themselves, and even reaching pop music.[12] Let us finally point out that a most satisfying and probably final answer to the question of the location of hell has been proposed by the televangelists Jack and Rexella Impe, who were granted the 2001 Ig Nobel Astrophysics Prize for their discovery that black holes fulfill all the technical requirements to be the location of hell.[13]

Science(s), Culture(s), Language(s), Writing(s)

Despite the examples we have presented of the theoretical plurality of physics, a stubborn inevitabilist could still argue, first, that the various mathematical formalizations of a given theory, as they are logically equivalent, can be considered to be one and the same, and second, that the co-existence of alternate theorizations is but temporary, and only one of them (at most) will ultimately survive. Plurality and contingency in that view would be but ephemeral aspects, due to disappear in the long run to leave room for a final and necessarily unique understanding. Let us accept this thesis for the sake of discussion, despite the rather ideal character it confers on inevitability. One strong argument still remains to be advocated by the pluralist-contingentist.

A movie by Alfred Hitchcock, *Torn Curtain*—admittedly not one of his best—tells a story of spying and science.[14] It features a strange scene, in which two physicists confront one another on some theoretical question. Their "discussion," if it may be so called, consists solely of one of them writing some equations on the blackboard, only to have the other angrily grab the eraser and wipe out the formulas to write new ones of his own, with neither ever uttering a single word. This picture of theoretical physics as an aphasic knowledge entirely consisting of a game with mathematical symbols, as common as it may be in popular representations, we know to be wrong, of course, and we have to acknowledge that, far from being mute, scientists are very talkative; physics is made of words. The notion of a purely formal and entirely rational expression of scientific knowledge is clearly a fantasy, already in the case of mathematics. Contrary to what Hilbert advocated, it is not true that, in elementary geometry, "one must be able to say at all times—instead of points, straight lines, and planes—tables, beer mugs, and chairs" (just try it . . .).

Physics, despite its intrinsic mathematization, which seems to make it more abstract than any other natural science, cannot be reduced to its mathematical formalism. Formulas cannot be understood and, for that matter, cannot be stated without words. The letters or other symbols entered into such formulas are only shorthand representatives of concepts that have no existence independent of language. The words we use to name these concepts are of crucial importance to their being grasped. This is so true that, without going back too far in antiquity, at the beginning of modern physics itself, the symbolic machinery was absent or very limited. When he discovered the law of free fall, Galileo never wrote anything like our familiar formula $z = \frac{1}{2} gt^2$ but had to express himself using a full sentence instead: "The spaces described by a body falling from rest with a uniformly accelerated motion are to each other as the squares of the time-intervals employed in traversing these distances" (Galileo 1914, 209).

Even with Descartes and Huygens, mathematical formalism was still rather limited and had to await Leibniz, Newton, and Varignon to start looking somewhat similar to the one commonly used today. Now one has to recall that in these times, most of the scientific vocabulary was borrowed from ordinary language and necessarily carried (and still carries) a heavy load of empirical representations, historical connotations, and conceptual associations. For example, just take the word "force," chosen by Newton to name a theoretical concept in mechanics. Although such technical terms are given a restricted and specific meaning within physics, this nevertheless cannot suffice to cut them from their deep vernacular roots in the fields of nonscientific practice (of course, "root"—in algebra—precisely is an example, as well as "field"). Later on, especially in the nineteenth century, it became customary to forge new technical terms by relying on Greek and Latin languages (think of "thermodynamics" or "electromagnetism"), with the purpose of separating this specialized jargon from the vernacular. But there cannot be an impenetrable barrier between scientific parlance and ordinary language. Indeed, many words initially created for and found in professional scientific discourse slowly leak out to find their way into common parlance, where they take on original meanings that cannot but return within the scientific discourse to give it new colors ("energy" and "entropy," "electricity" and "magnetism" are cases in point). In modern times, it has again become customary to use ordinary words in order to compensate for the highly abstract nature of our concepts and to endow them with some popular appeal, in what is probably more a communication and advertising tactics than a well thought-out epistemic strategy. "Big bang," "black holes," "quarks," "superstrings," and the like clearly illustrate the point. After all, one should not be surprised that physics also has its "spin" doctors.

My point here is that the vocabulary of science (and physics in particular), once we are convinced that it is not a secondary and external aspect of scientific knowledge, might be the most sensitive area to ascertain its deep contingent elements. For it is the common ground where historical, cultural, and personal contingencies meet with the production of scientific knowledge. Insofar as modern science developed in the specific context of precapitalist and Christian Europe and was thus founded on the linguistic background of Latin, how could one negate the impact of these historical and thus contingent facts on the way our science is today formulated, and hence understood?

One major example (although many others could be discussed) of the way some of our seemingly more deeply entrenched scientific notions are tributary to the surrounding culture is the idea of "natural laws." How come a word taken from the domain of law should apply to nature? A law is a rule that has to be proclaimed by some ruling power, whether divine or human. It is no surprise then that the idea entered science in the context of monotheism, through the Christian interpretation of Aristotle by Thomas Aquinas, whereas it was certainly foreign to most Greek thinkers; it is no mystery either that it developed after the Scientific Revolution in a period of strong monarchies. The paradox here is that, in human societies, a law is necessary only because of the all too obvious fact that its commandments are liable to be violated. A "law of nature," that is, a statement that is supposed to be obeyed always and absolutely is thus somewhat of a logical inconsistency. It seems, for instance, that Chinese science never had recourse to such a metaphor.

On a larger scale, a comparative study of the scientific vocabulary in many languages would show the large diversity of forms in which science is integrated in various cultures. Consider, for instance, how terms coined in the dominating occidental area, and more specifically from Greco-Latin roots, are translated in other linguistic areas through two different main processes:

—phonetic transcription, as is very common in Japanese for example (computer = *konpyuta*), with the ensuing loss of any significant relationship;
—adaptation of the original etymology, frequent in Chinese (particle = *lizi*, which means "grain" and is written with a character associated with rice), with the inverse drawback of emphasizing too heavy a vernacular load.[15]

One can pursue an investigation of the way culture influences science at its very heart by descending from the level of language and words to that of the writing system and its signs. Once more, let us take a single but crucial example: the atomic view of matter. When the idea was first conceived in antiquity, it strongly relied on an alphabetical metaphor, as clearly expressed by Lucretius (first century BCE):

> Thus easier 'tis to hold that many things
> Have primal bodies in common (as we see
> The single letters common to many words)
> Than aught exists without its origins.

The idea is that bodies are made of atoms exactly in the same manner as words are made of letters. This is a way to explain how the huge diversity of materials can be reduced to a small number of fundamental constituents and also how the emerging properties of the compound are not akin to those of the components. Many authors used the same metaphor (Hallyn 2000). It is to be stressed that the very term *elementa*, which is the one used by Lucretius (and not *atomos*!), in fact refers to the solid letters that served in Roman education to learn reading. Of course, this is not to say that an atomic theory cannot be conceived in another cultural context. But the understanding of the theory, and first of all its wording and its semantic associations, would have been quite different.

More generally, if we are to take seriously the deep Christian metaphor of the "Book of Nature" as the first volume of the Scriptures, looking at the world is reading it—but what is the writing in this Book? According to Galileo, it is written in "mathematical characters, which are geometrical figures," pointing to a curious ideographical conception. These geometrical forms will soon be replaced by literal symbols when algebra and analysis take over geometry in the mathematization of physics (Descartes, Leibniz, etc.). But it can be shown that the sources and status of these modern characters point to a very old story, establishing a neglected link between modern science (physics) and traditional mystics (kabbalah) (Lévy-Leblond 2005).

In fact, one may well think that only alphabetical writing, with the total arbitrariness of its signs, has allowed science as we know it to develop (de Kerkhove and Lumsden 1988). By disjoining the thing and the sign, the alphabet would have enabled the labor of abstracting, the detachment from sensate appearances, which constitutes the very basis of proper scientific knowledge. Permitting a shared access to common forms of writing, easy to teach, learn, and reproduce, the alphabet would have favored the collective exchange and expansion of this knowledge. Such is one of the possible elements of an answer to the great question of Joseph Needham, who wondered about the reason why science, in the modern acceptance of the word, was born in the West (Greece, Islam, Europe) and not in the East (China), despite the numerous priorities of the latter. There is some irony in the archaic notion of ideograms coming back in science through the window of writing, after having been thrown out at the door of thinking. For indeed, mathematical symbols in the equations of physics are much more than simple abbreviations and come to carry a heavy conceptual load. It is no surprise that a major role here has been played by Leibniz, the most inventive and fecund creator of mathematical signs; suffice it to remember his search for an "alphabet of human thoughts," the "universal characteristics" he was dreaming of. One also knows the acute and explicit interest he had in Chinese writing; he even went as far as advocating the use of Chinese ideograms, readable (so he thought) in any spoken language, as a universal written language for science. Indeed, the flexibility of contemporary Chinese in transcribing scientific concepts originating in Western languages may be commended. Nonetheless, it is a fact that science was born within an alphabetical environment, even though it later endowed its characters with more pictorial contents. Finally, these considerations indicate, at the very core of the modern formalism of mathematics and physics, the strong contingency at work in the very choice of the symbols we use, whether they are alphabetical (why "x" for an unknown quantity?) or typographically specific (why "+" for addition?).[16]

I hope to have shown through the various examples I have discussed, at several levels of generality, that the circumstances in which different scientific outlooks are born and the very order in which they surface are not dictated by purely internal conceptual reasons but depend on the historical and cultural context, thus supporting the idea that contingency is indeed at work in the building of science.

CHAPTER 15

Cultivating Contingency

A Case for Scientific Pluralism

HASOK CHANG

Contingency and Plurality

In addressing the issue of contingency in science I cannot help wishing for a reframing of the question itself. I will take Léna Soler's formulation of the question, inspired by Ian Hacking (2000a), as the most carefully crafted version available. What is at issue, according to Soler, is a choice between the two following positions: "contingentism claims that it is possible for there to be a science that is . . . as successful and progressive as ours but radically different in content"; in contrast, inevitabilism claims that "any science which is as successful and progressive as ours, and which has addressed the same questions as ours, would inevitably yield answers essentially similar to those that have been actually offered by our own science" (Soler 2008b, 230).[1] Here the question is taken as a descriptive one (albeit about counterfactual or future prospects); this question should be answered by reference to the nature of science, the nature of human beings and their social interactions, or the nature of the universe itself.

If we take the contingency question as a purely descriptive one, it is quite unanswerable. This is obvious to me, having followed the careful analysis given by Soler. Yes, it could be that the results of science are inevitable in the sense that all possible paths of development will ultimately converge into one, when we reach The End. But this kind of future is something we can only speculate about. Soler's thought experiment of "divided physics" is an interesting idea, but her own conclusion is that it "would not provide a decisive, inescapable argument for one or the other side," whatever the outcome (Soler 2008b, 240). The actual historical record of science up to the present day cannot give us the answer, either. On the whole, all that history seems to tell us is that scientists have developed various successful systems of knowledge, some of which have differed very much from each other. There are some convergent tendencies, and there are some divergent tendencies; from actual history as well as hypothetical history, contingentists and inevitabilists can each pick out their favorite cases and plead for more time to see inconvenient cases turn out differently in the end.

I think a major clue for shifting this debate into a more productive direction lies in the "able" in "inevitable," which gently suggests that the question is really about what we are able to do.[2] Inevitability is unavoidability, and whether something is unavoidable is not something we can tell without having made an attempt to avoid it—unless we understand the underlying situation so well that we can make confident in-principle predictions about it, which clearly does not seem to be the case when it comes to our understanding of the process of scientific development and of the deep structures of human cognition and the universe itself that shape it. So, given a particular result of science that seems strongly justified, the contingency question comes down to whether we are able to sustain a result that contradicts the existing one, with equally strong justification. The answer to that question may well depend on how hard we try, whether we are skilled enough, and whether we have sufficient and appropriate resources at our command. Therefore it is not quite right to say that any given result of science *is* either inevitable or not. Underlying any answer we might give to the question is a normative decision on how seriously we should try to seek alternatives. What we mean by contingency is partly a function of what we *allow* to be challenged and what we do not.

These considerations imply that the contingency question can be answered only through an *active pluralism*, which is an effort to *cultivate* multiple systems of knowledge in any given area of science. Inevitabilism could only be meaningfully vindicated by a distinct failure of active pluralism, after it has been tried in a serious way—in practice, inevitabilism is a negative doctrine. After the inconclusiveness of the thought experiment on "divided physics," Soler (2008b, 240) rightly recommends "examining the nature and the plausibility of variations within our own physics" and cites the work of James Cushing (1994) on Bohmian mechanics as an exemplary study. But it is not sufficient to turn our philosophical gaze upon past science and contemplate what could have turned out differently. A more active kind of pluralism is required, and there are two dimensions to that. First, if David Bohm himself had not taken the trouble of actually developing an alternative to standard quantum mechanics, there would have been no contingency there for Cushing to consider. Second, when the scientific community itself fails to produce pluralists like Bohm, it becomes necessary for the philosophers or historians interested in the contingency question to attempt to create alternatives ourselves. Less radically, we can engage in the recovery and rehabilitation of existing alternatives from the past of science that have been forgotten or unjustifiably dismissed.

It may seem wildly implausible to imagine that philosophers would have sufficient expertise to create scientific alternatives that scientists have not come up with themselves, or even to spot anything of value from history that scientists themselves have not picked up. Turn that thought on its head: at least, if you had a situation in which even nonexperts (such as philosophers) can come up with viable alternatives, then you would know for sure that the current scientific orthodoxy is not inevitable. Even if the likelihood of success is small, the returns would be great, which means that an attempt may be worth making. And it is not entirely ludicrous to think that a nonexpert may be able to see something that is in the collective blind spot of a homogeneous scientific community. We know that expert scientific communities have rejected valuable possibilities before, such as Avogadro's hypothesis of bi-atomic molecules of elementary substances, Semmelweis's ideas of contagion and chemical disinfection, and Wegener's theory of continental drift. And Thomas Kuhn reminds us that new ideas that set off scientific revolutions often come from relatively inexperienced scientists, who are either very young or recent arrivals from a different discipline. In any case, a commitment to explore scientific alternatives is required of any philosophers wishing to tackle the contingency question in a serious way. Without an active cultivation of alternatives, it will be all too easy to issue a meaningless armchair verdict in favor of inevitabilism based on a lack of imagination.

Pluralism provides more than the practical means of testing the contingency thesis. Taking pluralism seriously will change the entire spirit in which the contingency question is tackled and change its formulation subtly yet fundamentally. Soler's formulation of the question is clearly an approach from the side of inevitabilism, even as her answer is impartial. Radical contingency is considered to be "nonbenign," or even "harmful" in her formulation (Soler 2008b, 233, 231), and this reflects a dominant intuition among scientists. Many philosophers and scientists share an inevitabilist instinct, and I think that is due to a strongly monist scientific education that many of us have received. From that starting point, contingency is either feared as a threat to scientific knowledge or relished in a rebellious spirit. Soler's strategy is to start from this common way of framing the problem and then to show that the taken-for-granted inevitability is actually not so obvious.[3]

I have great sympathy for Soler's approach and admit the need to begin the discussion on broadly accepted terms, but my own approach is to reject inevitability as an implicit starting point of the discussion. Generally in my work I can break down the inevitabilist presumption by means of detailed historical case studies, although in the present chapter I will stay with an abstract philosophical argument. The kind of pluralism I advocate, which argues that plurality is good, should change the tenor of the debate fundamentally. Pluralists should be somewhat worried if scientists exhibit a complete consensus, rather than enter into a premature celebration of inevitability. Faced with a uniformity of opinion, we should be prompted to check whether it is not the result of an excessive herd instinct, institutional structures that suppress dissent, or an external drive by shared political or economic interests. If in some cases it turns out that we are just not capable of sustaining multiple accounts of the part of the world we are studying, then we should of course accept that as a practical fact, but with an anxious curiosity about why this is so. I believe that this sort of pluralism is a truly mature attitude about science (compared to the youthful bravado and ambition reaching for the "final theory" or the one "theory of everything"). The rest of this chapter will be devoted to the articulation and defense of pluralism in science.

Active Normative Epistemic Pluralism

It is time now to define what exactly I mean by "pluralism." I call my position *active normative epistemic pluralism*, the core of which is a belief that it is beneficial to have multiple systems of knowledge in each area of inquiry.[4] Scientific pluralism in general can be characterized more clearly by contrast to its antithesis, scientific monism, which is helpfully defined by Stephen

Kellert, Helen Longino, and Kenneth Waters (2006, x) in their recent edited collection on scientific pluralism. Note especially the following points: "1. the ultimate aim of a science is to establish a single, complete, and comprehensive account of the natural world (or the part of the world investigated by the science) based on a single set of fundamental principles; . . . 4. methods of inquiry are to be accepted on the basis of whether they can yield such an account; and 5. individual theories and models in science are to be evaluated in large part on the basis of whether they can provide" such an account.[5] In countering this monistic position, I would first of all submit that science should give accounts of the natural world that best serve whatever ultimate aims we may have (truth, practical utility, etc.); the monistic character of those accounts cannot in itself be our ultimate aim. On the contrary, as I will argue in some detail, I believe that the aims of science can be served better if we have multiple accounts of each given domain of nature.

I have designated my position "epistemic" pluralism because it does not primarily concern, or rest on, any specific beliefs about the fundamental ontology of nature, which I think is unknowable. Without denying the inevitable linkage between epistemology and metaphysics, I want to show that there are strong arguments for pluralism that do not rely on strong and specific views about what the world is really like. So, for example, while being quite inclined to accept Sandra Mitchell's (2003) view on the complexity of the biological domain, I do not want to tie the general arguments for pluralism too strongly to the special complexity of biology, which would make them inapplicable to much in the physical sciences. And I do not share Nancy Cartwright's (1999) metaphysical conviction about the "dappledness" of the universe, but I think the kind of pluralistic epistemology she advocates still has plenty of justification.

One important way in which I want to go further than some other pluralists is that the pluralism I advocate is unapologetically normative. I do have great sympathy with descriptive pluralism, which argues that in fact scientific practice has been more disunified than often imagined. But the descriptive thesis is not my main focus. Nor is it my main concern to interpret or explain the plurality that exists (or not) in scientific practice. The primary business I am in is not the so-called science of science but a critical engagement with science. To paraphrase Karl Marx (if one is still allowed to do such a thing), I might say: philosophers have interpreted science in various ways; the point, however, is to change it.

Therefore, going beyond passing value judgments, the pluralism I advocate is an active stance. A passive version of normative pluralism would simply point out the benefits of having multiple systems. An active pluralism actually engages in cultivating multiple systems of knowledge. This activism not only is essential for the purpose of tackling the contingency question as explained in the previous section but also has much more general implications about how we should practice history and philosophy of science, which I will touch on briefly at the end.

Arguments for Pluralism

A normative argument must begin with the clarification of the relevant aims and values in play. And I want to cast my axiological net widely, to argue that pluralism is more beneficial to science than monism, given any reasonable position regarding the fundamental aims and values of science. In this method of argument, I am inspired by Paul Feyerabend's (1975, 27) declaration that "anarchism helps to achieve progress in any of the senses [of progress] one cares to choose." But unlike Feyerabend, I want to do this in a systematic fashion, by surveying all the various things that one might think science should desire to achieve. This axiological survey will dictate the structure of this section of the chapter. (I am using "aims" and "values" almost synonymously here, both under the rubric of "axiology"; this is not ideal, but not so problematic if we take the achievement of something valued as an aim.)

Pluralism with Truth as the Ultimate Aim of Science

In the common realist conception, science has only one ultimate aim, and that is Truth ("with a capital T"), which is objective and univocal. This is what Bas van Fraassen (1980, 8) calls "the correct statement" of scientific realism. I want to show that pluralism is more productive than monism even for those who follow this version of scientific realism.

(a) Argument from Unpredictability

The most obvious difficulty with the search for Truth is that we can never be sure whether we have got it, or if we are even approaching it, especially about the unobservable portions of the universe. Larry Laudan's (1981) pessimistic meta-induction from the history of science is not a conclusive positive argument for anything, but it does point up the basic insecurity of our theoretical positions in science. Kyle Stanford's (2006) "problem of unconceived alternatives" has the same disturbing effect on any alleged security about the theory choices that scientists make. What both arguments point to is an unpredictability about the direction of scientific development, when it comes to ideas regarding the parts of nature that are not directly observable. This point was already emphasized by Kuhn (1970, 206–7): for example, in cosmology, there has been no clear direction of development as scientists moved from the closed and spherical Aristotelian universe to the open and infinite Newtonian universe, and then to the closed four-dimensional curved space-time of modern cosmology. Some realists have sought to address this worry by arguing that some elements of scientific knowledge are preserved even through revolutionary upheavals; I have argued elsewhere against this realist move, which I designated "preservative realism," and I will not go into the details of that argument here (see Chang 2003).[6]

Faced with an insurmountable unpredictability, what rational agents have to do is clear: hedge our bets.[7] Given that we do not know which line of inquiry will ultimately lead to the Truth, we should keep multiple lines open, instead of pursuing one line faithfully to its dead end, only then to try a different one. To put it in Bayesian terms, all theories with nonnegligible prior probabilities should be monitored for signs of life (that is, increases in posterior probabilities) as further evidence comes in.[8] It is most irrational to insist that only the theory with the highest probability at the moment should be preserved and all others killed off. The hedging strategy clearly allows, and encourages, the simultaneous development of mutually incompatible models and theories.

(b) Argument for Cross-Fertilization

For the next argument, we need to distinguish two different versions of pluralism. The weaker version I call "tolerant pluralism" and the stronger one "interactive pluralism."[9] The main feature of tolerant pluralism is quite simply to allow different systems to coexist, with respect and toleration by each side for the others; it is not required that the different systems have any interaction with each other; what is important is that we allow each system to exist and pursue its own potential.

Going further than tolerant pluralism, interactive pluralism seeks additional benefits from having different systems of knowledge interact with each other, rather than standing separately and delivering separate sets of contributions. If the slogan for tolerant pluralism is "Let a hundred flowers bloom," interactive pluralism says: "Yes, let them all bloom, and also cross-fertilize." At least one key aspect of interactive pluralism has been articulated in convincing detail in Sandra Mitchell's "integrative pluralism" (2003, esp. sec. 6.3). She gives an in-depth discussion of the theories of social insect communities (such as ants and bees). There are several competing theoretical models of such communities, but none are adequate for explaining actual cases. But each case can be dealt with successfully, by an ad hoc integration of various aspects of these models. Interestingly, Mitchell's integration has a resonance with Otto Neurath's view on the unity of science required at the point of action (so we can claim the Vienna Circle on the side of pluralism, too!). Neurath acknowledged that in one sense there was no unity present or required in science: there were all kinds of scientific disciplines with different methods and theories. However, when action needed to be taken (e.g., in economic planning), all relevant sciences had to be brought together, and in order for that orchestration to be possible, all sciences should be put in the same basic physicalist language (of objects located in space and time).[10]

I would add that even when different systems of knowledge are not being pulled together to achieve a specific aim, one system of knowledge can be helped in its development by the use of ideas and results taken from another system, even an apparently opposing one. Various historical examples illustrate this point: the long-running competition between the wave theory and the particle theory of light was a complex story of mutual stimulation as well as confrontation; Lavoisier would not have arrived at his new chemistry without co-opting results obtained by phlogiston theorists such as Priestley and Cavendish; and so on. Cross-fertilization can be helpful in many ways, but in the search for Truth the main point is the heuristics of discovery: we will increase our chances of discovering the true theory if we are not restricted in our sources for new ideas; inspiration for new ideas can come from anywhere, but one very important source is other systems of knowledge in the same area of inquiry.

Pluralism in Inclusive Uni-Axial Regimes

Not everyone thinks that "Truth with a capital T" is the ultimate aim of science; many antirealists have rejected Truth as a legitimate or productive aim of science. Other values have been proposed as providing the main aim of science; here I will consider just two prominent examples, namely, empirical adequacy and understanding.[11] What kinds of arguments for pluralism apply if we regard the achievement of one of these values as the main aim of science? There is one important contrast between Truth and these other values. Truth is exclusive, in the sense that if one system of knowledge attains it, it cannot be possessed by any alternative system that contradicts the successful one. Not all aims are exclusive in this sense. Both understanding and empirical adequacy are inclusive, which is why I group them together here. (Incidentally, another inclusive aim is approximate truth, partial truth, or truth within a perspective, the last as characterized by Ronald Giere [2006]. That is to say, in practice many realists are pursuing an inclusive aim.) Being exclusive should be distinguished clearly from being "uniaxial," which is a word I use in order to indicate a situation in which there is one overriding aim or value that is categorically more important than others. So, in that terminology, science governed by the search for Truth constitutes an exclusive uniaxial regime. In an *inclusive* uniaxial regime we can expect there will be further arguments for pluralism, because there is no demand for unity in the provider even though there is a unity in the desideratum.

In Search of Empirical Adequacy

Van Fraassen's constructive empiricism articulates an inclusive uniaxial regime of science. It is uniaxial because it privileges empirical adequacy as a value to override others, such as simplicity, elegance, scope, and explanatory power, which are relegated to the status of mere "pragmatic virtues" (see van Fraassen 1980, 87–89). It is inclusive because there can be multiple theories dealing with a given domain of nature that are all equally empirically adequate while they contradict each other in what they say about unobservables. Immediately, a plurality is allowed. But are there arguments for *pluralism* in this regime? I think there are many.

(a)–(b) Unpredictability and Cross-Fertilization

First of all, the argument from unpredictability applies equally well here, with some obvious adjustments. We do not know which line of inquiry will deliver the most empirically adequate theory, so it makes sense to pursue various lines. Likewise for the argument for cross-fertilization.

(c) Argument Concerning Coverage (Out of Present Necessity)

In addition, because empirical adequacy comes in degrees and parts much more readily than Truth does, there is an argument for pluralism for the here and now. If our best theory covers only some of the known observable phenomena in a domain, then we need to have other theories that cover the rest. It will not do to reject all theories while we wait for the single theory that will have complete coverage; that would be to privilege simplicity (or unity) over empirical adequacy, and for van Fraassen simplicity is a mere pragmatic virtue.

(d) Argument Concerning Progress

The monist will want to argue that we should still seek one theory that is fully empirically adequate, rather than happily multiply theories any more than strictly necessary. In a happy ending for constructive-empiricist science, with one theory that is adequate for all phenomena in the relevant domain, both the argument from unpredictability and the argument concerning coverage become inapplicable, and there is no need for a plurality of theories. However, any serious look at the history and current state of science indicates that such a dream of "the end of science" is not only dreary (nothing interesting left to do) but highly unlikely to be realized. Almost always, scientists' desire for increasing precision and scope will end up pushing any successful theory to failure by revealing fresh anomalies. Such a recognition lay behind Kuhn's assumption that a paradigm, no matter how successful in its heyday, would almost inevitably be replaced by another paradigm eventually. Success encourages ambition, and as our ambition grows, so does the scope for inadequacy. That is to say, if science is a progressive enterprise, it is almost inevitable that our current best theory will develop imperfections, and there will be scope for other theories to make valuable contributions to empirical adequacy.

(e) Argument from Observational Incommensurability

My discussion so far has been premised on the assumption that all competing systems of knowledge in a given domain will have the same set of observations to deal with. That is a highly debatable assumption. Whether or not there is an absolute impossibility of translation between competing paradigms, some degree of incommensurability in the realm of observations is common in science. Each paradigm (or system of knowledge, more generally) will tend to elicit and highlight its own distinct set of observations, to reveal and retain different facts about nature. So, not only is empirical adequacy a dynamic value as the observational basis grows in its extent and refinement, but it is also a system-dependent value as each system makes a different contribution to the observational basis. The benefit of plurality is clear, then. If we accept this mild degree of observational incommensurability, it becomes imperative that we keep multiple systems of knowledge, in order to retain all facts that can claim the status of observational statements and open up new possibilities for expanding the observational bases of science.

Two Notes on Arguments (c)–(e)

I would now like to make two general notes concerning the last three arguments. First, the applicability of these arguments is not restricted to van Fraassen's constructive empiricism. They should apply to any empiricist regime of science in which the ability to account for observations is chiefly prized. Second, underlying all three arguments is an assumption about the plenitude of nature, which is in fact compatible with scientific realism. Priestley (1790, 1:xviii–xix) argued that the subject matter of science was inexhaustible and gave a very nice visual image for it: "Every discovery brings to our view many things of which we had no intimation before. . . . The greater is the circle of light, the greater is the boundary of the darkness by which it is confined." As knowledge grows, so does ignorance. Priestley's notion was based on the infinity of God, but for nonbelievers his picture may simply be taken as a fact of life about science, or about the "mangle of practice," as Andrew Pickering (1995a) has put it. It does not need to be a highfalutin metaphysical doctrine, either. It just does seem that indefinite numbers and types of facts about nature are yet to be discovered, and this makes it likely that each system of knowledge would tap into a different part of that unexhausted reservoir and continue to tap into more of it for the foreseeable future. In van Fraassen's terms, empirical adequacy means being able to account for all *observable* phenomena, not just the so-far-observed phenomena. Then the task of science clearly includes making as many observations as possible, and there is a clear argument for pluralism in that task. If we are like the blind people feeling the elephant, it makes sense to involve more of us to reach different parts of the animal. Less metaphorically, and with a less realist backdrop, we might say that science is an enterprise of asking questions and articulating problems, and there is no apparent end to that process.

In Search of Understanding

So much for empirical adequacy. Although unfashionable among analytic philosophers, understanding is still a value often cited by scientists themselves as the ultimate aim of science, so it makes sense to give it at least some brief consideration here. Under the rubric of "understanding" I also include all senses of explanation that go beyond logical subsumption. If it is not reduced to Truth, understanding is at least possibly an inclusive aim. I will not address here the question of what exactly scientific understanding is.[12] For most meanings of "understanding" that seem to be operative in science, I think all of the arguments for pluralism given so far apply. And there are two further arguments.

(f) Argument from Subjectivity, or Human Variation

Understanding may be a more strongly inclusive aim than empirical adequacy. In the empirical-adequacy regime, if one has a fully empirically adequate theory in a given domain, it is not clear what one would gain by having another empirically adequate theory, except perhaps some convenience in certain applications. For example, witness the difficulty experienced by advocates of Bohmian quantum mechanics. But if we grant a subjective dimension to understanding at all, then we have to allow that different people will derive understanding from different types of systems of knowledge. As Pierre Duhem ([1906] 1962, 70–72) infamously put it, the English physicist could only understand something if a mechanical model could be made of it; the French physicist derived all the necessary understanding from formal mathematical systems, with no need for childish models. But if the overall aim of science is the greatest understanding by the greatest number of people (including even English people), then pluralism is the only viable method, since no single system is likely to be able to provide this sort of intuitive understanding to everyone.

(g) Argument for Multiple Accounts

At least according to some notions of understanding, having multiple ways of understanding the same set of phenomena is a positive thing in itself. It means having more windows on nature, giving us more enriched understanding. In fact modern physicists are quite used to this sort of situation. They all learn multiple formulations of classical mechanics (Newtonian, Lagrangian, and Hamiltonian) and multiple formulations of quantum mechanics (due to Heisenberg, Schrödinger, Dirac, and also Feynman). Each physicist not only chooses which formulation of the theory to use depending on the exact shape of the problem at hand but also derives a different kind of understanding from each formulation of the theory. Physicists normally ease their monist conscience by repeating the mantra that these formulations are all empirically equivalent to each other, but the equivalence proofs that the ordinary physicist may have once learned are quite superfluous to his practice and his pleasure of understanding even the same problem in a few different ways. A rare statement of a truly pluralist viewpoint on this matter is given by David Hull in his review of the collection on pluralism edited by Kellert, Longino, and Waters (2006). Wishing for a rectification of "the bias that we all seem to have with respect to multiplicity and variability," Hull notes: "In response to the usual objections raised to 'anything goes,' [Michael] Dickson remarks that a 'multiplicity of dynamics is not necessarily a bad thing' (Hull 2008, 57). Not necessarily a bad thing? It is not a bad thing at all. In fact, it is good. . . . Waters remarks that scientists must be 'tolerant of diversity' (210). Tolerant? Diversity deserves more than 'tolerance.'"[13]

Pluralism in Pluriaxial Regimes

So far I have considered uniaxial regimes of science, in which there is one overriding value that governs normative assessments. Now I want to open up my thinking further. There is no convincing reason to think that science has only one overriding value or aim. While Kuhn privileged problem-solving ability as the key value in some places, in other places he gave us the often-cited list of accuracy, simplicity, consistency, fruitfulness, and scope.[14] Van Fraassen gave us a whole list of pragmatic virtues, and I do not think he provided a convincing reason for regarding these as secondary in importance to empirical adequacy. Without trying to decide exactly which values provide key aims of science, I want to consider general arguments for pluralism in what I call "pluriaxial regimes," in which there are multiple values that drive scientific work. In a pluriaxial regime, all the above arguments for pluralism from the uniaxial regimes still apply, in relation to each value. In addition, there are two other arguments for pluralism that arise from the *multiplicity* of values and aims.

(h) Argument Regarding Divergent Needs

Once we grant that there are multiple human needs that science is called upon to satisfy, it is easy to recognize that we will most likely not be able to come up with *the* perfect system that satisfies all needs. Call it pessimism, but I do not think it is unwarranted pessimism. I would rather think of it as reasonable humility concerning human ingenuity, or a recognition of the complexity of life, or both.[15]

(i) The Lacuna Argument: Preservation of Values

The other argument arises from a subtle aspect of interactive pluralism, something I call the "lacuna effect." Imagine a situation in which two systems of knowledge have different aims, and one system does well in achieving the aims laid down by itself, and the other one does not succeed in its own aims. Common sense would dictate that the system that cannot even succeed in its own terms should be discarded. I want to argue the opposite: as long as we think the failing system's aims are worthwhile, then we should keep it around because its failures will serve as a reminder of valuable aims that we should strive to satisfy. If we simply discard the failing system, it will be easy to forget the unachieved aims. Just to give a quick example, I would argue that the dominance of special relativity has caused many physicists to forget that it may be a good thing to seek dynamic explanations of relativistic effects.

Risks of Pluralism

Having listed various benefits of plurality, I cannot complete my argument without considering some potential harms. Foremost in the minds of traditional philosophers and orthodox scientists will be the general worry that pluralism will make science descend into a relativist chaos in which the voices of reason and science will be drowned out by a cacophony of quacks, charlatans, religious fundamentalists, New Age mystics, and plain idiots. Worse yet, without social authority vested in proper science, how can we prevent the relativist chaos from settling into the nightmare of one crazy faction taking power and imposing itself on the rest of society? We are still less than a century away from the horrors of "Aryan physics" and Lysenkoism.

The first point to stress is that there are some real differences between pluralism and relativism. Pluralism takes a stance against absolutism, in a way that relativism actually cannot. A system of knowledge that denies the rights of other systems to exist would have to be banned in a pluralistic system of science.[16] This is just as a truly free society needs to impose constraints on individuals and groups to prevent them from restricting the freedom of others. (This parallel is a reminder of the ultimate political dimension of science, and of knowledge in general, which has been stressed by Nicholas Rescher [1993], for example.) Furthermore, pluralism requires that there should actually be many systems of knowledge simultaneously in a given domain. This is a point one should not have to make, but I must emphasize that the demand for *plurality* is an essential feature of *pluralism*. Curiously, plurality is not required by relativism, if it only insists on the equal treatment of any alternatives that *do* exist. If all members of a community actually agree on a particular system and neglect all other possibilities, there is nothing contrary to relativism in that. Pluralism is about the benefits of actually having multiple systems in coexistence. So, my slogan for pluralism is not "Anything goes," but "Many things go."

But how about the relativist chaos itself, which may be benevolent but still seriously counterproductive? Again, it is important to distinguish pluralism from relativism. To use a metaphor that Sandra Mitchell has put into my head: how does a pluralist decide who gets to come to the table? What are the criteria for determining which people and research programs are deserving of research funding, university posts, space in academic journals, the right to teach children, and even access to the public? This is an unavoidable question that we must take seriously, and often people become unhappy because pluralism in itself is not able to give an answer to it. But this unhappiness is based on a category mistake: pluralism is a doctrine about the number of places we should have at the table; it cannot be expected to answer a wholly different question, which is about the guest list. More specifically, pluralism is based on a gentle worry about eating (or talking) alone and recommends that it is good to have more than one place at the table. How many places? It does not make sense to lay down a strict number (and the metaphor also breaks down here because scientific research programs are not so clearly individuated as people)—any reasonable number over one is a good start, perhaps more interesting if it is not just two. If all this seems intolerably inadequate, it is good to remind ourselves that monism, in itself, does not specify the criteria of choice, either! Deciding that there will be only one place at the table does not determine who gets to sit there—"me, of course" is the usual unspoken presumption, but having a room full of tables for one is not only somewhat sad but no more productive than having one big table with an uncontrolled guest list. In the end, it should be plain that neither pluralism nor monism determines very much about what we should believe in science, or even how we should decide what to believe.

If the above worries are based on a misunderstanding of pluralism, the following is a genuine worry about pluralism proper. This objection, congenial to Kuhn's view of normal science, starts with an observation about human psychology: scientists can only focus down on esoteric questions if they are not unduly distracted; monism is the best mind-set for this activity. This is a valid point, at least about some people's psychology, and poses a difficulty for putting *interactive* pluralism into practice. However, I do not think that it is a universal and immutable feature of human psychology. Just as (some) people can learn to multitask and be multilingual, people can also learn to do focused work while entertaining other ways of thinking. Even in the arena of science and other complex practices, where the benefits of focus go well beyond a matter of simple psychology and into the necessity for training by immersion, I believe it is possible to expand the scientific mind more than is customary at present. At least we know that many research scientists do have the intellectual capacity to pursue very serious and absorbing avocations alongside their scientific work, whether it be political campaigning (Pauling), psychology (Pauli), history of science (Duhem, Partington), or what have you. Could the same kind of energy not be directed to exploring other conceptual possibilities in people's own field of expertise?

In any case, it is fine at least for *tolerant* pluralism if individuals or groups pursuing their own systems of knowledge are monists at heart, as long as no one prevents any others from pursuing their own schemes. That way all the benefits of Kuhnian normal science can be had within each paradigm, while allowing multiple paradigms.[17] But will scientists allow this sort of situation? Are they not disposed to regard any unorthodox system as unscientific and therefore shut it down? I suggest that this is less true than it might appear at first glance. Kuhnian extraordinary science is precisely the kind of situation in which multiple paradigms are called forth in order to deal with difficult scientific problems (those problems that precipitate a crisis). The flourishing of competing paradigms does pay dividends, resulting in the discovery of a paradigm that can resolve the crisis. Now, in the Kuhnian picture scientists at that point abandon all other paradigms and go right back to practicing monistic normal science. What I am arguing is that the scientific community could learn to be "extraordinary" on a more sustained basis, so that the benefits of pluralism do not stop flowing as soon as there is a reasonable degree of consensus on the hot topics of the day. There are reasonably undisruptive ways of doing this, and Kuhn himself does identify one clearly as a dominant trend in modern science: increasing specialization and the proliferation of subdisciplines. In Kuhn's own picture of science the developmental pattern of one dominant paradigm being replaced by another lives side by side with the pattern of "speciation," in which one paradigm splits into two or more, all of which continue to flourish.

One might also note that only the "hard sciences," particularly physics, chemistry, and experimental biology, exhibit such a high degree of unity and homogeneity within each field. Currently or in recent enough years most fields of the social sciences and various fields of the natural sciences such as evolutionary biology, ecology, and psychology have flourished while accommodating debates on fundamentals and different frameworks for handling important problems. Sometimes this has been achieved by the splitting of each field into very separate subdisciplines, and sometimes with a less clear separation. Economics provides an interesting case, as it now seems to be on the verge of fulfilling its "physics envy," at least in the pervasive use of mathematics and the near-universal neoclassical consensus in university departments. However, pockets of dissent are still alive and well enough, and we should be thankful for them: how stifling, impoverished, and even harmful the field of economics would be if absolutely everyone practiced it in the same orthodox way!

But can we really afford a pluralistic science? Scientific research requires a great deal of time, money, and talent, and it is not possible to pursue all plausible lines of inquiry; so don't resources need to be pooled into one line of inquiry, at least within each field? I have three layers of responses to this worry. First of all, in the modern era science is not so underresourced, despite the continual protests by scientists themselves. Surely there are enough resources to go beyond a strict monopoly; just how pluralistic we go is the question. And it may not take a great amount of resources to keep ideas and lines of inquiry alive (I remember James Lovelock, the author of the Gaia hypothesis, asking just 1 percent of the science budget to be given to all the unorthodox schemes). Second, there are also points of diminishing return: I think that modern scientists have actually tended to put too much investment into monopolistic lines of thinking—have we, for example, gotten a good return for our investment by putting so much of the best talent in theoretical physics in recent decades into string theory? Having too many people trying the same fashionable approach can be wasteful. Third and finally, it is a pessimistic fallacy to assume that the amount of resources that society devotes to science will remain the same; if we inspire people, we will increase the number of people going into science and even the amount of funding.

There is one last common objection that I must discuss: it may be all well for scientific research to be pluralistic, but at the point of application, when society needs to use scientific knowledge, should scientists not deliver one clear advice? And will that not be impossible on the basis of a pluralistic science? For example, if we have to decide on whether to invest a tremendous amount of money and change our whole lifestyle in order to control climate change, do we not need to have an agreed scientific answer about whether the climate change we observe now is being caused by carbon dioxide emissions from human activities? Without a monistic science, how can we have such useful and necessary answers? Although based on a genuine and serious worry, I think this objection misses the point. It will not help us in life if we make decisions on the basis of false certainty that comes from a narrow-minded monistic science. If there are some scientific results that are really inevitable, then all honest pluralistic attempts will not be able to deny them; on the other hand, if there is genuine contingency and uncertainty in the situation, active pluralism is the best way to find that out. If a cherished scientific result needs to be protected from refutation and doubt by monism, then such a result is not good enough for us to rely on absolutely. And there is always the worry that a trusted scientific consensus of today will turn out to be mistaken tomorrow.

It is true that each given action can only be one thing, so there is a stark need for "monism" there, but that is not the sense of monism we have been discussing here. The need to take action in the face of uncertainty is an unavoidable fact of life. When scientists disagree with each other (which they often do, even in today's heavily monistic climate), policymakers need to make judgments on the basis of conflicting advice, rather than wishing that science would just deliver a simple verdict so their lives would be easier. But how will nonexperts be able to make such judgments? They will, because they have to. Somehow we do not seem to have such trouble in trusting judges, or even juries, to rule on matters of life and death that involve scientific judgments. Perhaps our politicians and policymakers currently make science-based decisions without much appropriate training or procedure, and that is a significant worry. But this is a problem that we have to work to solve, not simply wish away with dreams of a monistic science. I suggest that policymakers at all levels from the United Nations to private foundations need to have advisers who can collect, evaluate, and synthesize all available scientific results that are relevant to the problems at hand, which may be mutually conflicting or incommensurable. And what better people to serve that function than pluralist philosophers of science with a reasonable degree of scientific training!

Action Points for History and Philosophy of Science

Having made arguments in favor of pluralism, I now want to consider briefly how to put it into practice within the field of history and philosophy of science. I focus my attention on my own field in which I can credibly do something, instead of preaching to the scientists about what they ought to do. The main point of action is to *proliferate*: to foster valuable alternatives to add to what is recognized as orthodox science. I believe that true pluralism is a program of knowledge building, not just knowledge evaluation. Therefore it is something for practicing scientists to engage in; however, it is also likely that scientists are already being as pluralistic as their professional constraints allow, and unlikely that they will be inclined to change what they do just because philosophers advocate something unconventional. I would like to propose some concrete ideas about what we philosophers and historians of science can do to help ensure that "many things go" in science, and how such attention to pluralism can change our own practices. I shall be brief here, since the points I am going to make have been elaborated further in another publication (Chang 2012, esp. ch. 5). Here I will try to focus on the points that are most relevant to how we deal with the contingency question.

Dismantling the Discourse on Theory Choice and Realism

A pluralist updating of the philosophical discourse on theory choice will help dispel a philosophical–psychological ideal of monism held by many scientists and counter an inherent bias toward inevitabilism present in philosophical discourse. It is a sufficiently widespread intuition that if we have the correct theory in place, all other (genuinely different) theories in that domain must be eliminated. Even admitting that they do not know whether they are in possession of the ultimate true theory, scientists still tend to think that if one of the competing theories is clearly better than the others, then the latter need to be eliminated. This notion is implicitly shared by many philosophers, which reinforces the scientists' inclinations. This monistic presumption about theory choice is underwritten by scientific realism: having signed up to the exclusive ideal of Truth, and sufficiently impressed by the success of modern science to assume that we must at least be on the correct road to Truth, realist philosophers and scientists have a strong intuition against giving any respect to alternatives to the current best theory.

Even among those who do not think science deals in "Truth," there is a widespread idea that scientists ought to work with only one theory at a time. The emblematic example here is Kuhn, with his insistence that a paradigm does and should enjoy a monopoly within a given field of science in its "normal" phases. Extraordinary science, in which competing paradigms coexist, is presented by Kuhn as a temporary and uncomfortable state that inevitably settles into another period of normal science. Imre Lakatos is the exception that proves the rule here: against Kuhn he maintains that there should always be multiple research programs in a field of science; however, this is only so that these programs can *compete* with each other, allowing scientists to choose the best (most progressive) one at the end of the process. Lakatos does not explain why there should be an "end" to the process of scientific research; that just comes as part of the common conceptual framework of theory choice.

This obsession with monistic choice tends to predetermine the contingency issue; if the normal task of scientific communities is to reach a consensus, then any appearance of contingency will automatically be seen as a problem to be solved, not something we can benefit from. But instead we can take "choice" simply as a matter of each scientist deciding which avenue of investigation to take, without implying that all other avenues are inferior and that inferior avenues should be closed off. It is even possible to maintain a respectable degree of realism within this pluralistic view on theory choice. Elsewhere I articulate and advocate a pluralistic realism, which locates the sense of objective reality in the resistance that nature offers to our epistemic activities. This is a doctrine of realism that does not rely on the correspondence theory of truth and that can grant reality simultaneously to the discoveries produced by different lines of inquiry (Chang 2012, ch. 4). Scientific rationality should not have to consist in every individual making a monistic choice and all individuals agreeing in that choice. It is very often rational for the scientific community to refrain from making a monistic choice between divergent systems of knowledge in a given field of study, and we philosophers may help scientists see that point more clearly, if we start talking about theory evaluation and realism in a pluralist way.

Tasks for Pluralist Historiography

Traditional historiography of science was dominated by a certain kind of triumphalism, presenting history from the viewpoint of the winners, especially at certain prominent junctures. This triumphalism is different from whiggism, which writes history as a progression toward the present, not toward some rather randomly chosen points of triumph in the past, such as Lavoisier's caloric theory of combustion.[18] This triumphalist historiography goes hand in hand with the dominant philosophical discourse on theory choice in reinforcing scientific monism. A pluralist reorientation of the historiography of science would have a profound impact; a great deal of work has already been done in that direction, but there is much more to be done.

Pluralism can be framed as two historiographical directives. First, pay particular attention to losing sides in past scientific debates, and do your best to construct and understand them as sensible alternatives that *unfortunately* got dropped. Historiographical pluralism is founded on a commitment to challenge the complacent triumphalist assumption that the winning side won because it was right (although a thorough investigation could in the end result in a verdict in favor of the winning side). I believe that pluralism is what provided a large part of the key insights in many recent classics in the history of science—Kuhn (1957) on the Copernican Revolution, Pickering (1984a) on elementary particle physics, Steven Shapin and Simon Shaffer (1985) on Boyle and Hobbes, Gerald Holton (1978) on Millikan and Ehrenhaft, Harry Collins (2004) on gravitational waves, Martin Rudwick (1985) on the "Great Devonian Controversy," and so on and so forth—though the authors of these works themselves have often not identified their method explicitly as pluralism. Second, turn away not only from the celebration of the winners but also from the focus on consensus points and explanations of closure. Pluralist historiography would counter the retrospective tidying-up tendency of other historians and most scientists, and even some sociologists. It would seek out and celebrate the rugged individualists and quirky subcommunities and take seriously those phases of scientific development in which no clear consensus emerges. There is much to be gained from a pluralist retelling of even those historical episodes that are widely considered to have been "done to death" already.

Complementary Science

The basic pluralist reorientation of philosophy and historiography suggested so far will prepare the ground for more activist work. Pluralism allows us to see that the judgment of inferiority does not and should not equal a death sentence. If we look back at history with that in mind, we will begin to see that there is still life in many of the "false" and "outdated" systems of knowledge. If an empirical system of knowledge once becomes well established for good reasons, it is difficult to see how it would suddenly become invalid or useless, if there is no genuine, metaphysical change in the very laws of nature. In fact scientists often do preserve and use systems of knowledge that are supposed to be invalid in an ultimate sense. Newtonian mechanics, with its absolute space and time, is still in use in most of its practical applications. Orbitals still form the basis of much work in chemistry, although they are not supposed to exist according to up-to-date quantum theory. Geometric optics still has its uses; classical wave optics even more so. It is of course acknowledged that the old theories do not apply well outside the domains in which they are well established, but it is also acknowledged in practice that they still function in their own right and the in-principle reductions to newer theories are often mere promissory notes.

One practical task for history and philosophy of science here starts with an act of conservation, giving due recognition to what does still survive and attempting to ensure that it does not become extinct. As well-informed and slightly removed observers of science, we can appoint ourselves as guardians of worthwhile systems of knowledge threatened with extinction. When we survey the history of science, we may find that some supposedly rejected past knowledge actually lives on in some form; in that case, we can highlight its survival. We may also find that some systems of knowledge were actually killed off prematurely; in that case, we can revive them. And what we preserve and revive, we can also develop further. The ultimate aim of the active normative epistemic pluralism that I advocate is to improve science by cultivating multiple systems of knowledge. The most active service that history and philosophy of science can perform in this connection, going beyond description and commentary, is to address *scientific* questions that are being ignored by scientists bound by monist traditions—sometimes out of necessity, sometimes through lack of imagination. History gives us a convenient starting point, if we approach it with sufficient philosophical acumen to discern elements of the past that became discarded or hidden without good reason. That same approach can be applied to current science, too. I have given the name of "complementary science" to my own brand of history and philosophy of science: using the intellectual tools and perspectives of history and philosophy to address scientific questions that are neglected by current specialist science.[19] What I did not quite see when I initially put that idea forward was that the project of complementary science was the expression of a thoroughgoing pluralism.

Pluralism and Contingency

Having outlined a program for scientific pluralism, I now return to the question of contingency. Pluralist practice in the history and philosophy of science will change the very way in which the contingency question is framed. An important preliminary step comes from history. The bulk of extant historiography of science that commands the attention of philosophers and scientists is still written with a predisposition toward inevitabilism. I predict that removing this bias will seriously weaken the common presumption that the actual development of science has been on an overall inevitabilist trajectory. Without the inevitabilist bias, we will be able to see that there have been a variety of developmental patterns in the historical development of science. If we are informed by such history, we will be more inclined to ask under which circumstance science would or should produce convergence rather than divergence in its results, instead of asking whether *all* results of science are inevitable.

As for the contingency question as presented by Soler, active pluralism will turn it into a question at least partly about our plans for the future development of science, rather than a matter to be determined entirely by factors beyond our control. When I say "our plans" and "our control" here, I mean "we" in a broad sense, for the future course of science will not be determined by research scientists alone. An equally important role will be played by science educators and those who determine the funding and other support structures for science. And I believe that philosophers and historians of science have an important role to play in the shaping of social and political attitudes toward science, as well as scientists' own conception of science. This is the deep reason why the contingency question is unanswerable: the answer depends on what we do! There are two ways in which a scientific result may become inevitable: first, we *make* it so, by shutting down all other alternatives; second, we try to establish different results and fail in all of our attempts. Active pluralism will increase the contingency present in science, by opposing the first course of action. It will encourage the second course of action, the outcome of which is itself contingent, as far as we can tell.

NOTES

Introduction

I am grateful to Katherina Kinzel and Ian J. Kidd for their useful comments about the contents and structure of this introduction. Many thanks also to Andrew Pickering and Peter Kaiser for their corrections and suggestions for improvement concerning the English language, and to Sjoerd Zwart for his help concerning bibliographic references.

1. The term "contingentism" was already introduced by Biagioli (1996), but in this paper, Biagioli does not oppose "contingentism" to inevitabilism. Instead, the contingentist position he defends is defined in contrast to relativism.

2. In this introduction, I use "science" as a generic term that, unless otherwise stated, is not restricted to the natural sciences but encompasses mathematics, logic, the human or social sciences, and potentially any cognitive enterprise that pretends to a scientific status—taking into account the evident fact that attributions of such a status can be an object of dispute.

3. For different ways of framing and vindicating the contingency and inevitability theses, see Soler (2008a, 2008b).

4. On the practice turn and its lessons, see Soler et al. (2014).

5. See also Shapin (1982, 194) for a similar point.

6. Steven Shapin and Simon Shaffer's famous book, *Leviathan and the Air-Pump* (1985), can also be viewed as conveying a contingency thesis about method. In their case, it is a thesis about what is often dignified as the most powerful exemplar of scientific method, namely, the *experimental* method. Many writings of David Bloor, although primarily couched in the "social" idiom rather than directly framed as an issue about contingency in science, can also be considered as pioneering efforts in support of the idea that methods and results of the sciences, mathematics included, are contingent, in the sense that they could have been otherwise, even profoundly different, without implying that the alternatives should be dismissed as human mistakes, subjective or irrational accomplishments, or the like (see Bloor [1976] 1991). On Bloor's contingentist claims about mathematics, see Jean Paul Van Bendegem's chapter 9 in this volume.

7. For an early analysis of the differences between Collins's and Pickering's conceptions of contingency in science, see Pickering (1987).

8. Note, however, that a number of recent writings, although not *specifically* focused on *science* studies, have endeavored to analyze the uses of "contingency" in the social and natural sciences. Several have attempted to carefully distinguish different senses of "contingency." Many of these works offer conceptual tools that prove transposable to the contingency issue *applied to science*. Exploiting these tools could help to impose some order on the multiple, often confusing problem formulations that are presently found in science studies. Some especially relevant works include Ballinger (2008, 2013); Beatty (2006); Ben-Menahem (1997, 2009); and Inkpen and Turner (2012).

9. For Hacking's other relevant writings on contingency, see Hacking (2006a, 2006b)—the first corresponds to a French short version and the second to a longer English version—as well as Hacking (2014; forthcoming).

10. After I had conducted some preliminary inquiries (see Soler 2006a and Soler 2006b, 60–70 and 163–68, both in French), a first international workshop was organized in 2006 in Nancy (France). It was the origin of the publication of a symposium in *Studies in History and Philosophy of Science* (Soler and Sankey 2008) that included papers by Allan Franklin (2008), Howard Sankey (2008), Emiliano Trizio (2008), and myself (Soler 2008a, 2008b).

11. PratiScienS stands for "Rethinking science from the standpoint of scientific practices." For further information about the PratiScienS project and the PratiScienS team, see http://poincare.univ-lorraine.fr/fr/operations/pratisciens/accueil-pratisciens. Two edited collections have been published in relation to this research program. They contain a number of passages related to the contingentist/inevitabilist issue: first Soler et al. (2012), in which Pickering (2012), Nickles (2012), and Soler (2012a, 2012b) are especially relevant; and second Soler et al. (2014), which is full of comments about the way the "practice turn" fostered attention to contingent aspects of science. See also Soler (2011) for connections between the contingentist/inevitabilist issue and the theme of "tacit knowledge" in science.

12. In the incomparably stimulating setting of Les Treilles in Provence (France). This is an occasion to thank the Fondation des Treilles warmly and to pay homage to Anne Gruner Schlumberger, the mother of the corresponding exceptional project (see http://www.les-treilles.com/?page_id=961).

13. Martin graduated from the University of Minnesota in 2013, and is currently a historian and philosopher of science at Colby College in Maine where he is a Faculty Fellow in Science, Technology, and Society.

14. On the contingency/inevitability of the life and earth sciences, see Radick (2003, 2005a), Radick and Jamieson (2013), and Bowler (2008, 2013). In relation to the issue of a possible field dependency of the contingentist/inevitabilist issue, see also Hacking (forthcoming), where the author suggests that the situation substantially differs in medicine and physics because different senses of "success" are at stake in each case.

15. Although within the historical and social sciences, counterfactuals *in general* (i.e., not specifically directed at *science*) have been the object of multiple analyses from diverse perspectives, often in the context of discussions about the determinist/indeterminist nature of historical processes involving human actions. See notably Byrne (2005), Carr (1961), Cowley (1999, 2005), Ferguson (1999), Hawthorn (1991), Weber ([1906] 1949), and Weinryb (2009). For an early attempt to apply counterfactual thinking to the history of quantum physics, see Hund (1966).

16. For an attempt to provide some kind of empirical support to plausible counterfactual histories of physics, see Pessoa (2001). The attempt starts from an analysis of the contents of scientific publications and uses computer programs to process this historical material.

17. Radick's symposium was in press when the Soler and Sankey (2008) symposium appeared—as recorded by Radick in the last note of his introduction, where he

kindly describes the publication of the latter as an "encouraging sign" given the little attention so far received by the topic of contingency in science (Radick 2008a, 551). None of the contributors to Soler and Sankey (2008) was, at the time, aware of Radick's forthcoming symposium—or, it should be confessed, even of Radick's antecedent contributions to the issue of counterfactual history and contingency (Radick 2003, 2005a, 2005b). Accordingly, and unfortunately, the corresponding references were not included in the bibliographies of Soler and Sankey (2008).

18. Outside of this book, see Radick (2005a) for stimulating systematic suggestions about features of this type, illustrated in the case of a counterfactual history of biology.

19. For general reflections on "alien science," see Rescher (2009), esp. chapter 3.

20. For a discussion of this type outside of this volume, see Arabatzis (2008), which also contains multiple relevant references.

21. See also Bloor ([1976] 1991), who points to some ways of "writing history" as tactics for rendering variations in science invisible (esp. 129–30).

22. Katherina became an associate member of the PratiScienS research group in 2011. This is the occasion to thank Martin Kusch for having invited me to give a talk on contingency in Vienna, in October 2013 (this talk was an abbreviated oral version of the text presented in chapter 1). Concerning the evidential support that histori-cal-social case studies can provide to philosophical positions, including contingentism versus inevitabilism, see Kinzel (forthcoming [a], forthcoming [b]).

23. All quotations are from the introduction of Kinzel's PhD dissertation (2014).

24. Concerning the initial resources of Kidd's work on contingency, I have been delighted to learn from Kidd that the 2008 edited volume in *Studies in History and Philosophy of Science* (Soler and Sankey 2008) was what persuaded him that contin-gency was a viable topic for philosophy of science.

25. In three recent essays, among which only one has been published so far, Kidd (2013; unpublished manuscript). For more about the contents of these papers and a discussion of aspects of them, see chapter 1, 000, 000, and 000 in this volume. In the third reference, Kidd (unpublished manuscript) discusses contingency in relation to Kyle Stanford's "unconceived alternatives" as developed in Stanford (2006).

26. On pluralism versus monism, and diverse forms of these two stances, see, for example, Kellert, Longino, and Waters (2006).

27. For more recent bibliographical references focused on the empirical equivalence of standard and Bohmian quantum theories and on the implications of such a situation in terms of contingency, see the references indicated in my chapter 1. As far as I can see, noth-ing of what has happened in quantum physics until 1994 requires the introduction of any substantial change regarding Cushing's general characterization of the situation in 1994.

28. Ludwik Fleck ([1935] 1979) is also often mentioned, albeit less frequently scru-tinized (see, e.g., Hacking, forthcoming).

29. It is worth noting that Hacking does not consider his 1992 paper as a contin-gentist argument or even as offering an obvious resource for contingentists. He said (private communication, April 2004) he was surprised to discover, when reading Pick-ering's introduction to the edited collection in which his 1992 article appeared (see Pickering 1992, 8–10), that the theme of contingency was evident in his article.

30. Path dependency and contingency harbor complex relations that await a systematic characterization. On path dependency in science, technology, and more generally in human history, see, for example, Garud and Karnøe (2001, in particular, Pinch's contribution) and Peacock (2009) for an analysis of differences between science and technology.

31. On the relation between contingency in science and contingency in the natural world as conceptualized by sciences using the evolutionary framework, see also Hacking (1999, 74), Radick (2003, 161–63; 2005a, 25–30). In an (unfortunately) unpublished talk of 1997, Pickering reconsiders and illuminates his contingentist picture of the *Mangle* in the light of the evolutionary scheme of thought (Pickering 1997).

32. For a critical reflection on mathematics as "the Realm of Necessity," see Bloor ([1976] 1991, 179), which provides innumerable proposals of great relevance and interest with respect to the contingentist/inevitabilist issue. See, in particular, chapters 5, 6 ("Can There Be an Alternative Mathematics?"), and the "Afterword" in the 1991 edition. For more recent works that deal with the contingency/inevitability of mathematics, see Buzaglo (2002) and Mancosu (2009).

33. See also Bloor ([1976] 1991), for example, 129–30, and 180–83.

34. This was also a concern of Bloor ([1976] 1991), see, in particular, 179–80, about "how people decide what is inside or outside mathematics."

Chapter 1. Why Contingentists Should Not Care about the Inevitabilist Demand to "Put-Up-or-Shut-Up"

1. Of course, contingentist/inevitabilist claims can also be directed toward *nonscientific* items (any extrascientific set of events, beliefs, actions, etc.) in human history. They can also include sets of factors conceived of as conditions of possibility for the emergence of Western science through the "scientific revolution" of the sixteenth to seventeenth centuries (on the contingency/inevitability of the scientific revolution, see Henry [2008] and Fuller [2008]). In this chapter, however, after some brief remarks about the historical emergence of the experimental method (46–47), I take for granted the existence of Western science as we know it (in particular, I take for granted that "science" goes hand in hand with an *experimental* practice and a *mathematical* practice).

2. But the corresponding alternative science could nevertheless be viewed as different from our science in ways that essentially matter. See Rouse's contribution, chapter 13, for a vindication of this thesis.

3. See, in particular, four sections (58–63 and 73–75). In these sections, the issue of scientific method enters into play via the issue of the *kinds of factors* involved in the choice of a theoretical option rather than another one, and of whether or not the factors involved are *compelling enough* to *universally* impose one *unique* option on scientists.

4. We could add scientific concepts to this repertory. Regarding the contingency/inevitability of the expansion of scientific concepts, Buzaglo (2002) provides relevant elements (more specifically focused on *mathematical* concepts).

5. An attempt to distinguish different types of contingency "depending on what parts of science that claim specifies [the claim that science is contingent]" can be found in a recent paper by Joseph D. Martin (2013). The section titled Restrictions

Concerning the Target of the Inevitabilist Thesis offers illustrations of conceptions that differ with respect to "what part of science" (here what part of *physics*) is taken to be contingent/inevitable. See also Yves Gingras's chapter 8 in this volume for a strong emphasis on the need to distinguish contingentist/inevitabilist theses according to the object under scrutiny.

6. In my paper (Soler 2008b), I took for granted that nobody—or more precisely, no professional analyst of science—would contest possibilities such as those conveyed by scenarios 1, 2, or 3. However, under further examination, it seems that this is not so simple. John Henry, for example, argued that both "positivist commentators of science" and, more surprisingly, "contextualist historians of science" ought, on pain of inconsistency, "to be committed to the view that counterfactual changes in the history of science would have made no significant difference to its historical development" (Henry 2008, 552). More specifically, contextualist social constructivist historians of science "cannot . . . hold that small changes in the actual history of science would have made a difference without simultaneously invalidating the historiography of science of the past half century or so. To do so would be to suggest that the wider social milieu of the sciences is not really a significant [i.e., 'causal'] factor in understanding their development." Provided they "do manage to come up with a historical explanation of all the factors involved in accounting for the burgeoning of science, beginning in the Renaissance, then there would be a strong tendency to suppose that, given all these factors, the rise of science was inevitable" (558). If Henry's reconstruction of contextualist historians' inevitabilist positions is correct, it might be used to question the plausibility, and even the very possibility, of at least scenarios 2 and 3—in opposition to my claim above that "nobody would contest" such possibilities. Henry concedes, however, that contextualist historians can, and usually do, endorse the following contingentist position without being inconsistent: "if the *entire cultural background* had been different," then, science might have been very different, and modern science may even never have emerged at all (552; emphasis added). He calls such counterfactual thinking "radically contingentist counterfactual" (559). Provided that contextualist historians do indeed accept such counterfactuals, at least scenario 1 would be widely assumed.

Similarly, Ian J. Kidd, in two recent papers that rely inter alia on (Henry 2008), reconstructed and systematically criticized positions that, if indeed held as such by the named authors, would call into question the claim that scenarios 1, 2, and 3 are uncontested genuine historical possibilities (Kidd forthcoming; 2013). In particular, some of the positions criticized by Kidd—termed "scientific imperialism"—claim the existence of "developmental teleologies," that is, predetermined inevitable pathways for science or for a particular discipline.

Whether or not some people are ready to contest the possibility of scenarios 1, 2, and/or 3, however, we can act—and I will in what follows for the sake of the discussion—as if the possibility of these scenarios was not contested. This is so, because even for an extreme inevitabilist who would be prepared to claim that human beings *had* to develop a successful physics and *had* to ask the questions physicists actually asked *and not others*, the real stake of the debate, from an *epistemic* point of view, would *remain, as for those who accept the possibility of scenarios 1 to 3*, the inevitability or contingency *of the answers* provided to the scientific questions asked in the context of the successful physics under discussion.

7. Jean Paul Van Bendegem makes a similar point in chapter 9 of this volume and provides illustrations in mathematics (see, in particular, 226–28). See also Joseph Rouse's chapter 13 (317), where a nice concise version of this point is stated. In a related vein, Bloor ([1976] 1991) already insisted on a number of historical or philosophical practices through which the possibility and reality of variations inside of mathematics can be rendered invisible or can be dismissed as mathematics worthy of the name (see, e.g., 129–30, 180–83). Although specifically focused on the case of mathematics, his remarks are mutatis mutandis applicable to any other scientific field. To dismiss a proposed alternative science as a science worthy of the name amounts to denying what I call below (51) the "genuine physics" condition (or generalizing, the "genuine science" condition).

8. For an illustration of such an inevitabilist claim relativized to the scientific questions that have been asked and investigated, which also use the case of the speed of light, see Gingras's chapter 8 in this volume (205): "If one decides, for some contingent reason, to measure the speed of light, then it is, by necessity, either finite or infinite."

9. Emilano Trizio (see chapter 4 in this volume) also formulates the question in terms of "same subject matter," following Hacking. For a discussion of the "same subject matter" condition, and insights about the special difficulties that this condition raises in the field of mathematics compared to physics, see Salanskis's contribution to this book, chapter 10 (241–49).

10. For these further qualifications, see 79–80.

11. The proposed definitions immediately suggest that an additional problematic point will sooner or later be involved in the discussion: namely, the conditions under which two scientific items can be viewed as "the same" (or "sufficiently similar," or "different but reconcilable," etc.), or should rather be considered as "truly different" items (or "incompatible," or "irreconcilable," or "irreducibly different" items). See Soler (2008b) for further developments on this issue (and also 71–73) for an illustration of possible disagreements at this level, in relation to the particular case of Bohm's quantum physics. This issue is strongly related to the compromise, mentioned above, between scientific alternatives that are too radically different and scientific alternatives that are too similar (see 49–50).

12. See Soler (2008a, 225–27) for more, and for suggestions of some inevitabilist representatives.

13. The framing in terms of "dialogic reconstruction" has been suggested by Emiliano Trizio. I am grateful to him for this suggestion, and more generally for his insightful comments on an antecedent version of this chapter.

14. "This kind of counterfactual history has a credibility handicap—we know how things did turn out but can only imagine how they might have turned out" (Shapin 2007). The subject of Shapin's paper is an alternative *technology.*

15. Systematic analyses of the power of counterfactuals, especially in relation to the history *of science*, are rare. A welcome example, however, is the symposium directed by Gregory Radick (2008b), "Counterfactuals and the Historian of Science," published in *Isis*, which contains interesting contributions and provides further references. See my introduction to this volume (9–10).

16. I borrow this phrase from Ian Kidd (forthcoming). In the same vein, Jean-Michel Salanskis writes (in a first longer version of his chapter 10 in this volume) that "the inevitabilist conviction is taken as the default value of reason."

17. For more about the reasons that inevitabilists think the burden of proof lies with contingentists, see 81–82). For a reflection on the sources of our inevitabilist intuitions, see Pickering's contribution to this book, chapter 3.

18. I thank Andy for his helpful clarifications about some aspects of his position that are involved in this section. It is also the occasion to express my gratitude to him for the time he spent to improve the English language of this chapter.

19. For more about "scientific symbioses," see 61–63.

20. For a structurally similar situation in mathematics, see Van Bendegem's analysis in chapter 9 of this volume, about Bloor's attempt to put up examples of actual alternative mathematics.

21. For characterizations of the symbiotic or robust-fit conception of science, see, for example, Hacking (1992), Pickering (1995a), Soler (2008c, 330–36). A clear synthetic account of the symbiotic framework as understood by Pickering, and its relation to contingency, is provided by Hacking (1999, 71–74).

22. The case of sciences that would involve completely different experimental bases and associated phenomena has been considered by Hacking and Pickering. Hacking (1992) talks of "literal incommensurability" and Pickering (1995a) of "machinic incommensurability." In Hacking's terms, the two sciences would be "incommensurable in the straightforward sense that there would be no body of instruments to make common measurements, because the instruments are peculiar to each stable science" (Hacking 1992, 31). See Soler (2008c) for a discussion.

23. Trizio also insists on similar important points when he discusses the condition—built into the formulation of the contingentist/inevitabilist issue as he frames it following Hacking—of two "equally successful sciences"; see Trizio's contribution to this volume, chapter 4 (149–50) and Trizio (2008, 256–57).

24. However, this intuition can be questioned. With respect to this issue, see the disagreement between Henry (2008) and Bowler (2008) in relation to the social explanation of the history of science. As Radick summarizes it: "For Henry, counterfactuals that suppose only small-scale changes . . . must be intellectually inert . . . . The mathematization of nature and other legacies of the Scientific Revolution were inevitable, given the wide distribution of the relevant background factors. For Bowler, however, a commitment to social explanation does not require so full blown a denial of the influence of the small" (Radick 2008a, 549). Bowler's particular thesis is that "if Darwin had not been there to write his *Origin of Species* the subsequent development of biology would have occurred along a line that steadily diverged from the sequence of events we actually experienced" (560), with some important "effect on the end-product" (567). His more general claim is that "the whole process [of the history of science] is open ended and small events . . . can have major consequences for all subsequent developments" (561). See also Bowler (2013).

25. See Cushing (1992) for a brief summary of relevant steps in the history of the causal quantum theory program (i.e., BQM-type attempts), including developments

before Bohm's publications in 1952 (embryonic versions of BQM-like frameworks were developed in 1926–27 by Erwin Madelung and Louis de Broglie, but the well-developed and coherent BQM was first introduced by Bohm in 1952, which is why SQM is often called the "de Broglie–Bohm theory" in the literature), reception of Bohm's publications, and some extensions after 1952.

26. Cushing writes that "the causal quantum theory program [BQM] either is entirely unknown to most scientists and philosophers concerned with foundational problems in quantum mechanics or has been badly presented to them" (Cushing 1994, xi). I had a recent opportunity to experience that BQM is indeed very poorly known to physicists, including those who made invaluable contributions to quantum mechanics and who feel deeply concerned with the interpretation of its formalism. This was specifically striking in one circumstance. During several years (2010–14), physicist and philosopher Bernard d'Espagnat organized in Paris, at the Institut de France under the auspices of the Académie des Sciences Morales et Politiques et du Collège de Physique et de Philosophie, a seminar intended to favor exchanges between physicists and philosophers around interpretational issues in relation to quantum physics, and he kindly invited me to participate. One meeting of the seminar was devoted to a presentation of BQM by Franck Laloë, followed by a general discussion. I was struck by the fact that BQM was so foreign to the participant physicists (among whom were famous figures such as Alain Aspect, Roger Balian, Édouard Brézin, Michel Le Bellac, and Jean-Michel Raimond) and struck as well by the rapidity with which, following the presentation, most physicists felt that BQM was valueless. For a transcription of the presentations and discussions of the seminar, see d'Espagnat and Zwirn (2014), and for more specically on BQM, see 211–60, and 42ff.

27. Whether "equivocal" or not, Cushing's use of "theory" is perfectly self-conscious. Cushing (1994) takes the trouble to give an explicit definition of "scientific theory" as he means it. "A scientific theory can be seen as having two distinct components: its formalism and its interpretation. These are conceptually separable, even if they are often entangled in practice." In modern physics, "a formalism means a set of equations and a set of calculation rules for making predictions that can be compared with experiments. . . . The physical interpretation refers to what the theory tells us about the underlying structure of these phenomena (i.e., the corresponding story about the furnitures of the world—an ontology). Hence, *one* formalism with *two* different interpretations counts as *two* different theories" (Cushing 1994, 9). For interesting remarks on the usual division of quantum physics into two supposedly "distinct cognitive objects," theory and interpretation and on some implications regarding the potential influence of a new unorthodox scientific proposal, depending on the fact that it is viewed as an interpretation of the theory rather than as the theory itself, see Pinch (1977, 176–84).

28. See Soler (2008b, 234, [d]), in which I stress a similar point and connect it to theory individuation and theory comparison in the history and philosophy of (our actual) science.

29. Simplicity is often presented by critics of BQM as one reason to prefer SQM. See below the quotation from Hervé Zwirn. Understandability is often presented by

advocates of BQM as one reason to value BQM. As Cushing stresses, "The quest for a more (nearly) understandable worldview can be a motivating factor in seeking another interpretation of quantum formalism" (Cushing 1994, 77). "Claims of increased understanding" (78) were put forward from the start by Bohm himself, in his seminal 1952 papers, as an asset of BQM.

30. A systematic classification of the different objections raised against BQM, as well as insightful possible replies to each, can be found in Passon (2005). This paper also provides multiple illustrations of the divergences between physicists concerning theory assessment on nonevidential grounds. See also Pinch (1977, 1979) for an earlier analysis of the negative reactions to Bohm's theory in the 1950s–1960s and additional illustrations.

31. Note, however, that not all inevitabilist-inclined minds reject counterfactuals. Some have attempted to build counterfactual histories of science that, *if indeed conceded as plausible,* favor inevitabilism rather than contingentism. For an early attempt of this kind applied to the beginning of the history of quantum physics, see Hund (1966).

32. "What is important is that there were precedents for such moves and that the necessary pieces were already there" (Cushing 1994, 175). "This [alternative] 'story' is neither ad hoc (in the sense of these causal models having as their sole justification an origin in successful results of a rival program) nor mere fancy, since all of these developments exist in the physics literature" (191–92).

33. There would be another possibility, corresponding to what I have called the "experiment of the divided physics." The experiment would consist in dividing the scientific community into two completely separate subcommunities and in examining what kind of science has resulted on each side after a "very long" time. A discussion of the possible results and epistemic force of an experiment of this type can be found in Soler (2008b, 235–40).

34. On the relations between contingentist/inevitabilist commitments on the one hand and ways of interpreting and narrating historical episodes on the other, see Fuller (2008).

35. In his contribution to this volume (chapter 8), Yves Gingras makes observations, as a historian and sociologist of science, that support these lines of thoughts. He writes: "The advantage of using real cases is that we can then analyze how real scientists do in fact react when faced with very different theories that pretend to cover the same phenomena. Such an approach also raises an interesting question: can scientists really accept being faced with two successful but incompatible sciences covering the same object? What we know about the history of science suggests that *they will in fact do everything they can to make these two theories compatible* even when they involve clearly incompatible ontologies like those of particles and waves" (207, emphasis added). Gingras goes on to provide several examples. However, he does not consider the status of this "fact" about "how scientists react when faced with 'different' theories of the same object": he does not discuss the issue of whether or not we should see such a fact as inevitable.

36. Concerning kinds of pluralist regimes, two cases would need to be distinguished with respect to the contingentist/inevitabilist issue: pluralism in the background of a *realist* conception of science and pluralism in the background of an *instrumentalist* conception of science. The first case corresponds to what Chang calls "Pluralism with Truth as the ultimate aim of science" (this volume, chapter 15, 364). In this case,

multiple scientific options are cultivated, but one of them is placed above all the others as *the* candidate for truth, thus conserving a superior, privileged status. Since "Truth is exclusive, in the sense that if one system of knowledge attains it, it cannot be possessed by any alternative system that contradicts the successful one" (367), a form of uniqueness commitment is maintained (though weaker than in the monist regime), that is, one option is still unique as far as its status (its superior value) is concerned. In such an exclusive pluralist regime, an inevitabilist-inclined philosopher can see the candidate for truth as the inevitable option, and all the other ones as contingent human useful means with respect to some human aims different from truth. The situation is notably different in an instrumentalist pluralist regime that would ignore the absolute aim of truth and would relativize the value of each scientific option to a multiplicity of not-exclusive and not-hierarchized aims—such as "understanding and empirical adequacy" (367, see Chang's chapter 15 for a discussion of this case). In such a pluralist regime, no need would be felt to elect one option as *the* best one (in an absolute sense). One option could be judged superior with respect to some aims but inferior with respect to others. It is clear that the first kind of pluralist regime is a much more fertile terrain for the culture of inevitabilist commitments than the second one.

37. See, for example, Feyerabend (1965, 1993), Chang (2009b, 2012), and for a more general perspective on pluralism in science, Kellert, Longino, and Waters (2006). In his contribution to this volume, Hasok Chang argues that "pluralism is more beneficial to science than monism, given any reasonable position regarding the fundamental aims and values of science" (chapter 15, 364).

38. On this point, see, for example, Pickering and Trower (1985) or Galison (1987).

39. I thank Ian Kidd for agreeing to send me his paper before publication. On the put-up-or-shut-up demand, see also, in a quite different framework, Joseph Rouse's stimulating suggestions that the demand could be viewed as already satisfied (this volume, end of chapter 13).

40. Some optimist scholars in science studies may think that this impossibility claim is too strong and too defeatist. Hasok Chang is a case in point. Considering the possibility that philosophers of science might be able to lend plausibility to alternative sciences, he writes: "Even if the likelihood of success is small, the returns would be great, which means that an attempt may be worth making," adding that "it is not entirely ludicrous to think that a nonexpert may be able to see something that is in the collective blind spot of a homogeneous scientific community." He concludes that "a commitment to explore scientific alternatives is required of any philosophers wishing to tackle the contingency question in a serious way. Without an active cultivation of alternatives, it will be all too easy to issue a meaningless armchair verdict in favor of inevitabilism based on a lack of imagination" (this volume, chapter 15, 361; see the same chapter, 377–81, for an overview of the kinds of activities in which historians and philosophers could engage to attempt to bring scientific alternatives to life or to support the credibility of already existing but devalued scientific alternatives).

41. Kidd has confirmed that it is (private communication).

42. Kidd's paper provides arguments for the position that "Inevitabilism should . . . not be the default stance within the philosophy of science" (sec. 2) and that "inevitabi-

list claims are hubristic and should be rejected" (i.e., "The inevitabilist is . . . guilty of epistemic hubris because they lack the cognitive powers to perform the epistemic tasks needed to establish that a given scientific result was inevitable," sec. 3).

43. Such determined critics could attempt to rely on the existing body of work on scientific pluralism and disunity, for example, by Helen Longino, John Dupré, David J. Stump, Peter Galison, Nancy Cartwright, and others, and more generally by a wider set of scholars, sometimes called "post-Kuhnians," who have argued that, in fact, science has never been as monistic as Kuhn suggested but, to the contrary, has always been pluralistic and disunified. I am not sure that Kuhn's view is indeed so monistic, but more important with respect to our purpose, regarding the issue of the degree to which our science is actually monistic, I think we should carefully distinguish two things: (1) the regulative ideal under which our science is practiced, which is strongly monistic—the search for a unique worldview, the will to eliminate competing ones, and so forth; and (2) the actual coexistence of a plurality of fields, programs, methods, theories, models, and so on.

Chapter 2. Some Remarks about the Definitions of Contingentism and Inevitabilism

1. Here are the main books and papers we refer to: Hacking 1999, 2000a, forthcoming; Pickering 1984a, 1995a; Sankey 2008; Soler 2006a, 2008a, 2008b; Trizio 2008. Here is the definition of Ian Hacking (1999) inspired by Andrew Pickering, *Constructing Quarks* (1984a): "in the case of physics, (a) physics (theoretical, experimental, material) could have developed in, for example, a non-quarky way, and by the detail and the standards that would have evolved with this alternative physics, could have been as successful as recent physics has been by *its* detailed standards. Moreover, (b) there is no sense in which these imagined alternative physics could be equivalent to present physics" (Hacking 1999, 78–79). We find in the same book of Hacking the following affirmation: "To sum up Pickering's doctrine: there could have been a research program as successful ('progressive') as that of high-energy physics in the 1970s, but with different theories, phenomenology, schematic descriptions of apparatus, and apparatus, and a different, and progressive, series of robust fits between these ingredients. Moreover— and this is something badly in need of clarification—the 'different' physics would not have been equivalent to present physics. Not logically incompatible with, just different" (72). Hacking also proposes a definition of inevitabilism: "If the results R of scientific investigation are correct, would any investigation of roughly the same subject matter, if successful at least implicitly contain or imply the same results?" (Hacking 2001, 61).

2. In the same paper, we find another definition, which constitutes, according to Léna Soler, a more rigorous formulation of the contingentism/inevitabilism opposition:
Contingentism will be defined as the following thesis:
Contingentism
a) More or less the same initial conditions obtain as those which have occurred in the history of our own science;
b) Nevertheless, the *possibility*, as "final" (subsequent or later) conditions, at least in the long run, of an alternative physics,

- As successful and progressive as ours,
- With yields irreducibly different from ours (notably which involves an ontology incompatible with ours)

Correlatively, inevitabilism will be defined as the following thesis:

Inevitabilism

a) *if* more or less the same initial conditions obtain as those which have occurred in the history of our own science;

b) *and* a successful and progressive physics has indeed been developed;

c) then, *inevitably*, as "final" (subsequent or later) conditions, at least in the long run:

•more or less the same results and the same ontology as our own,

•or different but reconcilable results and ontologies as our own (Soler 2008b, 233).

3. "My sticking points emphasize philosophical barriers, real issues on which clear and honorable thinkers may eternally disagree" (Hacking 1999, 68). The emphasis is on the endlessness of the debate.

4. Hacking himself underlines the fact that the C-I debate may be conceived without any ontological implications: "When we turn to the metaphysics of the schools, the contingency thesis appears to be consistent with any standard metaphysics. (So much the worse for the standards and the schools, you may say.) For example, contingency is consistent with the scholastic debating point of the 1980s called 'scientific realism.' Many versions of that doctrine state that physics aims at the truth, and if it succeeds, it tells the truth. If the physics refers to some type of unobservable entity, then, if the physics is true, entities of that type exist. Many social students of science reject any version of scientific realism. So do many philosophers, such as Bas van Fraassen (1980). But the contingency thesis itself is perfectly consistent with such scientific realism, and indeed anti-realists, such as van Fraassen, might dislike the contingency thesis wholeheartedly" (Hacking 1999, 80).

5. It is the appearance of this kind of necessity that we find in Greek tragedy and in the movies, about which Jean-Paul Sartre says that when he saw them in the cinema, the strict narrative economy gave him the feeling that the contingency was completely absent in the universe in which the characters evolved: "I thought about contingency, after watching a movie. I saw movies in which there was no contingency at all, and when I walked out of the cinema, I was finding contingency everywhere. It is the necessity expressed in the movies which made me aware of the contingency I found in the real life" (Beauvoir 1987, 321; our translation).

6. For instance, in Soler (2008b), the word "sciences" is replaced by "physics" from page 2 of the article.

7. Only very few contributions to the C-I debate include case studies borrowed from fields other than physics, more precisely: the social and human sciences in Trizio (2008); geophysics in Sankey (2008); the biomedical sciences (Hacking 2000a, forthcoming). Trizio (2008) briefly mentions the social and human sciences. He emphasizes that they do not seem to be pertinently concerned by the contingentism/ inevitabilism controversy for two reasons. On one hand, for some philosophers, such

sciences are not able to obtain results that are universally admitted, and on the other hand, it seems very difficult to define what the notion of "successful science" might be when we are talking about the social and human sciences. To summarize, "the question of contingency is not easily addressed to these sciences." Concerning contingency in geophysics, and more precisely the theories that granted continental drift, Sankey simply eludes the question when he asserts that "this appears to be a clear case in which contingent factors play a role in the development of science" (Sankey 2008, 262), because the investigating tools that allow "pivotal evidence" do not appear before a certain date. But it is not what is at stake with the contingentist thesis, which is mainly concerned with the contingency or not of the results, and the constitution of a specific ontology. In fact, we have here only a case of "benign contingentism" (Soler 2008b, 231), which nobody contests. For examples concerning the biomedical sciences considered by Ian Hacking, we will present his position in the following pages of this chapter.

8. This is the case in regard to Newtonian physics, which today does not appear inevitable—it has been supplemented, but not really replaced.

9. This remark is not really relevant if we consider the radical inevitabilist's (in the "very hard" version) point of view, because it does not conceive any alternative path for the development of science. Accordingly, it gives a unique answer to the above question: the results *are* inevitable. But one can be a radical inevitabilist about the development of some determined scientific disciplines or fields and, at the same time, adopt a mitigated position about the development of some others.

10. This is an explicit methodological requirement in some sociological studies, precisely those chosen by Hacking when he talks about contingentism—Latour and Woolgar (1979), Pickering (1984a). But the contingentist might say that this criticism is not relevant. Indeed, he might invoke the fact that Hacking does not appeal to logical possibilities. He only changes the order of what actually happened (e.g., as is the case in Cushing 1994).

11. Hacking's position is a moving target. In Hacking (1992), he seems more contingentist than in Hacking (1999) and of course than in Hacking (1983).

12. "en suivant la physique contemporaine, nous avons quitté la nature pour entrer dans une *fabrique de phénomènes*" (Bachelard 1951, 10).

Chapter 3. Science, Contingency, and Ontology

1. What follows grows out of my book *The Mangle of Practice* (Pickering 1995a), itself based on detailed documentation and analysis of scientific practice. I argued there that in order to offer a satisfactory analysis of scientific practice we need to move from what I called the representational idiom to a performative idiom. The former is centered on an image of science as primarily a body of knowledge and is epistemological in just that sense. The latter recognizes and centers itself on a recognition that scientists do things in a world that also does things, and that these doings are interrelated. In the laboratory, scientists set up apparatus precisely to see what the apparatus will do and react to that. This led me to an understanding of practice as a transformative and open-ended *dance of agency*, which is an ontological vision, inasmuch as it refers in

the first instance to worldly performances rather than our knowledge of entities. My argument was not that we should ignore the epistemic strata of scientific culture, but that we should see them as part and parcel of an overall performative rather than purely epistemic process. The object of the present chapter is to elaborate this ontological vision with a particular concern for the contingencies it implies. My hope is to open up a space for thinking about contingency ontologically, in the face of intuitions that point to inevitability. I realize that the inevitabilist might be moved to produce a string of counterarguments, but to respond to them I would first have to translate all the inevitabilist arguments I could think of into the performative idiom and then respond to each of them—an endless task, not to be undertaken here. If any inevitabilist wants to contest the picture I sketch out in specific ways, I am happy to continue the discussion in the future.

2. Some elaboration might be useful here. My notion of emergence is that the world can always surprise us by performing in ways that we do not expect. When we try to latch on to it in a new way, we have to *find out* what will happen. It just turns out that, for some materials, electrical resistance decreases continuously with temperature down to some cutoff, where, it turns out, resistance vanishes and the materials become superconducting. From one angle, this is part of the standard story about science as an endless venture into the unknown. But the focus in the standard story is not on the endless venture but on colonialization; science as retrospectively making the surprising not, in fact, surprising—as if we should have known in advance how matter would behave were it not for our lamentable ignorance. I argue against this sort of retrospective accounting in Pickering (1984a, 1995a). But what if we latch on to the world in the same way, and not in a new one? This question runs through what follows, but for now we can note that "the same" is itself a problematic notion. No action is ever exactly the same in all its details as an earlier version, an observation played out in practice in what Collins (1992) calls the experimenter's regress. (I am thinking here of the physical sciences and the inanimate world; change and novelty are much more obvious and readily grasped in relation to biology.)

3. This is related to Latour's (1993) notion of "purification"—now at the level of human and material performance rather than the level of human understandings. Latour's idea can be traced back at least as far as the discussion of "splitting" and "inversion" in Latour and Woolgar (1986). For more on making the world dual, see Pickering (2009).

4. Descartes lived in a world in which there were only a few freestanding machines as defined above. In that context it was clever of him to conceive of animals and brute matter on the model of the machine. Only since the Industrial Revolution have parts of the world become so saturated with machines that Descartes's thought becomes a truism.

5. There is an important point that perhaps bears emphasis here. An ontological version of the inevitabilist argument might be that even if there are, in principle, an indefinite number of these islands of stability, still, people like us—or physicists like our physicists—will inevitably light on certain of them. A shared cultural background (material, social, conceptual) inevitably singles out some subset of these islands. The

argument of this paragraph (set out at much greater lengths in Pickering 1984a, 1995a) denies this. To all intents and purposes, Morpurgo and Fairbank shared a common culture; the new physics was elaborated by much the same scientists as the ones that had previously elaborated the old physics. Of course, one can point to cultural differences: Morpurgo's work, up to his quark-search experiments, had been in theoretical physics, Fairbank was an expert in low-temperature experiments, and so on. But the inevitabilist argument can hardly refer to "people like Morpurgo" if it is to have any force.

6. I stress the need for a culturally situated goal here as the simplest way to insist in our constitutive entanglement in the definition of islands of dualist purity. Of course, one would have to supply the imaginary being with very much more than the chemicals and a goal to make this process plausible. Needed would be all sorts of skills and disciplined training, glassware, a climate not too different from ours, other chemicals, textiles, and so on. And all of the elements of this indefinite list would be liable to be mangled en route to mauve, if indeed that is where their route led. We should also recognize the explicitly social elements of such a process, along the lines of Fleck's discussion of a multiplicity of human actors collectively arriving at the Wasserman reaction in different forms of open-ended experimentation over a period of centuries. All this makes the picture more complicated, without affecting the point about decentering.

7. For more on the key notion of "islands of stability" see Pickering (2014).

Chapter 4. Scientific Realism and the Contingency of the History of Science

1. For a detailed analysis of the most interesting alternatives, see Soler (2008a).

2. See Allamel Raffin and Gangloff in this volume, chapter 2.

3. We can therefore introduce the distinction between weak and strong inevitabilism (and between strong and weak contingentism): *weak inevitabilism* is inevitabilism as defined by Hacking and is logically compatible with the multiplicity thesis. It is the claim that *our* history of science could not have led to alternative stabilized stages as successful as ours of the investigation of a given subject matter. *Strong inevitabilism*, instead, is incompatible with the multiplicity thesis; it implies that equally successful mutually incompatible stabilized stages of the scientific investigations of a given subject matter are impossible. Therefore, given a certain subject matter, one can only allow for mutually incompatible scientific accounts of it enjoying degrees of success that are sharply different from one another. According to this view, any historical trajectory leading to a theory incompatible with ours must lead either to a theory less successful than ours or to a theory more successful than ours that is either less advanced or more advanced. Simply put, according to strong inevitabilism, mutually incompatible successful scientific accounts of a given subject matter must form a series of increasing successfulness; hence they could be, potentially, different successive steps in the investigation of a subject matter. To go back to the previous example, if intelligent aliens have developed a particle physics as successful as ours, according to weak inevitabilism that physics might be one we could not possibly have come up with in the course of our historical trajectory; whereas according to strong

inevitabilism, an alien particle physics as successful as ours should necessarily have to look pretty much like our own. In this respect, I should also add that the note 5 of my 2008 article contains a mistake (Trizio 2008, 254), for it equates strong inevitabilism with the thesis that there is only one possible account of a given subject matter that could ever deserve to be called successful. The latter thesis would instead amount to a sort of "extreme" inevitabilism asserting that success does not come in degrees and that there is only one possible successful account of a given subject matter. As one can see, the maze of possible histories of science is quite intricate, even setting aside the complex problem of giving a satisfactory characterization of scientific success and a clear criterion for its evaluation.

4. I prefer this formulation to the slightly different one that takes a metaphysical realistic thesis as a component of the definition of scientific realism, as is done by Stathis Psillos (1999, 2000). Throughout this article, words such as *realism* or *antirealism* used without further specification refer to *scientific* realism and antirealism.

5. See Soler (2008b, 235–41) for a thorough analysis of the different possibilities of conflict between rival theories, conducted in the interesting framework of the "divided physics" thought experiment.

6. More recently, Kyle Stanford has developed a new detailed argument for the underdetermination of scientific theory by empirical evidence. According to Stanford "we have, throughout the history of scientific inquiry and in virtually every scientific field, repeatedly occupied an epistemic position in which we could conceive of only one or a few theories that were well confirmed by the available evidence, while subsequent inquiry would routinely (if not invariably) reveal further, radically distinct alternatives as well confirmed by the previously available evidence as those we were inclined to accept on the strength of that evidence" (Stanford 2006, 19). These historical facts are taken as the inferential basis for a new "induction over the history of science," whose conclusion is that "there typically are alternatives to our best theories equally well-confirmed by the evidence, even when we are unable to conceive of them at the time" (20). As a matter of fact, Stanford does not use this thesis to argue for contingentism because he is directly targeting scientific realism. For this reason, and even if Stanford's analyses could be exploited to make a case for contingentism, I will focus on Cushing's work.

7. The nonlocality of Bohmian mechanics is illustrated by the case of entangled states of a system of particles. In such states the velocity of a particle depends on the positions of other distant particles of the system.

8. Ironically, a theory such as Bohm's, whose existence is used by Cushing as a weapon against scientific realism, presents a picture of reality that is much more "realistic" than that of standard quantum mechanics. Of course then, when Cushing defends the rights of the "realistic" Bohmian mechanics to be acknowledged as a legitimate alternative, he is not thereby defending scientific realism. One thing is the realism *of* a scientific theory, quite another the realism *about* scientific theory.

9. Indeed there is some perversion in the way in which the paradoxical character of a theory is used for or against it depending on whether it is already in a dominant position or not. Much in the same way, the eccentric behavior of a celebrity is taken as a sign of genius, whereas that of an unknown man is judged as a pathetic weakness

of the mind.

10. For a clear statement of the extended Duhem thesis, see Hacking (1992, 30–31, 52–55).

11. This situation can also be described by means of the metaphor of symbiosis. A robust fit would be a situation in which there obtains a good symbiosis among the ingredients of experimental science. No such item, including empirical data, could thus have a life on it own, so to speak, for it can be valuable only in a community of "symbiotic" items. The symbiotic metaphor has been introduced and developed by Pickering (see especially 1995a). See also Soler (2008c).

12. Philosophical discussions of the various constructivist approaches can be found in Hacking (1999) and Kukla (2000).

13. In other passages of the book the expression "production of a world" and "production of a worldview" are used interchangeably, see, for instance (Pickering 1984a, 405, 407).

14. Cushing writes: "Successful theories can prove to be poor guides in providing deep ontological lessons about the nature of physical reality" (1994, 215). See also Cushing's subsequent reference to Quine's naturalistic account of underdetermination.

15. For a recent development of this position, see also Pickering (2012). For Kuhn's metaphysical antirealism see Kuhn (2000, 104): "The ways of being-in-the-world which a lexicon provides are not candidates for true/false." See also Kuhn (2000, 219–21).

16. Here incommensurability is intended in a sense stronger than the original Kuhnian one, as implying the impossibility of adjudicating between two theories. On Kuhn's own view about the relation between incommensurability and incomparability see Kuhn (2000, 33–57). For general analyses of Kuhnian incommensurability see Hoyningen-Huene (1993, 206–22), Soler (2004), and Trizio (2004).

17. For one of the most recent versions of Pickering's "ontological," agency-based contingentism, see also chapter 3 of this volume. A host of philosophical challenges that cannot be discussed here awaits whoever tries to abandon metaphysical realism, whether this is done adopting a representational or an agency-based framework. Inter alia, the efforts of thinkers like Pickering must at once give a precise sense to and convincing arguments for the thesis that the choices of physicists *produce a world* and not just a world*view*, for, clearly, this requires a sense of "production" and of "existence" that is not ordinary. Indeed, if scientists' choices had literally brought elementary particles into existence, the scientists' realism about them could hardly be criticized. Doing it would not differ much from saying that God was wrong in believing that the world exists, right after he created it. Hence, notions such as "existence of the world" and "existence in the world" must be entirely redefined. Moreover, Pickering must persuade us that his account of scientific practices is not entirely compatible with metaphysical realism. Are we sure that the world could not just be endowed with a fixed, inner structure that is rich, complex, and inaccessible enough to support a huge variety of performative engagements *in* and *with* it? More generally, what needs to be proved is that the choice between metaphysical realism and metaphysical antirealism

is not underdetermined by all the evidence that can ever be provided by empirical research *on* science. To paraphrase Cushing's skepticism about the ontological import of physical theory, science studies might just prove to be poor guides in providing deep lessons about the very *ontological status* of reality. For the purpose of this chapter, this is an important point because, in principle, contingentism and the multiplicity thesis can be stated in terms of the notion of robust fit without endorsing metaphysical antirealism.

18. I prefer this solution to my earlier view, according to which in cases like this the clause "equally successful" should simply be dropped (see Trizio 2008).

19. Scientific realism in general should also not be confused with what can be called the *realistic attitude* of scientists, which is the objectifying attitude inbuilt in the engagement in scientific research, by virtue of which the claims and theories resulting from the latter are in most cases implicitly intended as tentative descriptions of how things really are. By virtue of this attitude, criticism of a claim is only a way to argue for competing scientific claims, or for the need to conceive of them; it never becomes a criticism of the epistemic limitations of science as such.

20. One would like to say that philosophy of science and science studies are today, methodologically speaking, apart from a few exceptions, naturalistic, if many current research trends were not rather sociologistic or historicistic in character. Perhaps the best way to capture this state of affairs is to speak, as Andrea Woody (2014) has done, of a shift from a priori to the empirical, without prejudging what kind of empirical evidence is involved.

21. But again see Stanford (2006) for an attempt to foreground the antirealist consequences of the doctrine of underdetermination.

22. For instance, as Hasok Chang (2003) has shown, preservative realists face the problem of having to rely on controversial continuity claims about the series of past successful theories and on bold inferences from this alleged continuity to truth.

23. However, on this issue, see Soler (2006a).

24. To be precise, this position would reconcile scientific antirealism with *weak* inevitabilism, an inevitabilism compatible with the view that it is not reality as such that "determines" the course of successful research, but, rather, reality coupled with some specific historical (or perhaps even biological) conditions of research. Instead, it is really hard to try to give any plausibility to a position holding together strong inevitabilism (which implies a denial of the multiplicity thesis) and scientific antirealism. Let us add, in passing, that this discussion highlights the interest of the distinction between strong and weak inevitabilism. The former grounds the inevitability of our successful science in the idea that the *world* does not support rival accounts of it as successful as ours, whereas the latter is committed only to a claim concerning our historical trajectory.

25. An extreme form of social determinism would provide another form of inevitabilist antirealism. In this case, the whole burden of explaining the inevitability of science will fall on the social, historical, and cultural conditions surrounding the emergence of a scientific result. For this reason, it would be a form of weak inevitabilism, and hence compatible with the multiplicity thesis. Needless to say, such an extreme

form of social determinism is hard to defend for properly social and historical phenomena themselves, let alone for the innermost content of scientific achievements. I am indebted to Katherina Kinzel for reminding me of the doctrine of social determinism.

26. *Metaphysical* realism, as we already know, is compatible with contingentism. One should also mention that some constructivists, although they reject metaphysical realism, hold unorthodox forms of realism that are deeply intertwined with contingentism. Pickering himself has subsequently characterized his own position as *noncorrespondence* or *pragmatic realism* (Pickering 1989b, 279–82). According to this position it would make sense to call "reality" (or at least "reality for us") precisely the contingent outcome of the processes of material and intellectual negotiation with the world that lead to the emergence and stabilization of scientific results. Needless to say, these forms of pluralistic or even relativist views of reality somehow "stretch" the very notion of realism in such a way that it can cover most of the positions that are normally termed antirealist and constructivist (with the disadvantage that the only antirealists would then be the radical empiricists such as van Fraassen). Hence, in this chapter, I prefer to conform to the standard terminology, and I discuss forms of scientific realism that imply metaphysical realism and are incompatible with constructivism and relativism.

27. Although, to be sure, contingentism about X is not incompatible with the claim that X exists.

28. However, as we shall see shortly, this does not imply that they are allowed not to take into account the problem of the plausibility of the existence of alternatives.

29. See Chang's chapter 15 in this volume for an analysis of the relation between contingentism and pluralism, and Soler's chapter 1 for the implications of the monist ideology on the empirical evidence we may hope to get in support of the contingentist position.

30. An example of such a local, small-scale alternative account at the level of laboratory science is provided by Pickering's reconstruction of Giacomo Morpurgo's researches on quarks (Pickering 1989b, 1995a, ch. 3).

31. See Allamel-Raffin and Gangloff in this volume, chapter 2. For instance, it does seem easier to concede the contingency of particle physics than that of, say, cellular biology, for it is really hard to imagine how our science could have been as successful as ours without the notion of cell.

32. See, for instance, Pickering (2012, 323): "Within the culture of 1950s particle physics the world revealed itself to us in the shape of the bubble chamber. In the culture of the 1960s, it revealed itself to Morpurgo as having no free quarks. But I find it easy to imagine that different cultures could have elicited quite different machines and instruments and material performances from the world; and I can see no reason not to imagine that."

33. In other words, the conflict concerns scientific realism, not metaphysical realism.

Chapter 5. Contingency and Inevitability in Science

I wish to thank Léna Soler and the PractiScienS group for their agenda-setting endeavors on this topic. I also wish to express my gratitude to Les Treilles Foundation in France for so kindly hosting the International Workshop "Contingency: Science as It Could Have Been" (September 2009) in a wonderful intellectual environment. I wish to thank Henk Procee and Federica Russo for their valuable suggestions on the content of this chapter. This research is supported by a Vidi grant from the Dutch National Science Foundation (NWO).

Chapter 6. Contingency and "The Art of the Soluble"

1. See Pinch (1982) for alternative readings of Kuhn. Here the early Kuhn is referred to. The early Kuhn had not separated practice from theory but amalgamated both, making a paradigm something like a Wittgensteinian (1953) form of life (see also Winch 1958, 120–21).

2. See Collins (2004) for a discussion of the controversy over high fluxes of gravitational waves.

3. See Collins and Pinch ([1993] 1998, ch. 2) for a discussion of the dispute over the constancy of the velocity of light.

4. Irrespective of the logical possibilities we cannot have endless Changs and Ashmores resurrecting expired ideas and practices or science would stop dead—mired in its own past. Therefore we can, at best, only revive one or two dead ideas at a time and everything else has to remain pretty much the same. But this limit is logistical or sociological and applies equally to any creative social activity.

5. The author of this chapter receives them too; one, promoting an alternative, "vortex," theory of gravity, came on April 5, 2010, as this very passage was being written. See Collins (2014b) for extended discussion of another case.

6. See Collins ([1985] 1992) for the experimenter's regress. See Collins (2010) for tacit knowledge. See Kennefick (2000) for the theoretician's regress. See Gingras and Godin (2002) for an interesting discussion of the relationship between the experimenter's regress and the ideas of Sextus Empiricus and Montaigne, and see Collins (2002) for a response.

7. This chapter is written from a sociological perspective that accepts that what counts as truth is the outcome of debate and discussion within the scientific community. It has nothing to say about whether individual scientists should hold that view. Indeed, the author has argued elsewhere that science is best conducted under the opposite assumption, namely, that a single individual, in lone interaction with nature, can discover the truth and that the truth might be different from what everyone else in the scientific community believes (see, e.g., Collins 1982). One might also use the cheeseburger metaphor in reference to the entire short-term versus long-term argument—short termism solves the problem of contingency in a nourishing but ugly way.

8. Collins and Evans (2002, 2007) and Collins (2014a) argue strongly against this trend.

9. See Boyce (2006) for a discussion of the MMR case.

10. See Edwards and Sheptycki (2009) for the problems of technological populism in regard to criminology.

11. For a more exact working out of the relationship between the political and technical spheres under this model, and for its relationship to technological populism

and to other approaches to the policy dilemma, see Collins, Weinel, and Evans (2010).

12. The *Brent Spar* was disposed of on land in response to public outcries led by environmentalists. Everyone now agrees that this caused more environmental pollution than would have been caused by disposal at sea. For arguments around the sinking of the oil rig see Collins, Weinel, and Evans (2010).

13. The position presented in this part of the chapter is presented at book length under the heading of "Elective Modernism" in Collins and Evans (submitted).

14. For an analysis of tacit knowledge, see Collins, *Tacit and Explicit Knowledge* (2010).

15. *Rethinking Expertise* (Collins and Evans 2007, ch. 4) contains an account of the early experiments, although many more have been completed since then.

16. See Weinel (2007) for a discussion of the Mbeki affair.

Chapter 7. Contingency, Conditional Realism, and the Evolution of the Sciences

I want to thank Léna Soler and the other participants at the conference on contingency and inevitability at Les Treilles in Provence in September 2009. The conversation was as stimulating as the setting was bucolic. This final version of my contribution is much improved thanks to comments by Soler and other reviewers.

1. Léna Soler (2008a, 2008b) has been among the most prominent proponents of the importance of the contingentist/inevitabilist issue.

2. That advocates of forms of social constructivism have always been concerned with contingency in science is clear just from titles of works early and late. Thus an article by Harry Collins (1981) is subtitled: "Social Contingency with Methodological Propriety in Science," and James Cushing's (1994) book carries the subtitle: "Historical Contingency and the Copenhagen Hegemony." What is missing from this literature is an explicit concern with any notion of "inevitability" in science. The understood contrast with contingency has typically been scientific realism.

3. This section is a short, dogmatic statement of a view developed more fully in (Giere 2006, ch. 4). There it goes under the name of "perspectival realism." The most important feature of perspectival realism, however, is that it is conditional. And even in the more extended presentation I do not attempt to defend a deflationary understanding of ordinary truth. That is a very big subject all by itself.

4. An apparent exception can be found in Stephen Hawking's much quoted passage at the end of his *Brief History of Time*. There he opined that "if we find the answer to [why it is that we and the universe exist], it would be the ultimate triumph of human reason—for then we would know the mind of God" (Hawking 1988, 175). I have always assumed the references to God in this final section of the book to be metaphorical. And in his recent book (Hawking and Mlodinow 2010), he argues for the creation of the universe, indeed, many universes, without a creator.

5. Philosophers will recognize conditional realism as being similar to Putnam's "internal realism" (1981, ch. 3), but untangling the many differences would be difficult. One obvious difference is that Putnam associates his internal realism with a Peircean notion of truth as the result of inquiry in the long run rather than with a simple deflationary view. One might also see conditional realism as a realistic inter-

NOTES TO PAGES 192–194

pretation of Kuhn's views, where one conditionalizes on a paradigm. Here a major difference is that conditional realism does not imply linguistic incommensurability. In the conflict between stabilist and mobilist paradigms in geology, to be discussed in the final section of this chapter, there was never any linguistic incommensurability. Everyone understood what was being claimed, which is how some came to be so opposed to mobilism.

6. Note that this presumed fixity is not the same as metaphysical scientific realism. It is just the presumption that the natural world has some structure or other, independent of our attempts to represent it. It says nothing about how that structure is to be represented. In fact, the structure could change over time, just not too rapidly, otherwise we could not have evolved. But any such changes would be independent of our activities.

7. There are, of course, those who question whether human actions, which play such a large role in history, are fully causal. So here I may betray a commitment to a kind of naturalistic materialism. Unlike Descartes, I cannot imagine how anything outside the natural causal nexus could influence developments within that nexus.

8. The closest to a counterfactual history of science are the essays by Kenneth Pomeranz (2008) and Joel Mokyr (2008), which focus on technology.

9. In chapter 6 of this volume, Harry Collins provides an inside look at controversies within the history profession over the value of counterfactual history. It seems to me that the authors of this book decided not to defend counterfactual history, under that name, because of its bad reputation among historians. Instead, they join the opposition to counterfactual history and promote instead "virtual history." To a disinterested observer it seems that virtual history is just one way of doing counterfactual history in a methodologically responsible manner. The methodology of *Unmaking the West* (Tetlock, Lebow, and Parker 2006) seems to me equally responsible, given its subject. The "rebranding" as "virtual history" seems a strategic rhetorical strategy for gaining a hearing within the community of historians.

10. Howard Sankey (2008) has recently argued for the compatibility of scientific realism with contingency, using geology as an example. The big differences between my views and his are that Sankey maintains a stronger form of scientific realism and has more faith in the power of scientific methods.

11. Here I am drawing on my earlier extensive study of the 1960s revolution in geology (Giere 1988, ch. 8). In that study I was concerned to note contingencies that fit with an evolutionary model of this history, and there are many more than noted in this essay. Questions about the inevitability of scientific results and the possibility of counterfactual history, however, were not at that time even recognized as issues to be considered.

12. These engravings, attributed to one Antonio Snider, are reproduced in Marvin (1973, 43).

13. Note that we are here considering merely the formulation of hypotheses, not the formulation of a scientific consensus on the way the world is. But my sense of inevitability still applies. I will be claiming, in effect, that there are no plausible contingencies that would have prevented these hypotheses from being brought to public attention.

14. Any number of people must have verified this impression by cutting a map along the coastline of the Western Hemisphere and moving the severed piece across to match it up with the coastline on the other side. The matchup is quite good and, with a few plausible distortions, very good.

15. For the original diagrams, see Wegener (1915) or the more accessible English translations of the third (1924) or fourth (1966) English editions. I have reproduced Wegener's iconic three-stage presentation of the breakup of "Gondwanaland" in Giere (1999, fig 7.4). Due to the difficulties and cost of reproducing original graphics, including some from sources such as *Science*, I shall include none in this chapter. From here on I will refer the reader to graphics that appear in either Giere (1988) or Giere (1999) or both. This should suffice for present purposes, but I do apologize to readers who would prefer diagrams to descriptions. I too prefer diagrams.

16. Also in 1924, Wegener, together with Wladimir Köppen, published a related volume in German on paleoclimatology (Köppen and Wegener 1924). It is plausible that the growing interest in Wegener's work was significantly driven by his publication of a second edition in 1920 and a third in 1922, each of which was substantially revised to include new material and responses to criticisms. This information comes from Kurt Wegener's brief biography of his brother included in Wegener (1966, iii–v).

17. I discuss the views of the critics at length in Giere (1988, 234–41). In the end I try to explain why there was no revolution in geology in the 1920s. But I conclude by asking: "Could things have been different?" My answer is, "Of course." I go on to mention a number of counterfactual conditions that I speculate might even have made Wegener himself unnecessary. I do now wonder, however, whether my counterfactuals could provide a basis for a good counterfactual history. And of course I nowhere ask whether the later triumph of mobilism was "inevitable." I only try to explain why just about everyone involved eventually decided it was far better than stabilism.

18. For Holmes's dramatic illustration of his model, first published in 1929 and later in the final chapter of his influential 1944 textbook, see Giere (1988, fig. 8.7; 1999, fig. 7.7).

19. It would be interesting to explore why it took thirty years for the relevance of radioactive heating to the possible movement of continents to be recognized (and published). It is a little easier to understand why this idea was revived when it was another thirty years later. In general, it seems worth exploring contingencies that led to a plausible avenue of research *not* being pursued or even recognized.

20. Just this objection, and in these terms, was raised by Andy Pickering at the conference in Provence. Of course he raised other objections as well.

21. I think this was the conclusion also held by Catherine Allamel-Raffin and Jean-Luc Gangloff in their presentation in Provence (see chapter 2 in this volume).

Chapter 8. Necessity and Contingency in the Discovery of Electron Diffraction

I would like to thank Léna Soler for her close reading of the text, and the two referees for their suggestions.

1. For an analysis of a sociology of science implicit in Bachelard, see Gingras (2003).

2. There is a debate on the question of whether or not this proof really shows the equivalence of the two formulations. But the solution to that question does not intersect with the point discussed here, which is to look at how scientists react when faced with "different" theories of the same object. Knowing whether or not the demonstration of the equivalence is true is a problem in mathematical physics, whereas knowing that physicists acted *on the basis of* the equivalence of the two formalisms, as shown by Schrödinger (Pauli did not publish his demonstration) is a question in historical sociology, where one has to take into account that what is perceived as real has real consequences. On the debate over equivalence, see Muller (1997a, 1997b).

3. I limit myself to that subset simply because I do not want to discuss here the so-called primitive societies and their supposedly different logic. On this question, see Moody-Adams (1997).

4. For a presentation of this idea, again proposed independently by different physicists to interpret empirical data, see Rose (1928).

5. For details of his model, see Navarro (2010, 267–69).

6. Cushing (1994, 32–34) does not even mention Duane's work in his analysis of the wave-particle duality.

7. In their introduction to a collective book in honor of Landé, Yourgrau and van der Merwe write that "it is the fault of an inexplicable failure of physicists to take note" of Duane's contribution that explains why nobody tried to explain diffraction of electrons on the basis of a corpuscular view; see Yourgrau and van der Merwe (1979, xxxii).

8. For a discussion of the effects of mathematics on substantialists' conceptions of physics, see Gingras (2001, esp. 403–6).

Chapter 9. Contingency in Mathematics

This chapter is based on three other papers, Van Bendegem (2000, 2002, and 2008), where altogether four cases have been presented for alternative mathematics. The cases not dealt with in this chapter are another approach to infinitesimals (different from Bloor's version, as mentioned at the beginning of this chapter; see Van Bendegem 2002), and a form of vague mathematics, where statements such as "small numbers have few prime factors" can be rigorously proved (see Van Bendegem, 2000). The two cases presented here differ from the original version in the deletion of a number of confusing details, a shortcoming rather typical of first versions. This chapter has greatly benefited from the comments of the referees and I want to express my thanks to them. At least in my mind, this version is a definite improvement on the previous one. In addition, the papers by Soler (2008a, 2008b) have been a great help to me in focusing my ideas concerning concepts such as alternativity, unicity, inevitability, and contingency.

1. To be precise, the chapter following the chapter being discussed here presents a fifth case, but that case is tied strongly to the necessity and certainty of the underlying logic, which I am not addressing here. If I had done so, the chapter would have been less about the contingency of mathematics and more about the contingency of logic, and it is not clear how much would have been gained because it is not immediately clear that the contingency of logic necessarily implies the contingency of mathematics.

2. The second edition of Bloor's book has an afterword, in which Bloor replies

to his critics (1991, 163–85). The paragraph titled "Mathematics and the Realm of Necessity" (179–83) discusses mainly the notion of proof and the necessity that goes together with it. It therefore deals with the question of the necessity of the underlying logic rather than with the specific examples themselves, which would have been my criticism.

3. As remarked by one of the referees, it is not necessarily a disaster if this comparability does not succeed. If genuine alternatives can be found for (elementary) arithmetic, this is already sufficient to show that alternatives do indeed exist. It is not required to show that for every part of mathematics an alternative can exist.

4. Variations on this theme are possible. Questions that do not get an immediate answer can be added to a second list, "Things not to forget," and at every moment, both lists are checked. If a case on the second list can be connected to the first list, those items are relocated and erased from the second list. Another possibility is quite simply the "active" solution where the members start with an equation, determine its neighbors, their neighbors' neighbors, and so on.

5. The other source of inspiration is the theory of cellular automata, and especially the result showing that Turing machines can be easily translated into such automata.

6. As one of the referees remarked, the phrase "they know how to add" does not imply that they know what it means to add two numbers. Are they not merely performing calculations without understanding what it is they are doing? The brief answer is that this matter has indeed been left open in the above scenario, but that the story could easily be expanded to include justifications for what they are doing: "2 + 3 = 5" gets an empirical justification, and the acceptance of the neighbors can be justified either empirically or by the principle that adding or subtracting 1 on both sides of an equation generates a new equation. What remains important is that these justifications do not require any notion of proof.

7. For an excellent introduction and starting point, see http://logica.ugent.be/ad-log/al.html. The originator of this program is Diderik Batens.

Chapter 10. Freedom of Framework

1. Translations from French are by David Webb and Jean-Michel Salanskis, unless otherwise indicated.

2. *Agrégation* is a competitive examination officially meant to recruit high school teachers as civil servants but working, rather, as an occasion to determine who are the best scholars.

3. ZFC is the formal set theory of Zermelo-Fraenkel with Choice axiom, inside of which all of contemporary mathematics is supposed to happen.

Chapter 11. On the Contingency of What Counts as "Mathematics"

This chapter covers some but only some of the material in my festschrift essay "What Makes Mathematics Mathematics?" (Hacking 2009a). The present version of this chapter owes a good deal to a very careful anonymous referee, who is explicitly mentioned a couple of times below but is in effect omnipresent in revisions made to this chapter, especially in the historical asides.

1. Interestingly Mark Steiner (2005) takes this question to embrace the ones I have listed. In the case of the a priori, this is a change in focus but it makes obvious good sense.

2. This phrase is the title of J. E. Littlewood's (1953) charming potpourri of mathematical anecdotes and examples.

3. The referee protested here that "Detlefsen's opinion is very customary, in fact. Concerning Horsten's article, it is true that it does not mention Kant, but Kant is mentioned and his idea discussed in a large number of the texts mentioned in this article." True, but the two encyclopedia articles do explicitly present, if only on the surface, two very different conceptions of what the philosophy is about.

4. A pedantic remark. Earlier I quoted Russell: "The question which Kant put at the beginning of his philosophy, namely 'How is pure mathematics possible' is an interesting and difficult one." Actually Kant did not put it at the beginning of the first edition of the *Critique*. It was inserted only in the second edition, after it had been put forward in the *Prolegomena*.

5. The referee suggested citing these passages in support of Klein: *Philebus*, 56 d–e, *Thaetetus*, 195e–196a, and *Georgias*, 451a–c.

6. The assertion that the term "mixed mathematics" is original with Bacon is due to Brown (1991). In a personal communication, Lorraine Daston suggests that medieval texts, perhaps drawing on Aristotle's *Physics* II.2, which speaks to the mixture of matter and form, might be found that anticipate Bacon's usage.

7. Brown displays Bacon's divisions on a tree-diagram, and it is common to speak of Bacon's "tree of knowledge." Despite the work of major scholars who write as if Bacon had spoken of a tree of knowledge, this attribution is probably due to d'Alembert's *Discours préliminaire* (1751) for the *Encyclopédie*. D'Alembert really did give us a tree, upon which rested mixed mathematics, and he compared his tree to Bacon's classification of branches of knowledge. Hobbes, who also had a tree, did not acknowledge mixed mathematics.

8. Mersenne (1634), Question 38. This observation is due to Daston, e-mail of October 13, 2009.

9. Two late quotations from the *OED* speak of applying, but differ in their meaning. In 1706: "*Mixt Mathematicks*, are those Arts and Sciences which treat of the Properties of Quantity, apply'd to material Beings, or sensible Objects; as Astronomy, Geography, Navigation, Dialling [sundials], Surveying, Gauging &c." In 1834, in Coleridge: "We call those [sciences] *mixed* in which certain ideas of the mind are applied to the general properties of bodies."

10. The referee observed that one may prefer a more cautious and well-argued variant on this proposal, advanced in Guicciardini (2009, ch. 13).

11. *OED*: "Without foreign or extraneous admixture: free from anything not properly pertaining to it; simple, homogeneous, unmixed, unalloyed." Grimm's *Deutsches Wörterbuch*: "frei von fremdartigem, das entweder auf der Oberfläche haftet oder dem Stoffe beigemischt ist, die eigenart trübend."

12. Kant wrote "eigentilichen (empirischen) Physik," which Kemp Smith renders "(empirical) physics, properly so called." Like the English noun "physics," *Physik* in

Kant's time still meant natural science in general. Kant might have meant something more like "real (empirical) physics."

13. I had grouped Lagrange with Legendre and Laplace, but the referee wanted to emphasize that Lagrange paid heed to the distinction between pure and applied: "take the table of contents of Lagrange's *Théorie des fonctions analytiques*, 2nd edition"—but adds, "the point is then what 'application' meant for Lagrange." The "2nd edition" will be the reprint of the *Théorie* (1797) in the *Journal de l'École polytechnique* (neuvième cahier, tome II [1800]), where on page iii, we find a break in the table of contents, "Second Partie/*Application de la Théorie à la Géométrie et à la Mécanique.*" The geometrical applications are what we now call analytic geometry. The mechanical are what one now learns in a course on elementary Newtonian mechanics. I do not myself see this as a distinction between "pure" and "applied'" as we now understand those terms. Compare Gergonne's 1810 listing of the subjects of applied mathematics, as reproduced in my next section below.

14. The quotes are literal and ironic. Think of modern heroes such as Émile Durkheim in Bordeaux and Pierre Duhem in Lille, for example.

15. I owe to Vincent Guillin the fact that Mill's lecture notes on the lectures are reprinted in vol. 26 of his *Collected Works*, pp. 146ff. Gergonne comes out like a standard fin-de-siècle *ideologue*: Mill's eighteenth-century metaphysics.

16. The first edition was 1908. Wittgenstein wrote marginal notes on the 1941 printing. These will be analyzed in a forthcoming study by Juliet Floyd. My own copy, bought new in 1956, is the new edition of 1951. It is still in print, as the *Course of Pure Mathematics Centenary Edition* 2008.

17. "Return to an old refrain: what proof does to concepts," a paper read at the thirty-second Wittgenstein Symposium, Kirchberg, Austria, August 9-15, 2009, to appear in the annual proceedings.

18. I owe this information to Friedrich Stadler.

Chapter 12. The Science of Mind as It Could Have Been

1. According to James, "Introspective observation is what we have to rely on first and foremost and always. I regard th[e] belief [in introspection] as the most fundamental of all the postulates of Psychology" (James 1890, 185).

2. As described by Warren and Carmichael, "In scientific introspection, great care is necessary in the arrangement and simplification of the experimental setting and in the training of the individual who is to give the report" (1930, 58).

3. The references to Gupta (2004) and Shafii (1973) were kindly suggested by a referee.

4. Translations are by the authors unless otherwise stated.

5. In meditation, stabilizing attention is allowed by long sessions of concentration on a single felt or imagined process (such as breath or pictures); and contact with the manifold processes of mental life is realized not only by broadening the field of attention but also by dropping "all aim and objective" in full, open, nondirectional mindfulness. See, for example, Genoud (2009) and Wallace (1998).

6. For another exposition of the classical objections, see Petitmengin and Bitbol (2009) and Vermersch (1999).

7. In quantum mechanics, it is well-known (to the dismay of realist philosophers of science) that the project of objectifiying "properties" behind phenomena can hardly be worked out. Yet, one objectifies a universal anticipative structure that is nothing other than the *state vector*, which generates probabilistic predictions by means of Born's rule.

8. The enactive theory of knowledge advocated in Varela, Thompson, and Rosch (1991) and Thompson (2007) has many resources in store to weaken representational-ism.

9. Nisbett and Wilson's (1977) protocol consists in demanding that subjects choose quickly between two items, and then presenting them with the wrong item while asking them to explain why they chose it. The outcome is that 70 percent of subjects do not detect the cheating and candidly give a fancy explanation. This seems to justify strong diffidence toward introspecting one's own cognitive processes. In the experiment of Petitmengin et al. (2013), a careful explicitation of *how* the subjects initially made their choice is performed before the wrong item is presented to them. In this case, only 20 percent of the subjects do not detect the cheating. And even these residual mistakes can be provided with an experiential rationale.

Chapter 13. Laws, Scientific Practice, and the Contingency/Inevitability Question

1. Lange (2009, 22–23) reminds us that a nested counterfactual conditional $(p\square\!-\!>(q\square\!-\!>r)$ is not equivalent to a single counterfactual conditional conjoining their antecedents $(p\ \&\ q)\square\!-\!>r$, especially but not exclusively, because p and q might be inconsistent. Nor is that latter consideration beside the point here, since the inevitabilist may well regard any radically alternative historical trajectory of science to be inconsistent with its empirical success.

2. John Haugeland's (1998, 2013) arguments that the constitutive standards governing scientific practice must be laws importantly augments Lange's analysis, but would add too much complexity to introduce in this chapter. For discussion of the complementarity between Haugeland's and Lange's work, see Rouse (2015, ch. 8).

3. One anonymous referee pointed out that this thought experiment is complicated by the fact that real scientific communities already incorporate considerable differences among their members concerning these prospective judgments about the plausibility and significance of various hypotheses and research programs. Although those internal differences are real and important, there can still be clear overall differences in the preponderance of views, in where the burden of proof is taken to lie, in which options will likely receive funding under normal conditions of scarce resources, and so forth.

4. One anonymous referee constructively pointed out that these different historical trajectories converging at least temporarily on the same results might be conceived in different ways. One might imagine the differences to concern different theoretical hypotheses having been considered, different ways of arriving at, understanding, and using key concepts, different experiments contemplated and performed, with different instruments and skills for using them, and so forth. One could also imagine two trajectories that largely agreed in such narrowly scientific features of their history but quite significantly diverged in their broader cultural context in ways that led to quite different interpretations of the significance of what at one point could be identified as "the same"

results. I think that in both cases, we should expect the subsequent histories of science in these two traditions to diverge in important respects. When I wrote the chapter originally, however, I was primarily thinking of the first case, largely because I thought that there would be more widespread agreement that such differences in narrowly scientific practices would incline the two scientific communities in different research directions.

5. This inevitabilist response highlights an important logical feature of nested counterfactual conditionals. If we use the standard symbolism (such that $p\square\!\!-\!\!>q$ marks the counterfactual conditional "if p had been true, then q would have been true), then $p\square\!\!-\!\!>(q\square\!\!-\!\!>r)$ is not equivalent to $(p\ \&\ q)\square\!\!-\!\!>r$, because of the possibility that $p\ \&\ q$ is logically contradictory. And that is exactly what the inevitabilist will likely claim in the case of the nested counterfactual conditional that proposes an alternative, comparably successful science.

6. To those readers who find this counterfactual scenario implausible, I have three rejoinders. First, close ties between Christian theology and mathematical physics persisted well into the nineteenth century, even among influential physicists such as James Clerk Maxwell and William Thomson, Lord Kelvin (Smith and Wise 1989), whose work remains integral to contemporary physics. Second, the gradual separation of physics from physicists' theological commitments has much more to do with shifts in the conception and role of religion among educated elites in Europe and North America than it did with any specific developments in physics. Indeed, to some of a more thoroughgoing naturalist bent, myself included, this counterfactual hypothesis instead remains uncomfortably close to the actual history of physics; whereas cosmology and high-energy physics are no longer understood in specifically Christian-theological terms, the aspiration to understand the universe theologically, in the sense that physics aims to describe the universe from an absolute standpoint, independent of any entanglements within it, remains pervasive. Third, and most important, the historical plausibility of this scenario does not matter to its relevance to the contingency debates, on two grounds. On the one hand, this hypothesis is drawn in an especially striking way in order to highlight the possible significance of what are taken to be inferential consequences of scientific claims. The inferential consequences of any claim depend on what collateral commitments are conjoined with it, so there is ample room for variation here even for those who are committed to the same claims in a science. Even if this particular scenario were ruled out, it would be very difficult to rule out all other conceivably far-reaching differences in the interpretation of the significance of scientific claims resulting from different collateral commitments. On the other hand, when the issue concerns the contingency or *inevitability* of scientific understanding, historical implausibility is not a sufficiently disabling consideration to ground a relevant objection to informative alternative scenarios.

7. Pickering's subsequent work (1995a) embraces the conception of contingency as omnipresent, but that extension of his view may be in some tension with how his earlier work has been used in discussing the contingency question. One familiar way of inferring from the contingencies emphasized in micro-level studies of scientific practice to large-scale contingencies of ontology and methodology is to follow Pickering in arguing for the contingency of major turning points in the history of science, on

the presumption that these alternative histories would then continue on divergent trajectories. Yet consider the parallel to early efforts at weather modification, which were premised on the idea that intervention at just the right "tipping points" would allow small interventions to make large-scale differences in the long run. The recognition that such sensitive "tipping points" were pervasive in the movement of weather systems blocked any such effort to produce reliable large-scale differences via timely interventions (Gleick [1987] provides an accessible version of this history). In the absence of such abilities to predict stable, long-term historical trajectories from "decisive" events, one could only draw reliable inferences about macrohistorical outcomes from microhistorical contingency either if one knew the dynamics of the theory-change system with sufficient precision to run reliable simulations or if one had independent grounds for identifying attractors for the system. Since the issue in dispute is whether there is one, or more than one, stable attractor that would also generate comparable empirical success, Pickering's later position may well render the dispute irresolvable.

8. Radick (2005a) also explores the possibility of counterfactual histories of biology, although he takes up different possible loci of divergence than I do below (he focuses on early twentieth-century controversies between Mendelians and biometricians and the debates over Lysenkoism in the former Soviet Union, whereas I consider relations between genetics and embryology or developmental biology throughout the twentieth century). Thanks to Léna Soler for bringing Radick's paper to my attention. For more general discussion of the role of counterfactuals in thinking about the history of science, see Radick (2008b).

9. Beurton, Falk, and Rheinberger (2000) and Keller (2000) offer extensive discussion of the challenges confronting the gene concept, and some consideration of the alternative approaches that were set aside in the course of that history.

10. Giere (1999) and Cartwright (1999) are exemplary of philosophical suspicion of the importance of laws in scientific understanding. Beatty (1995) is the classical locus for the claim that biology is not a nomological science.

11. The crucial difference between laws and their subnomic counterparts thus turns on their relation to counterfactual conditionals. The subnomic counterpart to a law is a regularity (a regularity in what actually happens, although without the modal qualifier that these are "actual" occurrences). Taken as a law, however, the regularity extends to cover the conditions expressed by counterfactual antecedents as well. Lange (2009) then argues that subjunctive facts (the facts expressed by counterfactual or subjunctive conditionals) should be understood to be the "law-makers," in parallel to Plato's famous question about piety and the gods in the *Euthyphro*: it is the truth of the subjunctive facts that determines which other truths are laws, rather than the laws that determine which subjunctive conditionals are true. When the distinction between laws and subnomic facts is understood in this way, of course, the empiricist tradition concerning laws that stems from Hume must deny that there are any laws (instead acknowledging only subnomic regularities). Goodman's (1983) classic *Fact, Fiction and Forecast* powerfully argued that empirical confirmation cannot be understood without understanding it to extend counterfactually (e.g., canonically, that the empirical evidence for emeralds being green also must be understood to confirm

counterfactual claims such as, had this emerald first been discovered much later, it would still have been green).

12. The point also applies in reverse: in pursuing an inductive strategy, scientists thereby implicitly commit themselves to taking a hypothesis that expresses that strategy to be a law. Inductive strategies can vary in both their scope and content. Consider the inductive strategies Mendel might have pursued from his data about inheritance patterns in peas. One strategy (not a wise one!) would have been to limit his inductive inferences geographically: his experiments confirm that Mendelian inheritance patterns hold in Brno, or in the Austro-Hungarian Empire. Another might be to limit them taxonomically: the experiments justify inferences to Mendelian ratios in peas, or in plants. Clearly these would have been inappropriate scope restrictions to impose at the outset, in the absence of reason to impose those limits. Variation in content is nicely exemplified by the difference between Boyle's law and van der Waals's law as expressions of inductive strategies to pursue from initial data about the covariation of the pressure and volume of gases at a constant temperature. Depending on which strategy one adopts, the initial data drawn from gases at relatively low pressure would have different implications for what to expect at very high pressures and low volumes. These two forms of variance function together, of course; Boyle's law suggests a highly reliable inference rule, but only within a limited scope that excludes data from gases at very high pressure.

13. See the opening two sentences of this chapter as a reminder that the entire debate concerns the truth of a nested counterfactual conditional.

14. Lange (2000, 2007) often talks about the "interests" or "purposes" of a given discipline or inquiry. Yet we should not mistake such talk as indicating a dependence on what some community is interested in de facto. In Rouse (2002, 2015), I introduce the terminology of what is "at issue" and "at stake" in a given inquiry, to indicate that although the norms in question are indexed to a particular historically identifiable practice, they are not just up to the participants. What is at stake in a practice may be at odds with what some or even all of its participants think is at stake there. Moreover, once we recognize that scientific communities often disagree about these matters, we have to recognize in a similar way that what is at stake in a scientific practice may itself be at issue within the practice.

15. Strictly speaking, if the principle of centering holds, that is, if (p & q—> (p $\square$—> q)), the set of laws would be a *nonmaximal* stable set, since the set of all subnomic truths would also be stable (since any supposition under which one of those truths would not have held would have to be inconsistent with at least one member of that subnomically all-inclusive set).

16. Lange acknowledges that there might be different levels of nomological necessity within a given science. It may turn out, for example, that if one sets aside the specific force laws for gravity, electromagnetism, and the strong and weak force, there is still a subnomically stable set composed of the fundamental dynamic law, the law of composition of forces, conservation laws, and the relativistic transformation laws, which are independent of whatever force laws there are. Reasoning about how the world would have been had the force laws been different depends on such more limited stability. Could one then have an alternative physics, in which these are the only

laws, and the force laws were taken to be merely accidental truths? Yes and no. Yes, in the sense that this more restrictive physics would not make erroneous predictions, since it takes the force laws actually to hold. It would not, however, be as successful empirically as our actual physics, since it could not license empirically verifiable predictions based on inferences from the subjunctive invariance of the force laws.

17. Lange (2000) discusses this case in some detail.

18. Although I have not discussed this aspect of his view, Lange allows for the possibility of meta-laws (and perhaps higher levels in turn), based on their nonmaximal nomic stability, just as the first-order laws exhibit nonmaximal subnomic stability. Symmetries in physics are good examples of meta-laws. Each discipline (or interrelated group of disciplines) whose laws constitute a subnomically stable set would then have its own set of meta-laws, which in some cases may be the empty set. Presumably, however, whatever other forms of necessity also constitute subnomically stable sets (logical necessity, broadly logical necessity that includes metaphysical or moral necessity, etc.) are all proper subsets of the laws of any of these disciplines.

Chapter 14. On the Plurality of (Theoretical) Worlds

1. I cannot resist quoting here, even if only because of the role it has played in my own thinking, the remarkable 1962 novel by Philip K. Dick, *The Man in the High Castle* (Dick 1993), describing America in 1962, twenty years after it has lost the war and is occupied by Nazi Germany and imperial Japan.

2. As an example, let me quote an interesting paper by E. Brian Davies (2002); the author uses the fictional device of an imaginary Earth permanently covered with a cloud hiding the sky in order to investigate the role of astronomy (nonexistent in this world) in our own history of science; this scenario of course is reminiscent of the famous science-fiction story by Isaac Asimov, "Nightfall" (1941).

3. I have offered elsewhere an extended reconstruction of this fictitious history, including its formal and mathematical developments, mixing true references and invented characters (Lévy-Leblond 2003) based on the detailed historical studies by Arthur I. Miller (1981).

4. Recall that around 1920, Sommerfeld would remark to Einstein that the choice of the word "relativity" to name the new theory was rather unfortunate, in that it masked the essential aspects, that is the invariant (absolute) ones—to which Einstein acquiesced. The name "chronogeometry" instead would be in accordance with the deeper contents of the theory.

5. For a critical comment, see Lévy-Leblond (2007).

6. By the way, it is striking that today physicists, usually ignorant of the history of their discipline and particularly of Le Sage's theory, when exposed to it are usually at a loss to offer a refutation. Most often, they only think of the simplest counterarguments that had been already considered and victoriously answered by Le Sage himself.

7. As put forward by the character Sportin' Life, expressing his doubts about several biblical statements in *Porgy and Bess* (1935), George Gershwin (music) and Ira Gershwin (lyrics).

8. The idea has been dealt with in well-known literary works, such as Lewis Carroll's *Sylvie and Bruno Concluded*, or Jorge Luis Borges's *On Exactitude in Science*.

9. Even before the developments of modern philosophy of science, from Duhem to Feyerabend, it is worthwhile to point out that such a view had been put forward with great effect by Victor Hugo ([1864] 2014), chapter 3, titled "L'art et la science."

10. One could recall here that the young Galileo wrote a most interesting short treatise on the place and location of hell (Galileo 1587). However, this work should not be confused with the later investigations by Swinden, Whiston, and others, because Galileo's lessons had no theological content and were only a learned commentary on Dante's *Inferno*, with essentially literary purposes. Still, it is true that this work played an important role in the later scientific accomplishments of Galileo, most notably his studies on the strength of materials (Lévy-Leblond 2008; Peterson 2002).

11. See also a much circulated Internet piece, discussed at http://www.snopes.com/college/exam/hell.asp.

12. A Google search with the keywords "Venus" and "hell" will suffice to convince one of the banality of the metaphor in most descriptions of the planet. As to popular culture, see for instance, lyrics by Stan Ridgway, "Venus Is Hell" (2007).

13. See http://en.wikipedia.org/wiki/Jack_Van_Impe and http://improbable.com/ig/winners/#ig2001.

14. *Torn Curtain*, directed by Alfred Hitchcock, with Paul Newman and Julie Andrews. The movie, released in 1966, is a political thriller typical of the Cold War period.

15. I cannot resist mentioning the case of French, with its well-known linguistic self-concern, which in certain domains at least has victoriously resisted the Anglo-Saxon terminological dumping, particularly in computer sciences (*informatique* being much better), with terms like *ordinateur* (much better than computer), *logiciel* (much better than software), and so on.

16. See the essential work by Florian Cajori (1928).

Chapter 15. Cultivating Contingency: A Case for Scientific Pluralism

I would like to thank Léna Soler and her colleagues for inviting me to the wonderful conference at Les Treilles, which gave me the first chance to present the thoughts contained in this chapter in a systematic way. Léna also gave me a very helpful and detailed critique of a previous version of this chapter, which I hope to have accommodated adequately. Similar thanks go to two anonymous reviewers and to other participants at the conference, especially Ron Giere, Joe Rouse, and Yves Gingras. Seminar audiences at the University of Minnesota, the University of Cambridge, the University of Pittsburgh, Stanford University, University College London (UCL), and the British Society for the Philosophy of Science have given helpful comments on other versions of various parts of this chapter. The "PPP" (Pragmatism, Pluralism, and Phenomenology) reading group at UCL has been essential in the maturing of my thoughts, and I also thank Ken Waters, Antigone Nounou, Helen Longino, Sabina Leonelli, and Grant Fisher for various conversations that helped me to shape the direction of my thoughts.

1. Later in the same paper Soler glosses the "radically different" in the definition of contingentism as "irreducibly different" and notes that inevitabilism can accommodate differences that are "reconcilable" with our science (Soler 2008b, 233). For

a helpful summary of other formulations of the contingency question, see Soler (2008a).

2. Soler (2008b, 234) makes a similar point regarding the "able" in reconcilable/irreconcilable.

3. Private communication, April 23, 2010.

4. What is meant by "system," "area," or "inquiry" here is intentionally left vague. "System" is meant to be an inclusive term that can be substituted by more specific notions such as "paradigm" or "research program." In Chang (2012, 16) I give the following definition and elaborate on it: "A system of practice [in science] is formed by a coherent set of epistemic activities performed with a view to achieve certain aims." This definition can be extended outside science, too, although I will be focusing on science.

5. The rest of Kellert, Longino, and Waters's definition of scientific monism is as follows: "2. the nature of the world is such that it can, at least in principle, be completely described or explained by such an account; 3. there exist, at least in principle, methods of inquiry that if correctly pursued will yield such an account" (2006, x).

6. John Worrall's (1989) structural realism does not remove this worry; for example, his often-repeated case of structural preservation in optics concerns only rather phenomenological laws—mathematical equations governing variables that are quite observable.

7. See Mitchell (2003, 209), who traces this idea back to Beatty (1987) and Kitcher (1990).

8. Why Bayesians do not usually think like this is not clear to me.

9. This has some similarity to the distinction that Mitchell (2003, 208) makes between competitive and compatible pluralism.

10. On Neurath, see Cat, Cartwright, and Chang (1996).

11. There are other important examples, such as Kuhn's problem-solving ability and Popper's empirical content.

12. For many instructive views on the nature of scientific understanding, see de Regt, Leonelli, and Eigner (2009).

13. This quotation occurs in the last paragraph of Hull (2008).

14. See Kuhn (1977) for the latter list of values, and Kuhn (1970, 169, 205–6) for the emphasis on problem-solving ability.

15. Monism is a beautiful dream that is liable to turn into various nightmares; it is too dangerous for us humans, at least in the political arena.

16. In that regard I follow Feyerabend's early ideas, as interpreted by John Preston (1997, 139). Eric Oberheim (2006, 281) disputes Preston's interpretation of the early Feyerabend and has many other useful ideas about Feyerabend's pluralism.

17. In this same way, a democratic society can let various individuals and groups pursue all sorts of outrageous views and activities, as long as they do not actively prevent others from pursuing their own.

18. On the distinction between whiggism and triumphalism, see Chang (2009b).

19. The idea of complementary science is elaborated in Chang (2004, ch. 6).

BIBLIOGRAPHY

Alajouanine, Théophile, and François L'Hermitte. 1964. "Essai d'introspection de l'aphasie." *Revue neurologique* 11: 609–21.

Allison, Henry, Reinhard Brandt, Paul Guyer, Ralf Meerbote, Charles Dacre Parsons, Hoke Robinson, Jerome B. Schneewind, and Allen W. Wood, eds. 2002. *The Cambridge Edition of the Works of Immanuel Kant in Translation.* Cambridge: Cambridge University Press.

Almond, Philip C. 1994. *Heaven and Hell in Enlightenment England.* Cambridge: Cambridge University Press.

Althusser, Louis. 1974. *Cours de philosophie pour scientifiques, 1967–1968.* Reprinted in Louis Althusser. 2001. *Ecrits philosophiques et politiques,* vol. 2. LGF-Livre de Poche.

Anonymous. 1972. "Heaven Is Hotter than Hell." *Applied Optics* 11, no. 8: A14.

Arabatzis, Theodore. 2006. *Representing Electrons: A Biographical Approach to Theoretical Entities.* Chicago: University of Chicago Press.

Arabatzis, Theodore. 2008. "Causes and Contingencies in the History of Science: A Plea for a Pluralist Historiography." *Centaurus* 50: 32–36.

Arthur, W. Brian 1989. "Competing Technologies, Increasing Returns, and Lock-in by Historical Events." *Economic Journal* 99: 116–31.

Ascher, Marcia. 1994. *Ethnomathematics: A Multicultural View of Mathematical Ideas.* London: CRC Press.

Ashburner, Michael. 1993. "Epilogue." In *The Development of Drosophila Melanogaster,* edited by A. M. Arias and M. Bate, 1493–506. Plainview, NY: Cold Spring Harbor Laboratory Press.

Ashmore, Malcolm. 1993. "The Theatre of the Blind: Starring a Promethean Prankster, a Phoney Phenomenon, a Prism, a Pocket, and a Piece of Wood." *Social Studies of Science* 23, no. 1: 67–106.

Asimov, Isaac. 1941. "Nightfall." *Astounding Science Fiction,* September 1941; reprinted in Isaac Asimov, *Nightfall and Other Stories.* Garden City, NY: Doubleday, 1969.

Bachelard, Gaston. 1934. *Le nouvel esprit scientifique.* Paris: Presses Universitaires de France.

Bachelard, Gaston. 1951. *L'activité rationaliste de la physique contemporaine.* Paris: Presses Universitaires de France.

Bachelard, Gaston. 1953. *Le Matérialisme rationnel.* Paris: Presses Universitaires de France.

Ballinger, Clint. 2008. "Classifying Contingency in the Social Sciences: Diachronic, Synchronic, and Deterministic Contingency." Preprint: https://www.academia.edu/450967/Classifying_Contingency_In_the_Social_Sciences_Diachronic_Synchronic_and_Deterministic_Contingency.

Ballinger, Clint. 2013. "Contingency Is Just So." https://www.academia.edu/3769855/Contingency_is_just_so.

Barnes, Barry. 1991. "How Not to Do the Sociology of Knowledge." *Annals of Scholarship* 8, no. 3: 321–35.

Beatty, John. 1987. "Natural Selection and the Null Hypothesis." In *The Latest on the Best: Essays on Evolution and Optimality*, edited by John Dupré, 53–76. Cambridge, MA: MIT Press.

Beatty, John. 1995. "The Evolutionary Contingency Thesis." In *Theories and Rationality in the Biological Sciences*, edited by G. Wolters and J. Lennox, 45–81. Pittsburgh, PA: University of Pittsburgh Press.

Beatty, John. 2006. "Replaying Life's Tape." *Journal of Philosophy* 103, no. 7 (July): 336–62.

Beauvoir, Simone de. [1981] 1987. *La cérémonie des adieux suivie de Entretiens avec Jean-Paul Sartre*. Paris: Gallimard.

Ben-Menahem, Yemima. 1997. "Historical Contingency." *Ratio* 10, no. 2: 99–107.

Ben-Menahem, Yemina. 2009. "Historical Necessity and Contingency." In *A Companion to the Philosophy of History and Historiography*, edited by A. Tucker, 120–30. Oxford: Wiley-Blackwell.

Bergson, Henri. [1922] 2010. *Durée et simultanéité*. New critical edition by Elie During. Paris: Presses Universitaires de France.

Bergson, Henri. 1934. *La pensée et le mouvant*. Paris: Presses Universitaires de France.

Beurton, Peter, Raphael Falk, and Hans-Jörg Rheinberger, eds. 2000. *The Concept of the Gene in Development and Evolution*. Cambridge: Cambridge University Press.

Biagioli, Mario. 1996. "From Relativism to Contingentism." In *The Disunity of Science: Boundaries, Contexts, and Power*, edited by P. Galison and D. J. Stump, 189–206. Stanford, CA: Stanford University Press.

Binet, Alfred. 1903, *L'étude expérimentale de l'intelligence*. Paris : Costes.

Bitbol, Michel. 1996. *Physique et philosophie de l'esprit*. Paris: Flammarion.

Bitbol, Michel. 1998. "Some Steps Towards a Transcendental Deduction of Quantum Mechanics." *Philosophia naturalis* 35: 253–80.

Bitbol, Michel. 2000. *Schrödinger's Philosophy of Quantum Mechanics*. Dordrecht: Kluwer.

Bitbol, Michel. 2002. "Science as if Situation Mattered." *Phenomenology and the Cognitive Science* 1: 181–224.

Bitbol, Michel. 2006. "Une science de la conscience équitable: L'actualité de la neurophénoménologie de Francisco Varela." *Intellectica* 43: 135–57.

Bitbol, Michel. 2008a. "Consciousness, Situations, and the Measurement Problem of Quantum Mechanics." *NeuroQuantology* 6: 203–13.

Bitbol, Michel. 2008b. "Is Consciousness Primary?" *NeuroQuantology* 6: 53–72.

Bitbol, Michel. 2012. "Neurophenomenology, an Ongoing Practice of/in Consciousness." *Constructivist Foundations* 7, no. 3: 165–73.

Bitbol, Michel, and Claire Petitmengin. 2011. "On Pure Reflection (a Reply to Dan Zahavi)." *Journal of Consciousness Studies* 18: 24–37.

Bitbol, Michel, and Claire Petitmengin. 2013. "A Defense of Introspection from Within." *Constructivist Foundations* 8, no. 3: 269–79.

Blight, James G., Janet M. Lang, and David A. Welch. 2009. *Vietnam if Kennedy Had Lived: Virtual JFK*. Lanham, MD: Rowman and Littlefield.

Block, Ned. 2011. "Perceptual Consciousness Overflows Cognitive Access." *Trends in Cognitive Sciences* 15: 567–75.

Bloor, David. [1976] 1991. *Knowledge and Social Imagery.* Chicago: University of Chicago Press, 1991 (first edition: London: RKP).

Bode, Boyd H. 1913. "The Method of Introspection." *Journal of Philosophy, Psychology and Scientific Methods* 10: 85–91.

Bohm, David. 1951. *Quantum Theory.* Englewood Cliffs, NJ: Prentice Hall.

Bohm, David. 1952a. "A Suggested Interpretation of the Quantum Theory in Terms of 'Hidden' Variables, I and II." *Physical Review* 85: 166–79, 180–93.

Bohm, David. 1952b. "Reply to a Criticism of a Causal Re-Interpretation of the Quantum Theory." *Physical Review* 87: 389–90.

Bohr, Niels. 1934. *Atomic Theory and the Description of Nature.* Cambridge: Cambridge University Press.

Boon, Mieke. 2004. "Technological Instruments in Scientific Experimentation." *International Studies in the Philosophy of Science* 18, nos. 2 and 3: 221–30.

Boon, Mieke. 2009. "Understanding in the Engineering Sciences: Interpretative Structures." In *Scientific Understanding: Philosophical Perspectives,* edited by Henk W. de Regt, Sabina Leonelli, and Kai Eigner, 249–70. Pittsburgh, PA: University of Pittsburgh Press.

Boon, Mieke. 2011. "Two Styles of Reasoning in Scientific Practices: Experimental and Mathematical Traditions." *International Studies in the Philosophy of Science* 25, no. 3: 255–78.

Boon, Mieke. 2012a. "Understanding Scientific Practices: The Role of Robustness Notions." In *Characterizing the Robustness of Science: After the Practical Turn of the Philosophy of Science,* edited by Léna Soler, Emiliano Trizio, Thomas Nickles, and William Wimsatt, 289–315. Dordrecht: Springer, Boston Studies in the Philosophy of Science.

Boon, Mieke. 2012b. "Scientific Concepts in the Engineering Sciences: Epistemic Tools for Creating and Intervening with Phenomena." In *Scientific Concepts and Investigative Practice,* edited by U. Feest and F. Steinle, 219–43. Berlin: Walter De Gruyter, Berlin Studies in Knowledge Research.

Boon, Mieke, and Tarja Knuuttila. 2009. "Models as Epistemic Tools in Engineering Sciences: A Pragmatic Approach." In *Handbook of the Philosophy of Technological Sciences,* edited by Anthonie Meijers, 687–719. Amsterdam: Elsevier Science.

Boring, Edwin G. 1929. *A History of Experimental Psychology.* NJ: Appleton-Century.

Boring, Edwin G. 1953. "A History of Introspection." *Psychological Bulletin* 50: 169–89.

Born, Max. 1927. "Physical Aspects of Quantum Mechanics." *Nature* 119, no. 2992: 354–57.

Bourdieu, Pierre. 1998. *Practical Reason.* Stanford, CA: Stanford University Press.

Boutroux, Emile. 1902. *De la contingence des lois de la nature.* Paris: Félix Alcan.

Bowler, Peter J. 2008. "What Darwin Disturbed: The Biology That Might Have Been." *Isis* 99: 560–67.

Bowler, Peter J. 2013. *Darwin Deleted: Imagining a World without Darwin.* Chicago: University of Chicago Press.

Boyce, Tammy. 2006. "Journalism and Expertise." *Journalism Studies* 7, no. 6: 889–906.

Boyer, Carl B. 1991. *A History of Mathematics*. 2nd ed., revised by Uta C. Merzbach. New York: Wiley.

Bragg, William Lawrence, and George Porter, eds. 1970. *The Royal Institution Library of Science*. London: Applied Science Publishers.

Brahami, Frédéric. 2002. "Empirisme et scepticisme dans la philosophie des sciences en Grande-Bretagne aux XVII et XVIII siècles." In *Les philosophes et la science*, edited by Pierre Wagner, 301–48. Paris: Gallimard, Folio essais.

Brentano, Franz. 1995. *Psychology from an Empirical Standpoint*. London: Routledge.

Bricmont, Jean. 2007. "La mécanique quantique pour non-physiciens." http://www.mathematik.uni-uenchen.de/~bohmmech/BohmHome/files/2007_bricmont-mqlnew.pdf.

Brock, Adrian, Johann Louw, and Willem van Hoorn, eds. 2004. *Rediscovering the History of Psychology*. History and Philosophy of Psychology. New York: Kluwer.

Brown, Gary I. 1991. "The Evolution of the Term 'Mixed Mathematics.'" *Journal of the History of Ideas* 52: 81–102.

Burchfield, Joe D. 1990. *Lord Kelvin and the Age of the Earth*. Chicago: University of Chicago Press.

Burnyeat, Miles. 2000. "Plato on Why Mathematics Is Good for the Soul." In *Mathematics and Necessity: Essays in the History of Philosophy*, edited by Timothy Smiley, 1–82. Oxford: Oxford University Press for the British Academy.

Butterworth, Brian. 1999. *The Mathematical Brain*. London: Macmillan. U.S. edition: *What Counts: How Every Brain is Hardwired for Math*. New York: Free Press.

Buzaglo, Meir. 2002. *The Logic of Concept Expansion*. Cambridge: Cambridge University Press.

Byrne, Ruth M. J. 2005. *The Rational Imagination: How People Create Alternatives to Reality*. Cambridge, MA: MIT Press.

Cajori, Florian. [1928–1929] 1993. *The History of Mathematical Notations*. Mineola, NY: Dover.

Canguilhem, Georges. 1989. *Études d'histoire et de philosophie des sciences*. Paris: Vrin.

Cardano, Girolamo. 1968. *The Great Art or The Rules of Algebra*, translated and edited by T. Richard Witmer. Cambridge, MA: MIT Press.

Carr, Edward Hallett. 1961. *What Is History?* London: Macmillan.

Cartwright, Nancy. 1983. *How the Laws of Physics Lie*. Oxford: Oxford University Press.

Cartwright, Nancy. 1989. *Natures Capacities and Their Measurement*. Oxford: Oxford University Press.

Cartwright, Nancy. 1999. *The Dappled World: A Study of the Boundaries of Science*. Cambridge: Cambridge University Press.

Castaneda, Carlos. 1968. *The Teachings of Don Juan: A Yaqui Way of Knowledge*. Harmondsworth: Penguin.

Cat, Jordi, Nancy Cartwright, and Hasok Chang. 1996. "Otto Neurath: Politics and the Unity of Science." In *The Disunity of Science*, edited by Peter Galison and David Stump, 347–69. Stanford, CA: Stanford University Press.

Catellin, Sylvie. 2014. *Sérendipité: Du conte au concept*. Paris: Seuil.

Cavaillès, Jean. 1994. *Œuvres complètes de Philosophie des sciences*. Paris: Hermann.

Chang, Hasok. 2003. "Preservative Realism and Its Discontents: Revisiting Caloric." *Philosophy of Science* 70: 902–12.

Chang, Hasok. 2004. *Inventing Temperature: Measurement and Scientific Progress.* Oxford: Oxford University Press.

Chang, Hasok. 2009a. "Ontological Principles and the Intelligibility of Epistemic Activities." In *Scientific Understanding: Philosophical Perspectives*, edited by Henk W. de Regt, Sabina Leonelli, and Kai Eigner, 64–82. Pittsburgh, PA: University of Pittsburgh Press.

Chang, Hasok. 2009b. "We Have Never Been Whiggish (about Phlogiston)." *Centaurus* 51: 239–64.

Chang, Hasok. 2011. "Beyond Case Studies: History as Philosophy." In *Integrating History and Philosophy of Science: Problems and Prospects*, edited by Seymour Mauskopf and Tad Schmaltz, 109–24. Durham, NC: Duke University Press.

Chang, Hasok. 2012. *Is Water H$_2$0? Evidence, Realism, and Pluralism.* Dordrecht: Springer.

Changeux, Jean–Pierre, and Alain Connes. 1989. *Matière à pensée.* Paris: Odile Jacob.

Changeux, Jean-Pierre, and Alain Connes. [1989] 1995. *Conversations on Mind: Matter and Mathematics*, edited and translated by M. B. DeBevoise. Princeton, NJ: Princeton University Press.

Churchland, Patricia S. 1986. *Neurophilosophy: Toward a Unified Science of the Mind-Brain.* Cambridge, MA: MIT Press.

Coghill, Robert C., John G. McHaffie, and Ye-Fen Yen. 2003. "Neural Correlates of Interindividual Differences in the Subjective Experience of Pain." *Proceedings of the National Academy of Sciences of the USA* 100, no. 14: 8538–42.

Cohen, Michael A., and Daniel Dennett. 2011. "Consciousness Cannot Be Separated from Function." *Trends in Cognitive Sciences* 15, 358–63.

Collins, Harry. 1974. "The TEA Set: Tacit Knowledge and Scientific Networks." *Science Studies* 4: 165–86.

Collins, Harry. 1981. "The Role of the Core-set in Modern Science: Social Contingency with Methodological Property in Science." *History of Science* 19, no. 1: 6–19.

Collins, Harry. 1982. "Special Relativism: The Natural Attitude." *Social Studies of Science* 12: 139–43.

Collins, Harry. 1985. *Changing Order: Replication and Induction in Scientific Practice.* Beverly Hills, CA: Sage.

Collins, Harry. 1992. *Changing Order: Replication and Induction in Scientific Practice*, 2nd ed. Chicago: University of Chicago Press.

Collins, Harry. 2002. "The Experimenter's Regress as Philosophical Sociology." *Studies in History and Philosophy of Science* 33: 153–60.

Collins, Harry. 2004. *Gravity's Shadow: The Search for Gravitational Waves.* Chicago: University of Chicago Press.

Collins, Harry. 2010. *Tacit and Explicit Knowledge.* Chicago: University of Chicago Press.

Collins, Harry. 2011. "Language and Practice." *Social Studies of Science* 41, no. 2: 271–300. DOI 10.1177/0306312711399665.

Collins, Harry. 2014a. *Are We All Scientific Experts Now?* Cambridge, MA: Polity Press.

Collins, Harry. 2014b. "Rejecting Knowledge Claims inside and outside Science." *Social Studies of Science* 1, no. 44 (October): 786–92. DOI: 10.1177/0306312714536011.

Collins, Harry, and Robert Evans. 2002. "The Third Wave of Science Studies: Studies of Expertise and Experience." *Social Studies of Science* 32, no. 2: 235–96.

Collins, Harry, and Robert Evans. 2007. *Rethinking Expertise.* Chicago: University of Chicago Press.

Collins, Harry, and Robert Evans. Submitted. "Elective Modernism."

Collins, Harry, and Trevor J. Pinch. 1993. *The Golem: What Everyone Should Know about Science.* Cambridge: Cambridge University Press. [New edition, 1998].

Collins, Harry, Martin Weinel, and Robert Evans. 2010. "The Politics and Policy of the Third Wave: New Technologies and Society." *Critical Policy Studies* 4, no. 2,:185–201.

Compton, Arthur H. 1923. "The Quantum Integral and Diffraction by a Crystal." *Proceedings of the National Academy of Sciences* 9, 359–62.

Comte, Auguste. [1830–1842] 2001. *Cours de philosophie positive.* BookSurge.

Cooper, David. 2002. *The Measure of Things: Humanism, Humility, and Mystery.* Oxford: Clarendon Press.

Corallo, G., Jerome Sackur, Stanislas Dehaene, and M. Sigman. 2008. "Limits on Introspection: Distorted Subjective Time during the Dual-Task Bottleneck." *Psychological Science* 19: 1110–17.

Costall, Alan 2006. "'Introspectionism' and the Mythical Origins of Scientific Psychology." *Consciousness and Cognition* 15: 634–54.

Cournot, Antoine Augustin. [1851] 1956. *Essay on the Foundations of Our Knowledge.* New York: Liberal Press.

Cowley, Robert, ed. 1999. *What If? The World's Foremost Military Historians Imagine What Might Have Been.* New York: Simon and Schuster.

Cowley, Robert, ed. 2005. *What If? 2: Eminent Historians Imagine What Might Have Been.* New York: Berkley Publishing Group.

Culp, Sylvia. 1994. "Defending Robustness: The Bacterial Mesosome as a Test Case." In *PSA 1994: Proceedings of the 1994 Biennial Meeting of the Philosophy of Science Association,* vol. 1, edited by David Hull, Micky Forbes, and Richard M. Burian, 46–57. East Lansing, MI: Philosophy of Science Association.

Culp, Sylvia. 1995. "Objectivity in Experimental Inquiry: Breaking Data-Technique Circles." *Philosophy of Science* 62: 430–50.

Cushing, James T. 1992. "Historical Contingency and Theory Selection in Science." *Proceedings of the Biennial Meeting of the Philosophy of Science Association,* volume 1: *Contributed Papers,* 446–57. Chicago: University of Chicago Press.

Cushing, James T. 1994. *Quantum Mechanics: Historical Contingency and the Copenhagen Hegemony.* Chicago: University of Chicago Press.

D'Ambrosio, Ubiratàn. 2006. *Ethnomathematics.* Rotterdam: Sense.

Danziger, Kurt. 1980. "The History of Introspection Reconsidered." *Journal of the History of the Behavioral Sciences* 16: 241–62.

Danziger, Kurt. 1994. *Constructing the Subject: Historical Origins of Psychological Research.* Cambridge: Cambridge University Press.

Darwin, George H. 1916. "Introduction to Dynamical Astronomy." In *Scientific Papers*, vol. 5, 9–15. Cambridge: Cambridge University Press.

Davies, E. Brian. 2002. "The Role of Astronomy in the History of Science." arXiv: physics/0207043.

Davisson, Clinton J. 1928. "Are Electrons Waves?" *Journal of the Franklin Institute* 206 (May): 597–623.

Davisson, Clinton J., and Lester H. Germer. 1927a. "The Scattering of Electrons by a Single Crystal of Nickel." *Nature* 119, no. 2998: 558–59.

Davisson, Clinton. J., and Lester H. Germer. 1927b. "Diffraction of Electrons by a Crystal of Nickel." *Physical Review* 30, no. 6 (December): 705–40.

Davisson, Clinton J., and Charles H. Kunsman. 1921. "The Scattering of Electrons by Nickel." *Science* 52, no. 1404 (November 25): 522–24.

Davisson, Clinton J., and Charles H. Kunsman. 1923. "The Scattering of Low Speed Electrons by Platinum and Magnesium." *Physical Review* 22: 242–58.

Dehaene, Stanislas. 1997. *The Number Sense: How the Mind Creates Mathematics*. London: Allen Lane/Penguin Press.

de Kerkhove, Derick, and Charles J. Lumsden, eds. 1988. *The Alphabet and the Brain*. Berlin: Springer.

Deleuze, Gilles, and Félix Guattari. 1987. *A Thousand Plateaus: Capitalism and Schizophrenia*. Minneapolis: University of Minnesota Press.

Dennett, Daniel 1992. *Consciousness Explained*. New York: Back Bay Books.

Depraz, Natalie. 2008. *Lire Husserl en phénoménologue*. Paris: Presses Universitaires de France.

Depraz, Natalie, Francisco Varela, and Pierre Vermersch. 2003. *On Becoming Aware*. Amsterdam: John Benjamins.

de Regt, Henk, Sabina Leonelli, and Kai Eigner, eds. 2009. *Scientific Understanding: Philosophical Perspectives*. Pittsburgh, PA: University of Pittsburgh Press.

Deser, Stanley. 1970a. "Self-Interaction and Gauge Invariance." *General Relativity and Gravitation* 1, no. 9. arXiv :gr-qc/04110223.

Deser, Stanley. 1970b. "Gravity from Self-Interaction Redux." arXiv:gr-qc/0910.2975.

d'Espagnat, Bernard, and Hervé Zwirn. 2014. *Le monde quantique: Les débats philosophiques de la physique quantique*. Paris: Editions Matériologiques, Collection Science et Philosophie.

Detlefsen, Michael. 1998. "Mathematics, Foundations of." In *Routledge Encyclopedia of Philosophy*, edited by E. Craig, 6:181–91. London: Routledge.

Dick, Philip K. [1962] 1992. *The Man in the High Castle*. New York: Vintage Books.

Dirac, Paul A. M. 1930. *The Principles of Quantum Mechanics*. Cambridge: Cambridge University Press.

Duane, William. 1923. "The Transfer in Quanta of Radiation Momentum to Matter." *Proceedings of the National Academy of Sciences* 9: 158–64.

Duhem, Pierre. [1906] 1962. *The Aim and Structure of Physical Theory*, translated by Philip P. Wiener. New York: Atheneum.

Dürr, Detlef, et al. 2004. "Bohmian Mechanics and Quantum Field Theory." *Physical Review Letters*, 93: 1–4.

Dürr, Detlef, et al. 2005. "Bell-Type Quantum Field Theories." *Journal of Physics A: Mathematical and General* 38: R1–R43.

Edelman, Gerald M., and Giulio Tononi. 2001. *A Universe of Consciousness: How Matter Becomes Imagination.* New York: Basic Books.

Edwards, Adam, and James Sheptycki. 2009. "Third Wave Criminology: Guns, Crime and Social Order." *Criminology and Criminal Justice* 9, no. 3: 379–97.

Elgin, Catherine. 1991. "Understanding in Art and Science." In *Philosophy and the Arts,* edited by P. French, T. E Uehling Jr., and H. K. Wettstein, 196–208, Midwest Studies in Philosophy, vol. 16. Notre Dame, IN: University of Notre Dame Press.

English, Horace B. 1920. "In Aid of Introspection." *American Journal of Psychology* 32: 406–10.

Epstein, Paul S., and Paul Ehrenfest. 1924. "The Quantum Theory of the Fraunhofer Diffraction." *Proceedings of the National Academy of Sciences* 10: 133–39.

Epstein, Paul S., and Paul Ehrenfest. 1927. "Remarks on the Quantum Theory of Diffraction." *Proceedings of the National Academy of Sciences* 13: 400–408.

Ericsson, K. Anders, and Herbert A. Simon. 1984. *Protocol Analysis: Verbal Reports as Data.* Cambridge, MA: MIT Press.

Falk, Raphael. 2000. "The Gene: A Concept in Tension." In Beurton, Falk, and Rheinberger, *The Concept of the Gene,* 317–48.

Ferguson, Niall. 1999. "Introduction." In *Virtual History: Alternatives and Counterfactuals,* edited by Niall Ferguson, 1–90. New York: Basic Books (first published, London: Picador, 1997).

Feyerabend, Paul. 1965. "Problems of Empiricism." In *Beyond the Edge of Certainty: Essays in Contemporary Science and Philosophy,* edited by R. G. Colodny, 145–260. Englewood Cliffs, NJ: Prentice Hall.

Feyerabend, Paul. 1975. *Against Method.* London: New Left Books.

Feyerabend, Paul. 1993. *Against Method,* 3rd ed. London: Verso.

Feynman, Richard P. 1965. *The Character of Physical Law.* London: BBC.

Findlay, John N. 1948. "Recommendations Regarding the Language of Introspection." *Philosophy and Phenomenological Research* 9: 212–36.

Flajoliet, Alain 2006. "Husserl et Messer." *Expliciter: Journal de l'association GREX (Groupe de recherche sur l'explicitation),* no. 66: 1–32.

Flanagan, Owen J. 1993. *Consciousness Reconsidered.* Cambridge, MA: MIT Press.

Fleck, Ludwik. [1935] 1979. *Genesis and Development of a Scientific Fact,* translated by Frederick Bradley and Thaddeus J. Trenn. Chicago: University of Chicago Press.

Floridi, Luciano. 2011. "A Defence of Constructionism: Philosophy as Conceptual Engineering." *Metaphilosophy* 42, no. 3: 282–303.

Fock, Vladimir. 1959. *Theory of Space, Time and Gravitation.* Oxford: Pergamon.

Fontenelle, Bernard Le Bovier de. 1686. *Entretiens sur la pluralité des mondes.* English translation by Henry A. Hargreaves, *Conversations on the Plurality of Worlds.* Berkeley: University of California Press, 1990.

Foucault, Michel. 1985. "La vie: l'expérience et la science." *Revue de métaphysique et de morale* 90: 3–14.

Franklin, Allan. 1986. *The Neglect of Experiment.* Cambridge: Cambridge University Press.

Franklin, Allan. 1990. *Experiment Right or Wrong*. Cambridge: Cambridge University Press.

Franklin, Allan. 1998. "Experiment in Physics." In *Stanford Encyclopedia of Philosophy*. http://plato.stanford.edu/entries/physics-experiment/.

Franklin, Allan. 2008. "Is Failure an Option? Contingency and Refutation." *Studies in History and Philosophy of Science* 39: 242–52.

French, Steven. 2008. "Genuine Possibilities in the Scientific Past and How to Spot Them." *Isis* 99, no. 3: 568–75.

Freud, Sigmund. 1976. *The Complete Psychological Works*, vol. 15. Norton.

Fuller, Steve. 2008. "The Normative Turn: Counterfactuals and a Philosophical Historiography of Science." *Isis* 99, no. 3: 576–84.

Galileo, Galilei. 1587. *Due Lezioni all'Accademia Fiorentina circa la figura, sito e grandezza dell'Inferno di Dante*. Florence.

Galileo, Galilei. 1638. *Discorsi e dimostrazioni matematiche intorno a due nuove scienze*, Elzevir, 1638. In English *Dialogues Concerning Two New Sciences*, translated by Henry Crew and Alfono de Savio, Macmillan, 1914. http://files.libertyfund.org/files/753/0416_Bk.pdf.

Galison, Peter. 1987. *How Experiments End*. Chicago: University of Chicago Press.

Galison, Peter. 1995. "Context and Constraints." In *Scientific Practice, Theories and Stories of Doing Physics*, edited by J. Buchwald, 13–41. Chicago: University of Chicago Press.

Galison, Peter. 1997. *Image and Logic: A Material Culture of Microphysics*. Chicago: University of Chicago Press.

Gallagher, Shaun, and Dan Zahavi. 2007. *The Phenomenological Mind*. London: Routledge.

Garfield, Jay L. 1989. "The Myth of Jones and the Mirror of Nature: Reflections on Introspection." *Philosophy and Phenomenological Research* 50: 1–26.

Garud, Raghu, and Peter Karnøe, eds. 2001. *Path Dependence and Creation*. Mahwah, NJ: Lawrence Erlbaum.

Gehrenbeck, Richard K. 1978. "Electron Diffraction: Fifty Years Ago." *Physics Today* (January): 34–41.

Gendlin, Eugen 1962. *Experiencing and the Creation of Meaning*. Evanston, IL: Northwestern University Press.

Genoud, Charles. 2009. "On the Cultivation of Presence in Meditation." *Journal of Consciousness Studies* 16: 117–28.

Gergonne, Joseph-Diez. 1810. "Prospectus." *Annales de mathématiques pures et appliquées* 1: i–iv.

Germer, Lester Halbert. 1964. "Low-Energy Electron Diffraction." *Physics Today* 17 (July): 19–22.

Giere, Ronald N. 1988. *Explaining Science: A Cognitive Approach*. Chicago: University of Chicago Press.

Giere, Ronald N. 1999. *Science without Laws*. Chicago: University of Chicago Press.

Giere, Ronald N. 2006. *Scientific Perspectivism*. Chicago: University of Chicago Press.

Gilbert, G. Nigel, and Michael Mulkay. 1984. *Opening Pandora's Box. A Sociological Analysis of Scientific Discourse*. Cambridge: Cambridge University Press.

Gingras, Yves. 2001. "What Did Mathematics Do to Physics." *History of Science* 39: 383–416.

Gingras, Yves. 2003. "Mathématisation et exclusion: socioanalyse de la formation des cités savantes." In Wunenburger, Jean-Jacques, *Bachelard et l'épistémologie française.* Paris: Presses Universitaires de France, 115–152.

Gingras, Yves. 2015. "The Creative Power of Formal Analogies in Physics: The Case of Albert Einstein." *Science and Education.* doi: 10.1007/s11191–014–9739–1.

Gingras, Yves, and Silvan S. Schweber. 1986. "Constraints on Construction." *Social Studies of Science* 16, no. 2: 372–83.

Glashow, Sheldon. 1992. "The Death of Science!?" In *The End of Science? Attack and Defence,* edited by Richard J. Elvee, 23–32. Book Series: Nobel Conference, vol. 25. Lanham, MD: University Press of America.

Gleick, James. 1987. *Chaos: The Making of a New Science.* New York: Penguin.

Godin, Benoît, and Yves Gingras. 2002. "The Experimenters' Regress: From Skepticism to Argumentation." *Studies in History and Philosophy of Science* 33: 137–52.

Goldman, Alvin. 1992. "In Defense of the Simulation Theory." *Mind and Language* 7: 104–19.

Goldman, Alvin. 2001. "Epistemology and the Evidential Status of Introspective Reports." In *Trusting the Subject?* edited by A. Jack and A. Roepstorff, 2:1–16. Exeter, UK: Imprint Academic.

Gooding, David C., Michael E. Gorman, Ryan D. Tweney, and Alexandra P. Kincannon, eds. 2005. *Scientific and Technological Thinking.* Mahwah, NJ: Lawrence Erlbaum.

Goodman, Nelson. 1983. *Fact, Fiction and Forecast.* 4th ed. Cambridge, MA: Harvard University Press.

Gopnik, Alison, and Andrew N. Meltzoff. 1994. "Minds, Bodies and Persons: Young Children's Understanding of the Self and Others as Reflected in Imitation and 'Theory of Mind' Research." In *Self-Awareness in Animals and Humans,* edited by S. Parker and R. Mitchell, 166–86. Cambridge: Cambridge University Press.

Gould, Stephen Jay. 1989. *Wonderful Life: The Burgess Shale and the Nature of History.* New York: Norton.

Guicciardini, Niccolò. 2009. *Isaac Newton on Mathematical Certainty and Method.* Cambridge, MA: MIT Press.

Gupta, Bina. 2004. "Advaita Vedanta and Husserl's Phenomenology." *Husserl Studies* 20: 119–34.

Hacking, Ian. 1979. "What Is Logic?" *Journal of Philosophy* 86: 285–319.

Hacking, Ian. 1983. *Representing and Intervening.* Cambridge: Cambridge University Press.

Hacking, Ian. 1992. "The Self-Vindication of the Laboratory Sciences." In *Science as Practice and Culture,* edited by A. Pickering, 29–64. Chicago: University of Chicago Press.

Hacking, Ian. 1999. *The Social Construction of What?* Cambridge, MA: Harvard University Press.

Hacking, Ian. 2000a. "How Inevitable Are the Results of Successful Science?" *Philosophy of Science* 67, supplement PSA 1998: 58–71.

Hacking, Ian. 2000b. "What Mathematics Has Done to Some and Only Some Philosophers." In *Mathematics and Necessity*, edited by Timothy J. Smiley, 83–138. London: British Academy.

Hacking, Ian. 2006a. "La philosophie de l'expérience: illustrations de l'ultrafroid." *Tracés*, no. 11: 195–228.

Hacking, Ian. 2006b. *Another New World Is Being Constructed Right Now: The Ultracold*. Preprint 316, Max-Planck-Institut für Wissenschaftsgechichte.

Hacking, Ian. 2009a. "What Makes Mathematics Mathematics?" In *The Force of Argument: Essays in Honour of Timothy Smiley*, edited by Jonathan Lear and Alex Oliver, 82–106. London: Routledge.

Hacking, Ian. 2009b. *Scientific Reason*. Taipei: National Taiwan University Press.

Hacking, Ian. 2014. *Why Is There Philosophy of Mathematics At All?* Cambridge: Cambridge University Press.

Hacking, Ian. Forthcoming. "Le Zeitgenössischen Begriff chez Ludwik Fleck." In *Actes de la conférence internationale Genèse et Développement d'un Fait Scientifique. 1935: Retour sur les Fondements, la Fécondité et l'Actualité de la Pensée de Ludwik Fleck, 1896–1961*. February 2006. Paris: Ecole des Mines.

Hallyn, Fernand. 2000. "Atoms and Letters." In *Metaphor and Analogy in the Sciences*, edited by F. Hallyn, 53–69. New York: Springer.

Hanson, Norwood R. 1958. *Patterns of Discovery: An Inquiry into the Conceptual Foundations of Knowledge*. Cambridge: Cambridge University Press.

Hardy, Godfrey Harold. 2008. *The Course of Pure Mathematics Centenary Edition*, 10th ed. Cambridge: Cambridge Mathematical Library, Cambridge University Press.

Harré, Rom. 2003. "The Materiality of Instruments in a Metaphysics for Experiments." In *The Philosophy of Scientific Experimentation*, edited by H. Radder, 19–38. Pittsburgh, PA: University of Pittsburgh Press.

Hassin, Ran R., James S. Uleman, and John A. Bargh, eds. 2006. *The New Unconscious*. New York: Oxford University Press.

Haugeland, John. 1998. *Having Thought*. Cambridge, MA: Harvard University Press.

Haugeland, John. 2013. *Dasein Disclosed*. Cambridge, MA: Harvard University Press.

Hawking, Stephen W. 1988. *A Brief History of Time: From the Big Bang to Black Holes*. Toronto: Bantam Books.

Hawking, Stephen W., and Leonard Mlodinow. 2010. *The Grand Design*. New York: Random House.

Hawthorn, Geoffrey. 1991. *Plausible Worlds: Possibility and Understanding in History and the Social Sciences*, Cambridge: Cambridge University Press.

Haynes, William M., and David R. Lide. 2014. *CRC Handbook of Chemistry and Physics: A Ready-Reference Book of Chemical and Physical Data*, 95th ed. Boca Raton, FL: CRC Press.

Healey, Tim. 1979. "A Refutation of the Proof that Heaven Is Hotter than Hell." *Journal of Irreproducible Results* 25, no. 4: 17–18.

Heidegger, Martin. 1977. "The Question Concerning Technology." In *The Question concerning Technology and Other Essays*, translated by William Lovitt, 3–35. New York: Harper and Row.

Heisenberg, Werner. 1949. *The Physical Principles of the Quantum Theory.* Mineola, NY: Dover.

Hendricks, Marion. 2009. "Experiencing Level: An Instance of Developing a Variable from a First Person Process so It Can Be Reliably Measured and Taught." *Journal of Consciousness Studies* 16: 129–55.

Henry, John. 2008. "Ideology, Inevitability, and the Scientific Revolution." *Isis* 99: 552–59.

Hickey, Thomas J. 2013. *History of Twentieth-Century Philosophy of Science*, Book 5. www.philsci.com.

Høffding, Harald. 1905. *The Problems of Philosophy.* London: Macmillan.

Holland, John G., and Burrhus Frederick Skinner. 1961. *Analysis of Behavior.* New York: McGraw-Hill.

Holton, Gerald. 1978. "Subelectrons, Presuppositions, and the Millikan–Ehrenhaft Dispute." *Historical Studies in the Physical Sciences* 9: 161–224.

Horsten, Leon. 2007. "Philosophy of Mathematics." In *The Stanford Encyclopedia of Philosophy*, edited by E. N. Zalta. http://plato.stanford.edu/archives/win2007/entries/philosophy-mathematics/.

Hoyningen-Huene, Paul. 1989. *Die Wissenschaftsphilosophie Thomas S. Kuhns: Rekonstruktion und Grundlagenprobleme.* Braunschweig: Vieweg. English translation by A. T. Levine. 1993. *Reconstructing Scientific Revolutions.* Chicago: University of Chicago Press.

Hugo, Victor. [1864] 2014. *William Shakespeare.* Paris: Garnier-Flammarion.

Hull, David L. 2008. "Review of Stephen H. Kellert, Helen E. Longino, and C. Kenneth Waters, eds., *Scientific Pluralism.*" *Notre Dame Philosophical Reviews*, May 11. http://ndpr.nd.edu/review.cfm?id=12963/.

Hume, David. 1962. *A Treatise of Human Nature.* London: Fontana/Collins.

Humphrey, George. 1951. *Thinking.* London: Methuen.

Hund, Friedrich. 1966. "Paths to Quantum Theory Historically Viewed." *Physics Today* 19, no. 8: 23–29.

Hurlburt, Russ T. 1990. *Sampling Normal and Schizophrenic Inner Experience.* New York: Plenum Press.

Hurlburt, Russ T. 1993. *Sampling Inner Experience in Disturbed Affect.* New York: Plenum Press.

Hurlburt, Russ T., and Christopher L. Heavey. 2001. "Telling What We Know: Describing Inner Experience." *Trends in Cognitive Sciences* 5: 400–403.

Hurlburt, Russ T., and Christopher L. Heavey. 2006. *Exploring Inner Experience.* Amsterdam: John Benjamins.

Husserl, Edmund. [1936] 1970. *Der Ursprung der Geometrie als intentional-historisches Problem.* Translated as "The Origin of Geometry," an appendix in *The Crisis of European Sciences and Transcendental Phenomenology: An Introduction to Phenomenological Philosophy.* Evanston, IL: Northwestern University Press.

Husserl, Edmund. 1972. *Erste Philosophie. 1923/4. Zweiter Teil: Theorie der phänomenologischen Reduktion*, edited by R. Boehm and Martinus Nijhoff. In French, *Philosophie première*, translated by Arion L. Kelkel. Paris: Presses Universitaires de France.

Husserl, Edmund. 1983. *Ideas Pertaining to a Pure Phenomenology and to a Phenome-*

nological Philosophy, vol. 1: General Introduction to a Pure Phenomenology. Berlin: Springer.

Husserl, Edmund. 2002. *Husserliana: Edmund Husserl Gesammelte* Werke, XXXIV, *Zur phänomenologischen Reduktion.* Dordrecht: Kluwer. French translation by Jean-François Pestureau and Marc Richir, *De la réduction phénoménologique, textes posthumes (1926–1935).* Grenoble: Jérome Millon.

Inkpen, Rob, and Derek Turner. 2012. "The Topography of Historical Contingency." *Journal of the Philosophy of History* 6: 1–19.

Jack, Anthony, and Andreas Roepstorff A. 2002. "The 'Measurement Problem' for Experience: Damaging Flaw or Intriguing Puzzle?" *Trends in Cognitive Sciences* 6: 372–74.

James, William. 1890. *The Principles of Psychology.* London: Holt.

James, William. 1907. *Pragmatism.* New York: Dover.

James, William. 1976. *Essays in Radical Empiricism.* Cambridge, MA: Harvard University Press.

Jean, Raymond. 1975. *Mesure et intégration.* Paris: Presses universitaires de France.

Johansson, Petter, Lars Hall, Sverker Sikström, and Andreas Olsson. 2005. "Failure to Detect Mismatches between Intention and Outcome in a Simple Decision Task." *Science* 310: 116–19.

Johansson, Petter, Lars Hall, Sverker Sikström, Betty Tärning, and Andreas Lind. 2006. "How Something Can Be Said about Telling More Than We Can Know: On Choice Blindness and Introspection." *Consciousness and Cognition* 15: 673–92.

Kant, Immanuel. [1783] 1953. *Prolegomena to Any Future Metaphysics That Will Be Able to Present Itself as a Science,* translated by Peter G. Lucas. Manchester, UK: Manchester University Press.

Kant, Immanuel. [1786] 2002. *Metaphysical Foundations of Natural Science.* In *Theoretical Philosophy after 1781,* edited by H. Allison, R. Brandt, P. Guyer, R. Meerbote, C. D. Parsons, H. Robinson, J. B. Schneewind, and A. W. Wood. *The Cambridge Edition of the Works of Immanuel Kant in Translation.* Cambridge: Cambridge University Press.

Kant, Immanuel. [1787] 1929. *Critique of Pure Reason,* translated by N. Kemp Smith. London: Macmillan.

Kant, Immanuel. [1800] 1988. *Logic,* translated by Robert S. Hartman. New York: Dover.

Kant, Immanuel. [1760–1790] 1997. *Lectures on Metaphysics,* translated by K. Ameriks and S. Naragon. Cambridge: Cambridge University Press.

Keller, Evelyn Fox. 2000. *Century of the Gene.* Cambridge, MA: Harvard University Press.

Kellert, Stephen H., Helen E. Longino, and C. Kenneth Waters, eds. 2006. *Scientific Pluralism.* Minnesota Studies in the Philosophy of Science, vol. 19. Minneapolis: University of Minnesota Press.

Kennefick, Daniel. 2000. "Star Crushing: Theoretical Practice and the Theoretician's Regress." *Social Studies of Science,* 30, no. 1: 5–40.

Kidd, Ian James. 2013. "Historical Contingency and the Impact of Scientific Imperialism." *International Studies in the Philosophy of Science* 27, no. 3: 317–26.

Kidd, Ian James. Forthcoming. "Inevitability, Contingency, and Epistemic Humility." *Studies in History and Philosophy of Science.*

Kidd, Ian James. Unpublished manuscript. "Epistemic Humility and Unconceived Alternatives."

Kinzel, Katherina. 2014. "Could the Results of Our Science Have Been Different? Contingency and Inevitability in the Philosophy and Historiography of Science." PhD thesis, University of Vienna, Austria.

Kinzel, Katherina. Forthcoming a. "Pluralism in Historiography: A Case Study of Case Studies." In *The Philosophy of Historical Case Studies,* edited by Tilman Sauer and Raphael Scholl. Boston Studies in the Philosophy and History of Science. Dordrecht: Springer.

Kinzel, Katherina. Forthcoming b. "Narrative and Evidence. How Can Case-Studies from the History of Science Support Claims in the Philosophy of Science?" *Studies in History and Philosophy of Science Part A.*

Kitcher, Philip. 1983. *The Nature of Mathematical Knowledge.* Oxford: Oxford University Press.

Kitcher, Philip. 1990. "The Division of Cognitive Labor." *Journal of Philosophy* 87, no. 1: 5–22.

Kitcher, Philip. 2001. *Science, Truth and Democracy.* Oxford: Oxford University Press.

Klasen, Martin, Mikhail Zvyagintsev, René Weber, Klaus Mathiak, and Krystyna Mathiak. 2008. "Think Aloud during fMRI: Neuronal Correlates of Subjective Experience in Video Games." In *Fun and Games,* edited by P. Markopoulos, B. de Ruyter, W. A. Ijsselsteijn, and D. Rowland. 132–38. Heidelberg: Springer.

Klein, Jacob. [1934] 1968. *Greek Mathematical Thought and the Origin of Algebra,* translated by E. Brann. Cambridge, MA: MIT Press.

Knuuttila, Tarja, and Mieke Boon. 2011. "How Do Models Give Us Knowledge? The Case of Carnot's Ideal Heat Engine." *European Journal Philosophy of Science* 1, no. 3: 309–34.

Kohler, Robert E. 1994. *Lords of the Fly: Drosophila Genetics and the Experimental Life.* Chicago: University of Chicago Press.

Köppen, Wladimir, and Alfred Wegener. 1924. *Die Klimate der geologischen Vorzeit.* Berlin: Gebrüder Bornträger.

Kosso, Peter. 1988. "Dimensions of Observability." *British Journal for the Philosophy of Science* 39: 449–67.

Kosso, Peter. 1989. *Observation and Observability in the Physical Sciences.* Dordrecht: Kluwer.

Kouider, Sid, Vincent de Gardelle, Jerome Sackur, and Emmanuel Dupoux. 2010. "How Rich Is Consciousness? The Partial Awareness Hypothesis." *Trends in Cognitive Sciences* 14: 301–7.

Krieger, Martin H. 1987. "The Physicist's Toolkit." *American Journal of Physics* 55, 1033–38.

Kripke, Saul. 1982. *Wittgenstein on Rules and Private Language.* Cambridge, MA: Harvard University Press.

Kroker, Kenton. 2003. "The Progress of Introspection in America, 1896–1938." *Studies in History and Philosophy of Biological and Biomedical Sciences* 34: 77–108.

Kuhn, Thomas S. 1957. *The Copernican Revolution*. Cambridge, MA: Harvard University Press.

Kuhn, Thomas S. 1970. *The Structure of Scientific Revolutions*, 2nd ed. Chicago: University of Chicago Press.

Kuhn, Thomas S. 1977. "Objectivity, Value Judgment, and Theory Choice." In *The Essential Tension: Selected Studies in Scientific Tradition and Change*, 320–39. Chicago: University of Chicago Press.

Kuhn, Thomas S. 2000. *The Road since Structure*. Chicago: University of Chicago Press.

Kukla, André. 2000. *Social Constructivism and the Philosophy of Science*. New York: Routledge.

Lakatos, Imre. 1976. *Proofs and Refutations: The Logic of Mathematical Discovery*. Cambridge: Cambridge University Press.

Laloë, Franck. 2012. *Do We Really Understand Quantum Mechanics?* Cambridge: Cambridge University Press.

Landé, Alfred. 1955. *Foundations of Quantum Theory*. New Haven: Yale University Press.

Landé, Alfred. 1965. *New Foundations of Quantum Mechanics*. Cambridge: Cambridge University Press.

Lange, Marc. 2000. *Natural Laws in Scientific Practice*. Oxford: Oxford University Press.

Lange, Marc. 2007. "Laws and Theories." In *A Companion to the Philosophy of Biology*, edited by S. Sarkar and A. Plutynski, 489–505. Cambridge: Cambridge University Press.

Lange, Marc. 2009. *Laws and Lawmakers*. Oxford: Oxford University Press.

Latour, Bruno. 1993. *We Have Never Been Modern*. Cambridge, MA: Harvard University Press.

Latour, Bruno. 2008. "The Netz-Works of Greek Deductions." *Social Studies of Science* 38: 441–59.

Latour, Bruno, and Steve Woolgar. 1979. *Laboratory Life*, 1st ed. Beverly Hills, CA: Sage; 2nd ed. Princeton, NJ: Princeton University Press, 1986.

Laudan, Larry. 1981. "A Confutation of Convergent Realism." *Philosophy of Science* 48, 19–48. Reprinted in *The Philosophy of Science*, edited by R. Boyd, P. Gasper, and J. D. Trout, 223–45. Cambridge, MA: MIT Press, 1991.

Lepeltier, Thomas. 2010. "Quand fallait-il abandonner la théorie de l'état stationnaire?" *Revue des questions scientifiques* 181, no. 4: 513–22.

Le Sage, George-Louis. 1782. "Lucrèce newtonien." In *Nouveaux mémoires de l'Académie royale des sciences et belles lettres*, année 1782, vol. 13: 404–32. English translation by C. G. Abbot. http://en.wikisource.org/wiki/The_Le_Sage_Theory _of_Gravitation#THE_NEWTONIAN_LUCRETIUS.

Lévy-Leblond, Jean-Marc. 1990. "Did the Big Bang Begin?" *American Journal of Physics* 58: 156–59.

Lévy-Leblond, Jean-Marc. 1992. "Why Does Physics Need Mathematics?" In *The Scientific Enterprise*, edited by Edna Ullmann-Margarit, 145–61. Boston Studies in the Philosophy of Science 146. Dordrecht: Kluwer.

Lévy-Leblond, Jean-Marc. 2001a. "Science's Fiction." *Nature* 413, 573.

Lévy-Leblond, Jean-Marc. 2001b. "De la relativité à la chronogéométrie: Pour en finir avec le 'second postulat' et autres fossiles." Colloque de Cargèse, Le Temps. http://o.castera.free.fr/pdf/Chronogeometrie.pdf.

Lévy-Leblond, Jean-Marc. 2003. "What if Einstein Had Not Been There?" *Proceedings of the Twenty-Fourth International Colloquium on Group Theoretical Methods in Physics*, edited by Jean-Pierre Gazeau et al. Paris, July 2002. London: Institute of Physics. http://o.castera.free.fr/pdf/Chronogeometrie.pdf.

Lévy-Leblond, Jean-Marc. 2005. "Figures and Characters in the Great Book of Nature." In *The Visual Mind II*, edited by Michele Emmer, 623–46. Cambridge, MA: MIT Press.

Lévy-Leblond, Jean-Marc. 2007. "Le boulet d'Einstein et les boulettes de Bergson." In *Annales bergsoniennes III: Bergson et la science*, 280–83. Paris: Presses Universitaires de France.

Lévy-Leblond, Jean-Marc. 2008. "Postface" to the French translation: Galilée, *Leçons sur l'Enfer de Dante*, translated by Lucette Degryse. Paris: Fayard.

Littlewood, John E. 1953. *A Mathematician's Miscellany*. London: Methuen.

Lloyd, Geoffrey. 1990. *Demystifying Mentalities*. Cambridge: Cambridge University Press.

Lucretius. 1916. *De Natura Rerum*. English translation by William E. Leonard, *Of the Nature of Things*. http://www.gutenberg.org/dirs/7/8/785/785.txt.

Lutz, Antoine. 2002. "Towards a Neurophenomenology as an Account of Generative Passages: A First Empirical Case Study." *Phenomenology and the Cognitive Science* 1, 133–67.

Lyons, William. 1986. *The Disappearance of Introspection*. Cambridge, MA: MIT Press.

Mac Lane, Saunders. 1986. *Mathematics: Form and Function*. New York: Springer.

Mach, Ernst. [1897] 1984. *The Analysis of Sensations*. Chicago: Open Court.

Macherey, Pierre. 2008. "L. S. Althusser et le concept de philosophie spontanée des savants." http://stl.recherche.univ-lille3.fr/seminaires/philosophie/macherey/macherey20072008/macherey21052008.html.

Mancosu, Paolo. 2009. "Measuring the Size of Infinite Collections of Natural Numbers: Was Cantor's Theory of Infinite Number Inevitable?" *Review of Symbolic Logic* 2: 612–46.

Martin, Joseph D. 2013. "Is the Contingentist/Inevitabilist Debate a Matter of Degrees?" *Philosophy of Science* 80, no. 5 (December): 919–30.

Marvin, Ursula B. 1973. *Continental Drift: The Evolution of a Concept*. Washington, DC: Smithsonian Institution Press.

Massimi, Michela. 2008. "Why There Are No Ready-Made Phenomena: What Philosophers of Science Should Learn from Kant." *Royal Institute of Philosophy Supplement* 83, no. 63, 1–35.

Medawar, Peter B. 1967. *The Art of the Soluble*. London: Methuen.

Mehra, Jagdish, and Helmut Rechenberg. 1982. *The Historical Development of Quantum Theory*, vol. 1, part 2. New York: Springer, 1982.

Merleau-Ponty, Maurice. 1953. *Éloge de la philosophie*. Paris: Gallimard.

Mersenne, Marin. 1634. *Questions harmoniques*. Paris: Jacques Villery. Facsimile edition. Stuttgart: Fromann, 1972.

Merton, Robert King, and Elinor Barber. 2006. *The Travels and Adventures of Serendipity: A Study in Sociological Semantics and the Sociology of Science*. Princeton, NJ: Princeton University Press.

Michel, Alain. 1992. *Constitution de la théorie moderne de l'intégration*. Paris: Vrin.

Mill, John Stuart. 1981. *Autobiography*, volume 1: *The Collected Works of John Stuart Mill*. Toronto: University of Toronto Press.

Miller, Arthur. 1981. *Albert Einstein's Special History of Relativity: Emergence (1905) and Early Interpretation (1905–1911)*. Reading, MA: Addison-Wesley.

Milo, Daniel S., and Alain Bourreau. 1991. *Alter-histoire: essais d'histoire expérimentale*. Paris: Les Belles Lettres.

Mira-Perez, Jorge, and Jose Viña. 1998. "Bible Used to Reexamine if Heaven Is Hotter than Hell." *Physics Today* 51, no. 7: 96.

Mitchell, Sandra D. 2003. *Biological Complexity and Integrative Pluralism*. Cambridge: Cambridge University Press.

Mokyr, Joel. 2008. "King Kong and Cold Fusion: Counterfactual Analysis and the History of Technology." In Tetlock, Lebow, and Parker, *Unmaking the West*, 277–322.

Moody-Adams, Michele. 1997. *Fieldwork in Familiar Places*. Cambridge, MA: Harvard University Press.

Moore, Jared S., and Herbert Gurnee. 1933. *The Foundations of Psychology*. Princeton, NJ: Princeton University Press.

Muller, F. A. 1997a. "The Equivalence Myth of Quantum Mechanics: Part 1." *Studies in History and Philosophy of Modern Physics* 28 (March): 35–61.

Muller, F. A. 1997b. "The Equivalence Myth of Quantum Mechanics: Part 2." *Studies in History and Philosophy of Modern Physics* 28 (June): 219–47.

Nahmias, Eddy A. 2002. "Verbal Reports on the Contents of Consciousness: Reconsidering Introspectionist Methodology." *Psyche* 8, no. 21 (October).

Nassau, Kurt. 1972. *Applied Optics* 11, no. 12: A14.

Natorp, Paul. 2007. *Psychologie générale selon la méthode critique*, translated by E. Dufour and J. Servois. Paris: Vrin.

Nauriyal, Dinesh K., Michael Drummond, and Y. B. Lal. 2010. *Buddhist Thought and Applied Psychological Research*. New York: Routledge.

Navarro, Jaume. 2010. "Electron Diffraction chez Thomson: Early Responses to Quantum Physics in Britain." *British Journal of History of Science* 43, 245–75.

Nelson, Edward. 1977. "Internal Set Theory." *Bulletin of the American Mathematical Society* 83, no. 6 (November): 1165–98.

Nelson, Edward. 1986. *Predicative Arithmetic*. Princeton, NJ: Princeton University Press.

Nelson, Thomas O. 1996. "Consciousness and Metacognition." *American Psychologist* 51: 102–16.

Nersessian, Nancy J. 2008. *Creating Scientific Concepts*. Cambridge, MA: MIT Press.

Netz, Reviel. 1999. *The Shaping of Deduction in Greek Mathematics: A Study in Cognitive History*. Cambridge: Cambridge University Press.

Netz, Reviel. 2002. "Counter Culture: Towards a History of Greek Numeracy." *History of Science* 40: 321–52.

Nickles, Thomas. 2012. "Dynamic Robustness and Design in Nature and Artifact." In Soler et al., *Characterizing the Robustness of Science*, 329–60. Dordrecht: Springer.

Nigro, Georgia, and Ulric Neisser. 1983. "Point of View in Personal Memories." *Cognitive Psychology* 15: 467–82.

Nisbett, Richard E., and Timothy D. Wilson. 1977. "Telling More Than We Can Know: Verbal Reports on Mental Processes." *Psychological Review* 84: 231–59.

Oberheim, Eric. 2006. *Feyerabend's Philosophy*. Berlin: Walter de Gruyter.

Ogden, Richard M. 1911. "Imageless Thought: Résumé and Critique." *Psychological Bulletin* 8: 194.

Olavo, Leopoldino Da Silva Filho. 1999. "Foundations of Quantum Mechanics (II): Equilibrium, Bohr–Sommerfeld Rules and Duality." *Physica* A 271, 260–302.

Overgaard, Morten, ed. 2006. "Introspection in Science." *Consciousness and Cognition* 15, no. 4: 629–33.

Overgaard, Morten, Mila Koivisto, Thomas A. Sørensen, Signe Vangkilde, and Antti Revonsuo. 2006. "The Electrophysiology of Introspection." *Consciousness and Cognition* 15: 662–72.

Passon, Oliver. 2005. "Why Isn't Every Physicist a Bohmian?" arXiv:quant-ph/0412119v2.

Peacock, Marc. 2009. "Path-Dependence in the Production of Scientific Knowledge." *Social Epistemology* 23, no. 2: 105–24.

Pessoa, Osvaldo Jr. 2001. "Counterfactual Histories: The Beginning of Quantum Physics." *Philosophy of Science* 68 (Suppl.): 519–30.

Peterson, Mark. 2002. "Galileo's Discovery of Scaling Laws." *American Journal of Physics* 70, no. 3 (February), 575–80. arXiv:phyics/0110031.

Petitmengin, Claire. 2006. "Describing One's Subjective Experience in the Second Person: An Interview Method for the Science of Consciousness." *Phenomenology and the Cognitive Science* 5: 229–69.

Petitmengin, Claire. 2007. "Towards the Source of Thoughts: The Gestural and Transmodal Dimension of Lived Experience." *Journal of Consciousness Studies* 14: 54–82.

Petitmengin, Claire, ed. 2009. *Ten Years of Viewing from Within: The Legacy of Francisco Varela*. Exeter: Imprint Academics.

Petitmengin, Claire, and Michel Bitbol. 2009. "The Validity of First-Person Descriptions as Authenticity and Coherence." *Journal of Consciousness Studies* 16: 363–404.

Petitmengin, Claire, Michel Bitbol, Jean-Michel Nissou, Bernard Pachoud, Hélène Curallucci, Michel Cermolacce, and Jean Vion-Dury. 2009. "Listening from Within." *Journal of Consciousness Studies* 16: 252–84.

Petitmengin, Claire, Vincent Navarro, and Michel Baulac. 2006. "Seizure Anticipation: Are Neuro-phenomenological Approaches Able to Detect Preictal Symptoms?" *Epilepsy and Behavior* 9: 298–306.

Petitmengin, Claire, Anne Remillieux, Béatrice Cahour, and Shirley Thomas. 2013. "A Gap in Nisbett and Wilson Findings? A First-Person Access to Our Cognitive Processes." *Consciousness and Cognition* 22: 654–69.

Piaget, Jean. 1985. *Equilibration of Cognitive Structures*. Chicago: University of Chicago Press.

Piccinini, Gualtiero. 2003. "Data from Introspective Reports." *Journal of Consciousness Studies* 10: 141–56.

Pickering, Andrew. 1984a. *Constructing Quarks: A Sociological History of Particle Physics.* Chicago: University of Chicago Press.

Pickering, Andrew. 1984b. "Against Putting the Phenomena First: The Discovery of the Weak Neutral Current." *Studies in the History and Philosophy of Science* 15: 85–117.

Pickering, Andrew. 1987. "Forms of Life: Science, Contingency and Harry Collins." *British Journal for the History of Science* 20: 213–21.

Pickering, Andrew. 1989a. "Editing and Epistemology: Three Accounts of the Discovery of the Weak Neutral Current." *Knowledge and Society: Studies in the Sociology of Science Past and Present* 8: 217–32.

Pickering, Andrew. 1989b. "Living in the Material World: On Realism and Experimental Practice." In *The Uses of Experiment*, edited by David Gooding et al., 275–97. Cambridge: Cambridge University Press.

Pickering, Andrew. 1995a. *The Mangle of Practice: Time, Agency, and Science.* Chicago: University of Chicago Press.

Pickering, Andrew. 1995b. "Beyond Constraint: The Temporality of Practice and the Historicity of Knowledge." In *Scientific Practice: Theories and Stories of Doing Physics*, edited by J. D. Buchwald, 42–55. Chicago: University of Chicago Press.

Pickering, Andrew. 1997. "The Mangle as an Evolutionary Theory of Science and Technology Studies." Physics/History of Science Colloquium, University of Utrecht, February 11.

Pickering, Andrew. 2001. "Reading the Structure." *Perspectives on Science* 9: 499–510.

Pickering, Andrew. 2005. "Decentering Sociology: Synthetic Dyes and Social Theory." *Perspectives on Science* 13, 352–405.

Pickering, Andrew. 2008. "New Ontologies." In *The Mangle in Practice: Science, Society and Becoming*, edited by Andrew Pickering and K. Guzik, 1–13. Durham, NC: Duke University Press.

Pickering, Andrew. 2009. "The Politics of Theory: Producing Another World, with Some Thoughts on Latour." *Journal of Cultural Economy* 2: 199–214.

Pickering, Andrew. 2010. *The Cybernetic Brain: Sketches of Another Future.* Chicago: University of Chicago Press.

Pickering, Andrew. 2012. "The Robustness of Science and the Dance of Agency." In Soler et al., *Characterizing the Robustness of Science.* Dordrecht: Springer, 317–28.

Pickering, Andrew. 2014. "Reflections on the Dance of Agency: Islands of Stability, Science as Performance." Presented at the History of Science Seminar, Uppsala University, Sweden, June 3.

Pickering, Andrew, ed. 1992. *Science as Practice and Culture.* Chicago: University of Chicago Press.

Pickering, Andrew, and Peter W. Trower. 1985. "Sociological Problems of High Energy Physics." *Nature* 318 (November 21), 243–45.

Pinch, Trevor J. 1977. "What Does a Proof Do If It Does Not Prove?" In *The Social Production of Scientific Knowledge*, vol. 1: *Sociology of the Sciences*, edited by Everett Mendelsohn, Peter Weingart, and Richard D. Whitley, 171–215. Dordrecht: Reidel.

Pinch, Trevor J. 1979. "The Hidden-Variables Controversy in Quantum Physics." *Physics Education* 14: 48–52.

Pinch, Trevor J. 1982. "Kuhn: The Conservative and the Radical Interpretations." *4S Newsletter* 7, no. 1: 10–25. Reprinted as a "Historic Paper." *Social Studies of Science* 27 (1997): 465–82.

Pinch, Trevor J. 2001. "Why Do You Go to a Piano Store to Buy a Synthesizer? Path Dependence and the Social Construction of Technology." In Garud and Karnøe, *Path Dependence and Creation*, 381–99.

Pomeranz, Kenneth. 2008. "Without Coal? Colonies? Calculus? Counterfactuals and Industrialization in Europe and China." In Tetlock, Lebow, and Parker, *Unmaking the West*, 241–76.

Preston, John. 1997. *Feyerabend: Philosophy, Science and Society*. Cambridge: Polity Press.

Price, Donald D., and Marat Aydede. 2005. "The Experimental Use of Introspection in the Scientific Study of Pain and Its Integration with Third-Person Methodologies: The Experiential-Phenomenological Approach." In *Pain: New Essays on Its Nature and the Methodology of Its Studies*, edited by Murat Aydede, 123–36. Cambridge, MA: MIT Press.

Priestley, Joseph. 1790. *Experiments and Observations on Different Kinds of Air, and Other Branches of Natural Philosophy, Connected with the Subject*. 3 vols. Formerly 6 vols., abridged, methodized, and expanded. Birmingham: Thomas Pearson.

Psillos, Stathis. 1999. *Scientific Realism: How Science Tracks Truth*. London: Routledge.

Psillos, Stathis. 2000. "The Present State of the Scientific Realism Debate." *British Journal for the Philosophy of Science* 51: 705–28. Reprinted in *Philosophy of Science Today*, edited by P. Clark and K. Hawley. Oxford: Clarendon Press, 2003.

Putnam, Hilary. 1981. *Reason, Truth, and History*. Cambridge: Cambridge University Press.

Quine, Willard Van. 1960. *Word and Object*. Cambridge, MA: MIT Press.

Radick, Gregory. 2003. "Is the Theory of Natural Selection Independent of Its History?" In *The Cambridge Companion to Darwin*, edited by Jonathan Hodge and Gregory Radick, 143–67. Cambridge: Cambridge University Press.

Radick, Gregory. 2005a. "Other Histories, Other Biologies." In *Philosophy, Biology, and Life*, edited by Anthony O'Hear, 21–47. Cambridge: Cambridge University Press.

Radick, Gregory. 2005b. "What If . . . ? Introduction to a Feature on the History of Science Counterfactuals." *New Scientist*, August 20. http://www.newscientist.com/channel/opinion/mg18725131.500-what-if-exploring-alternative-scientific-pasts.html.

Radick, Gregory. 2008a. "Why What If?" *Isis* 99: 547–51.

Radick, Gregory, ed. 2008b. "Counterfactuals and the Historian of Science." *Isis* 99: 547–84.

Radick, Gregory, and Annie Jamieson. 2013. "Putting Mendel in His Place: How Curriculum Reform in Genetics and Counterfactual History of Science Can Work Together." In *The Philosophy of Biology: A Companion for Educators*, edited by K. Kampourakis, 577–95. Dordrecht: Springer.

Raman, Vankata, and Paul Forman. 1969. "Why Was It Schrödinger Who Developed de Broglie's Ideas?" *Historical Studies in the Physical Sciences* 1: 291–314.

Rescher, Nicholas. 1993. *Pluralism: Against the Demand for Consensus*. Oxford: Clarendon Press.

Rescher, Nicholas. 2009. *Unknowability: An Inquiry into the Limits of Knowledge*. New York: Lexington Books.

Robbins, Philip. 2004. "Knowing Me, Knowing You: Theory of Mind and the Machinery of Introspection." *Journal of Consciousness Studies* 11: 129–43.

Robinson, Abraham. 1966. *Non-standard Analysis*. Amsterdam: North-Holland.

Robinson, Abraham. 1979. *Selected Papers of Abraham Robinson*, vol. 2. Amsterdam: North-Holland.

Rosa, R. 1979. "Electron Interference: Landé's Approach Upset by a Recent Elegant Experiment." *Lettere al Nuovo Cimento* 24 (April 21): 549–50.

Rose, C. D. 1928. "The Reflexion of Electrons from Aluminum Crystal." *Philosophical Magazine* 6, no. 37: 726–28.

Rosenthal, David M. 2005. *Consciousness and Mind*. Oxford: Oxford University Press.

Rothstein, Jerome. 1962. "Wiggleworm Physics." *Physics Today* 15 (September): 28–38.

Rouse, Joseph. 2002. *How Scientific Practices Matter*. Chicago: University of Chicago Press.

Rouse, Joseph. 2011. "Articulating the World: Experimental Systems and Conceptual Understanding." *International Studies in the Philosophy of Science* 25, no. 3: 243–55.

Rouse, Joseph. 2015. *Articulating the World*. Chicago: University of Chicago Press.

Roustang, François. 2009. *Le secret de Socrate pour changer la vie*. Paris: Odile Jacob.

Rudwick, Martin J. S. 1985. *The Great Devonian Controversy*. Chicago: University of Chicago Press.

Russell, Bertrand. [1912] 1946. *The Problems of Philosophy*. London: Oxford University Press (Home University Library).

Russell, Bertrand. 1921. *The Analysis of Mind*. London: Allen and Unwin.

Russo, Arturo. 1981. "Fundamental Research at Bell Laboratories: The Discovery of Electron Diffraction." *Historical Studies in the Physical Sciences* 12: 117–60.

Sackur, Jerome. 2009. "L'introspection en psychologie expérimentale." *Revue d'histoire des sciences* 62: 5–28.

Salanskis, Jean-Michel. 1999. *Le constructivisme non standard*. Lille: Presses du Septentrion.

Sankey, Howard. 2008. "Scientific Realism and the Inevitability of Science." *Studies in History and Philosophy of Science* 39: 259–64.

Sartre, Jean-Paul. [1943] 1973. *Being and Nothingness*. New York: Washington Square Press.

Sartre, Jean-Paul. [1936] 2000. *La Transcendance de l'ego*. Paris: Vrin.

Schmid-Schönbein, Christiane. 1998. "Improvement of Seizure Control by Psychological Methods in Patients with Intractable Epilepsies." *Seizure* 7: 261–70.

Schooler, Jonathan W. 2002. "Re-representing Consciousness: Dissociations between Experience and Meta-consciousness." *Trends in Cognitive Sciences* 6: 339–44.

Searle, John R. 2001. *Rationality in Action*. Cambridge, MA: MIT Press.

Shafii, Mohammad. 1973. "Silence in the Service of Ego: Psychoanalytic Study of Meditation." *International Journal of Psychoanalysis* 54: 431–43.

Shanon, Benny. 1984. "The Case for Introspection." *Cognition and Brain Theory* 7: 167–80.

Shapin, Steven. 1982. "History of Science and Its Sociological Reconstructions." *History of Science* 20: 157–211.

Shapin, Steven. 2007. "What Else Is New? How Uses, Not Innovations, Drive Human Technology." *New Yorker*, May 14. http://www.virtualrhetoric.com/onlineclass/moodledata/12/Readings/Shapin.pdf.

Shapin, Steven, and Simon Schaffer. 1985. *Leviathan and the Air-Pump: Hobbes, Boyle, and the Experimental Life*. Princeton, NJ: Princeton University Press.

Shubin, Mikhail A., and Alexander Zvonkin. 1984. "Non-standard Analysis and Singular Perturbations of Ordinary Differential Equations." *Russian Mathematical Surveys* 39, no. 2: 69–131.

Siegmund-Schultze, Reinhard. 2004. "A Non-Conformist Longing for Unity in the Fractures of Modernity: Towards a Scientific Biography of Richard von Mises, 1883–1953." *Science in Context* 17: 333–70.

Silverman, Michael, and Arien Mack. 2006. "Change Blindness and Priming: When It Does and Does Not Occur." *Consciousness and Cognition* 15: 409–22.

Skinner, Charles. 1935. *Readings in Psychology*. New York: Farrar and Rinehart.

Smith, Crosbie, and M. Norton Wise. 1989. *Energy and Empire*. Cambridge: Cambridge University Press.

Snobelen, Steven S. 2004. *Newton Heresy and His Science*. http://www.isaac-newton.org/.

Soler, Léna. 2004. "The Incommensurability Problem: Evolution, Current Approaches and Recent Issues." *Philosophia Scientiae*, 8, no. 1: 1–38.

Soler, Léna. 2006a. "Contingence ou inévitabilité des résultats de notre science?" *Philosophiques* 33, no. 2: 363–78.

Soler, Léna, ed. 2006b. *Philosophie de la physique: dialogue à plusieurs voix autour de controverses contemporaines et classiques (entre Michel Bitbol, Bernard d'Espagnat, Pascal Engel, Paul Gochet, Léna Soler et Hervé Zwirn)*. Paris: L'Harmattan.

Soler, Léna. 2008a. "Are the Results of Our Science Contingent or Inevitable?" *Studies in History and Philosophy of Science*, 39: 221–29.

Soler, Léna. 2008b. "Revealing the Analytical Structure and Some Intrinsic Major Difficulties of the Contingentist/Inevitabilist Issue." *Studies in History and Philosophy of Science* 39: 230–41.

Soler, Léna. 2008c. "The Incommensurability of Experimental Practices: The Incommensurability *of What?* An Incommensurability *of the Third-Type?*" In *Rethinking Scientific Change and Theory Comparison: Stabilities, Ruptures, Incommensurabilities?*, edited by Léna Soler, Howard Sankey, and Paul Hoyningen, 299–340. Dordrecht: Springer, Boston Studies for Philosophy of Science.

Soler, Léna. 2009. "The Convergence of Transcendental Philosophy and Quantum Physics: Grete Henry-Hermann's 1935 Pioneering Proposal." In *Constituting Objectivity: Transcendental Perspectives on Modern Physics*, edited by Michel Bitbol, Pierre Kerszberg, and Jean Petitot, 325–40. Dordrecht: Springer.

Soler, Léna. 2011. "Tacit Aspects of Experimental Practices: Analytical Tools and Epistemological Consequences." *European Journal for the Philosophy of Science* 1, no. 3: 394–433.

Soler, Léna. 2012a. "The Solidity of Scientific Achievements: Structure of the Problem, Difficulties, Philosophical Implications." In Soler et al., *Characterizing the Robustness of Science*, 1–60. Dordrecht: Springer.

Soler, Léna. 2012b. "Robustness of Results and Robustness of Derivations: The Internal Architecture of a Solid Experimental Proof." In Soler et al., *Characterizing the Robustness of Science*, 227–66. Dordrecht: Springer.

Soler, Léna, and Howard Sankey, eds. 2008. "Are the Results of Our Science Contingent or Inevitable? A Symposium Devoted to the Contingency Issue." *Studies in History and Philosophy of Science* 39: 220–64.

Soler, Léna, Emiliano Trizio, Thomas Nickles, and William Wimsatt, eds. 2012. *Characterizing the Robustness of Science: After the Practice Turn in Philosophy of Science*. Boston Studies in the Philosophy of Science, 292. Dordrecht: Springer.

Soler, Léna, Sjoerd Zwart, Michael Lynch, and Vincent Israël-Jost, eds. 2014. *Science after the Practice Turn in the Philosophy, History, and Social Studies of Science*. Studies in the Philosophy of Science. New York: Routledge.

Solomon, Miriam. 1995. "The Pragmatic Turn in Naturalistic Philosophy of Science." *Perspectives on Science* 3, no. 2: 206–30.

Solomon, Miriam. 2001. *Social Empiricism*. Cambridge MA: MIT Press.

Sperling, George. 1960. "The Information Available in Brief Visual Presentations." *Psychological Monographs* 74, no. 9: 1–29.

Stanford, P. Kyle. 2006. *Exceeding Our Grasp: Science, History and the Problem of Unconceived Alternatives*. Oxford: Oxford University Press.

Steiner, Mark. 2005. "Mathematics: Application and Applicability." In *The Oxford Handbook of Philosophy of Mathematics and Logic*, edited by Stewart Shapiro, 625–50. Oxford: Oxford University Press.

Stern, Daniel. 1985. *The Interpersonal World of the Infant*. New York: Basic Books.

Swinden, Tobias. [1714] 1727. *An Enquiry into the Nature and Place of Hell*, 2nd ed. London: Thomas Astley.

Ten Elshof, Gregg. 2005. *Introspection Vindicated: An Essay in Defense of the Perceptual Model of Self Knowledge*. Farnham: Ashgate.

Ten Hoor, Marten. 1932. "A Critical Analysis of the Concept of Introspection." *Journal of Philosophy* 29: 322–31.

Tetlock, Philip E., Richard Lebow, and Geoffrey Parker, eds. 2006. *Unmaking the West: "What-If" Scenarios That Rewrite World History*. Ann Arbor: University of Michigan Press.

Thompson, Evan. 2007. *Mind in Life*. Cambridge, MA: Harvard University Press.

Thompson, Evan. 2014. *Waking, Dreaming, Being: New Light on the Self and Consciousness from Neuroscience, Meditation, and Philosophy*. New York: Columbia University Press.

Thomson, George Paget. 1928. "The Waves of an Electron." *Nature* 122, no. 3069: 279–82.

Thomson, George Paget. [1931] 1970. "The Optics of Electrons: Discourse at the Royal Institution." December 4. Reprinted in Bragg and Porter, eds. *The Royal Institution Library of Science*, vol. 9, 357–64.

Thomson, George Paget. 1961. "Early Work in Electron Diffraction." *American Journal of Physics* 29: 821–25.

Thomson, George Paget. 1968. "The Early History of Electron Diffraction." *Contemporary Physics* 9: 1–15.

Thomson, George Paget, and Alexander Reid. 1927. "Diffraction of Cathode Rays by a Thin Film." *Nature* 119, no. 3007 (June 18): 890.

Titchener, Edward B. 1909. *Lectures on the Experimental Psychology of Thought Processes.* London: Macmillan.

Titchener, Edward B. 1910. *A Textbook of Psychology.* New York: Scholars' Facsimiles and Reprints.

Titchener, Edward B. 1912. "The Schema of Introspection." *American Journal of Psychology* 23: 485–508.

Titchener, Edward B. 1914. "Psychology: Science or Technology?" *Popular Science Monthly* 84: 39–51.

Titchener, Edward B. 1916. *A Textbook of Psychology.* London: Macmillan.

Trizio, Emiliano. 2004. "Incommensurability and Laboratory Science." *Philosophia Scientiae*, 8, no. 1: 235–67.

Trizio, Emiliano. 2008. "How Many Sciences for One World? Contingency and the Success of Science." *Studies in History and Philosophy of Science* 39: 253–58.

Van Bendegem, Jean Paul. 2000. "Alternative Mathematics: The Vague Way." In *Festschrift in Honour of Newton C. A. da Costa on the Occasion of His Seventieth Birthday. Synthese*, edited by Décio Krause, Steven French, and Francisco A. Doria vol. 125, nos. 1–2, 19–31.

Van Bendegem, Jean Paul. 2002. "Inconsistencies in the History of Mathematics: The Case of Infinitesimals." In *Inconsistency in Science*, edited by Joke Meheus, 43–57. Origins: Studies in the Sources of Scientific Creativity, vol. 2. Dordrecht: Kluwer Academic.

Van Bendegem, Jean Paul. 2008. "'What-if' Stories in Mathematics: An Alternative Route to Complex Numbers." In *Linguista Sum: Mélanges offerts à Marc Dominicy à l'occasion de son soixantième anniversaire*, edited by Emmanuelle Danblon, Mikhail Kissine, Fabienne Martin, Christine Michaux, and Svetlana Vogeleer, 391–402. Paris: Harmattan.

Van der Gracht, Willem A. J. M. van Waterschoot, ed. 1928. *Theory of Continental Drift.* Tulsa, OK: American Association of Petroleum Geologists.

Van Fraassen, Bas. 1980. *The Scientific Image.* Oxford: Oxford University Press.

Van Fraassen, Bas. 2008. *Scientific Representation.* Oxford: Oxford University Press.

Van Vliet, C. M. 2010. "Linear Momentum Quantization in Periodic Structures II." *Physica* A 389, 1585–93.

Varela, Francisco J. 1996. "Neurophenomenology." *Journal of Consciousness Studies* 3: 330–49.

Varela, Francisco J., and Jonathan Shear, eds. 1999. *The View from Within.* Exeter: Imprint Academics.

Varela, Francisco J., Evan Thompson, and Eleanor Rosch. 1991. *Embodied Mind.* Cambridge, MA: MIT Press.

Vermersch, Pierre 1994. *L'entretien d'explicitation.* Paris: Éditions ESF.

Vermersch, Pierre 1999. "Introspection as Practice." *Journal of Consciousness Studies* 6, 15–42.

Vermersch, Pierre. 2011. "Husserl the Great Unrecognized Psychologist!" *Journal of Consciousness Studies* 18: 20–23.

Wallace, B. Alan. 1998. *The Bridge of Quiescence.* Chicago: Open Court Press.

Wallace, B. Alan. 2000. *The Taboo of Subjectivity.* Oxford: Oxford University Press.

Wallace, B. Alan. 2006a. *Contemplative Science.* New York: Columbia University Press.

Wallace, B. Alan. 2006b. *The Attention Revolution.* Somerville, MA: Wisdom.

Warren, Donna D. 1999. "Externalism and Causality: Simulation and the Prospects for a Reconciliation." *Mind and Language* 14: 154–76.

Warren, Howard C., and Leonard Carmichael. 1930. *Elements of Human Psychology.* Boston: Houghton Mifflin.

Watson, John B. 1913. "Psychology as the Behaviorist Views It." *Psychological Review* 20: 158–77.

Watt, Henry J. 1905. "Experimentelle Beiträge zu einer Theorie des Denkens." *Archiv für die gesamte Psychologie* 4: 289–436.

Weber, Marcel. 2007. "Redesigning the Fruit Fly." In *Science without Laws, Model Systems, Cases, Exemplary Narratives,* edited by Angela N. H. Creager, Elizabeth Lunbeck, and M. Norton Wise, 23–45. Durham, NC: Duke University Press.

Weber, Max. [1906] 1949. "Critical Studies in the Logic of the Cultural Sciences." In *The Methodology of the Social Sciences,* edited and translated by E. A. Shils and H. A. Finch, 13–188. New York: Free Press.

Wegener, Alfred. 1915. *Die Enstehung der Kontinente and Ozeane.* Braunschweig, Germany: F. Vieweg, 2nd ed. 1920, 3rd ed. 1922, 4th ed. 1929.

Wegener, Alfred. 1924. *The Origin of Continents and Oceans,* translated by J. G. A. Skerl from the 3rd German ed. London: Methuen.

Wegener, Alfred. 1966. *The Origin of Continents and Oceans,* translated by J. Biram from the 4th German ed. Mineola, NY: Dover.

Weinberg, Steven. 1996a. "Sokal's Hoax." *New York Review of Books* (August 8): 11–15.

Weinberg, Steven. 1996b. "Reply." *New York Review of Books* (October 3): 55–56.

Weinberg, Steven. 2001. "Physics and History." In *The One Culture: A Conversation about Science,* edited by J. A. Labinger and H. M. Collins, 116–27. Chicago: University of Chicago Press.

Weinel, Martin. 2007. "Primary Source Knowledge and Technical Decision-Making: Mbeki and the AZT Debate." *Studies in History and Philosophy of Science* 38, no. 4: 748–60.

Weinryb, Elazar. 2009. "Historiographic Counterfactuals." In *A Companion to the Philosophy of History and Historiography,* edited by A. Tucker, 109–19. Oxford: Wiley-Blackwell.

Whiston, William. 1717. *Astronomical Principles of Religion, Natural and Reveal'd.* London: J. Senex and W. Taylor.

Whitehead, Alfred North. 1925. *Science and the Modern World.* New York: Macmillan.

Wigner, Eugene. 1960. "The Unreasonable Effectiveness of Mathematics in the Natural Sciences." *Communications in Pure and Applied Mathematics* 13: 1–14.

Wilson, Mark. 2006. *Wandering Significance.* Oxford: Oxford University Press.

Wimsatt, William. 1981. "Robustness, Reliability and Overdetermination." In *Scien-*

tific Inquiry and the Social Sciences, edited by M. B. Brewer and B. E. Collins, 124–63. San Francisco: Jossey-Bass.

Winch, Peter G. 1958. *The Idea of a Social Science*. London: Routledge and Kegan Paul.

Wittgenstein, Ludwig. 1953. *Philosophical Investigations*. Oxford: Blackwell.

Wittgenstein, Ludwig. 1978. *Remarks on the Foundations of Mathematics*, 3rd ed., edited by G. H. von Wright, R. Rhees, and G. E. M. Anscombe; translated by G. E. M. Anscombe. Oxford: Blackwell.

Wittgenstein, Ludwig. 1980. *Philosophical Remarks*. Chicago: University of Chicago Press.

Wolf, R. 1862. *Biographien zur Kulturgeschichte der Schweiz*. Orell, Füssli, 4:173–92.

Woodward, James. 2005. *Making Things Happen: A Theory of Causal Explanation*. New York, Oxford University Press.

Woodworth, Robert Sessions. 1906. "Imageless Thought." *Journal of Philosophy, Psychology and Scientific Methods* 3: 701–8.

Woody, Andrea. 2014. "Chemistry's Periodic Law: Rethinking Representation and Explanation after the Turn to Practice." In Soler et al., *Science after the Practice Turn*, 123–150.

Worms, Frédéric. 2009. *La philosophie en France au XXe siècle: Moments*. Paris: Gallimard.

Worrall, John. 1989. "Structural Realism: The Best of Both Worlds?" *Dialectica* 43, nos. 1–2: 99–124. Reprinted in *The Philosophy of Science*, edited by David Papineau, 139–65. Oxford: Oxford University Press.

Wroblewski, Andrzej K. 2002. "Development of Science: Determined or Arbitrary?" In *In the Scope of Logic, Methodology and Philosophy of Science*, edited by Peter Gärdenfors, Jan Wolenski, Katarzyna Kijania-Placek, 1:3–13. Dordrecht: Kluwer.

Wundt, Wilhelm. 1901. *Lectures on Human and Animal Psychology*. London: Swan Sonnenshein.

Wundt, Wilhelm. 1910. *Principles of Physiological Psychology*. London: Swan Sonnenschein.

Wunenburger, Jean-Jacques, ed. 2003. *Bachelard et l'épistémologie française*. Paris: Presses Universitaires de France.

Yourgrau, Wolfgang, and Stanley Mandelstam. [1968] 1979. *Variational Principles in Dynamics and Quantum Theory*. Mineola, NY: Dover.

Yourgrau, Wolfgang, and A. van der Merwe, eds. 1979. *Perspectives in Quantum Theory: Essays in Honor of Alfred Landé*. Mineola, NY: Dover.

Zwirn, Hervé. 2000. *Les limites de la connaissance*. Paris: Odile Jacob.

CONTRIBUTORS

Catherine Allamel-Raffin is associate professor of philosophy and history of science at the Université de Strasbourg, France. She is a member of the Institut de Recherches Interdisciplinaires sur les Sciences et la Technologie (Strasbourg), and a member of the Laboratoire d'Histoire des Sciences et de Philosophie–Archives Henri Poincaré (Nancy, France). She has recently worked on the production and functions of images in scientific investigation processes, especially in astrophysics, material physics, and pharmacology. Recent publications include "The Meaning of a Scientific Image: Case Study in Nanoscience," *Nanoethics* 5, no. 2 (2011); "Interpreting Artworks, Interpreting Scientific Images," *Leonardo* 48, no. 1 (2015).

Michel Bitbol is a researcher at the Centre National de la Recherche Scientifique, based at the Husserl Archive (Ecole Normale Supérieure, Paris). He successively received an MD and a PhD in physics, and a habilitation in philosophy. After a start in scientific research, he turned to philosophy, editing texts by Erwin Schrödinger and formulating a neo-Kantian philosophy of quantum mechanics. He then studied the relations between the philosophy of physics and the philosophy of mind, working in close collaboration with Francisco Varela, and is currently developing a phenomenological critique of naturalist theories of consciousness.

Mieke Boon is full professor of philosophy of science in practice at the University of Twente. She received her PhD (with honor) in chemical engineering from the Technical University of Delft, and studied philosophy at the University of Leiden. In 2003, she was awarded a Vidi research grant from the Dutch National Science Foundation (NWO) for developing a Philosophy of Science for the Engineering Sciences. Currently, she continues this line of research on an NWO Aspasia grant. She is a cofounder of the Society for Philosophy of Science in Practice. Her publications can be found at the UT Repository: http://www.utwente.nl/bms/wijsb/organization/boon/publication%20list.html.

Hasok Chang is the Hans Rausing Professor of History and Philosophy of Science at the University of Cambridge. He received his degrees from Caltech and Stanford, and has taught at University College London. He is the author of *Is Water H₂O? Evidence, Realism and Pluralism* (2012), and *Inventing Temperature: Measurement and Scientific Progress* (2004). He is a cofounder of the Society for Philosophy of Science in Practice, and the International Committee for Integrated History and Philosophy of Science.

Harry Collins is Distinguished Research Professor and directs the Centre for the Study of Knowledge, Expertise and Science at Cardiff University. He is a Fellow of the British Academy and winner of the Bernal Prize for Social Studies of Science. His eighteen books cover sociology of scientific knowledge, artificial intelligence, the nature of expertise, and tacit knowledge. He is continuing his research on the sociology of gravitational wave detection, expertise, fringe science, science and democracy, technology in sport, and a new technique—the "Imitation Game"—for exploring expertise and comparing the extent to which minority groups are integrated into societies.

Jean-Luc Gangloff teaches philosophy in high school. He is member of the Institut de Recherches Interdisciplinaires sur les Sciences et la Technologie (Strasbourg, France), and of the Laboratoire d'Histoire des Sciences et de Philosophie–Archives Henri Poincaré (Nancy, France). He is presently interested in issues related to the interactions between science and fiction. A recent publication (in collaboration with Catherine Allamel-Raffin) is "Robustness and Scientific Images," in *Characterizing the Robustness of Sciences: After the Practical Turn in Philosophy of Science*, edited by Léna Soler et al. (Springer, 2012).

Ronald N. Giere is professor of philosophy emeritus and a member, as well as a former director, of the Center for Philosophy of Science at the University of Minnesota. He is the author of *Understanding Scientific Reasoning* (5th ed., 2006), *Explaining Science: A Cognitive Approach* (1988), *Science without Laws* (1999), and *Scientific Perspectivism* (2006). He is a past president of the Philosophy of Science Association. He retired to Bloomington, Indiana, in 2011, where he is an adjunct professor in the Department of History and Philosophy of Science, his home department from 1966 to 1987. Web site: http://www.tc.umn.edu/~giere; e-mail: giere@umn.edu.

Yves Gingras is professor in the History Department and Canada Research Chair in History and Sociology of Science at the Université du Québec in Montreal. His research covers the mathematization of the sciences, the uses of formal analogies, the transformation of the universities, and the dynamic of scientific disciplines. His most recent books are *Sociologie des sciences* (Paris: Presses universitaires de France, 2013), *Les dérives de l'évaluation de la recherche: Du bon usage de la bibliométrie* (Paris : Raisons d'agir, 2014). He is also the editor of *Controverses: accords et désaccords en sciences humaines et sociales* (Paris: CNRS Éditions, 2014).

Ian Hacking is professor emeritus at the University of Toronto and the Collège de France (Paris). He is the author of several influential books and numerous articles pertaining to various areas of philosophy, logic, and history of science. They include *The Emergence of Probability* (1975), *Representing and Intervening* (1983), *Rewriting the Soul: Multiple Personality and the Sciences of Memory* (1995), *The Social Construction of What?* (1999), *Historical Ontology* (2004), and *Why Is There a Philosophy of Mathematics at All?* (2014). He has taught philosophy at Princeton University, the University of British Columbia, Stanford University, University of Toronto, and the Collège de France.

Jean-Marc Lévy-Leblond is physicist and philosopher of science, and professor emeritus at the Université de Nice, France. His research work covers theoretical physics (invariance principles and group theory, quantum theory, space-time structure), as well as philosophy of science (foundational problems of quantum theory and relativity theory). He is editor of several scientific series with the publisher Seuil (Paris), the founder and editor of the quarterly *Alliage (Culture-Science-Technique)*, and the author of several books on the social, political, and cultural issues of modern science.

Claire Petitmengin completed a PhD under the direction of Francisco Varela at the Ecole Polytechnique (Paris), on the lived experience that accompanies the emergence of an intuition. She is professor at the Institut Mines-Télécom and member of the Archives Husserl (Ecole Normale Supérieure, Paris). Her research focuses on the usually unrecognized microdynamics of lived experience and "first-person" methods enabling us to become aware of and describe lived experience. She studies the epistemological conditions of these methods and their educational, therapeutic, artistic, and technological applications, as well as the process of mutual enrichment of "first-person" and "third-person" analyses in the context of neurophenomenological projects.

Andrew Pickering is professor of sociology and philosophy at the University of Exeter, UK. He is a leading figure in the field of science and technology studies, and is the author of *Constructing Quarks, The Mangle of Practice,* and, most recently, *The Cybernetic Brain: Sketches of Another Future.* His current research explores questions of agency in life, art, environmental management, and traditional Chinese philosophy.

Joseph Rouse is the Hedding Professor of Moral Science in the Philosophy Department and chair of the Science in Society Program at Wesleyan University. He is the author of *Articulating the World* (University of Chicago Press, 2015), *How Scientific Practices Matter* (University of Chicago Press, 2002), *Engaging Science* (Cornell University Press, 1996), *Knowledge and Power* (Cornell University Press, 1987), and editor of John Haugeland's posthumous *Dasein Disclosed* (Harvard University Press, 2013).

Jean-Michel Salanskis is professor of philosophy of science, logic and epistemology at Université Paris Ouest Nanterre La Défense. He has worked in the fields of philosophy of mathematics, phenomenology, contemporary philosophy, and philosophy of Jewish tradition. He is the author of 21 books, has edited or coedited 10 collections, and published around 150 papers. In the field of philosophy of mathematics, his work includes *Le constructivisme non standard* (Lille: Presses Universitaires du Septentrion, 1999), *Philosophie des mathématiques* (Paris: Vrin, 2008), and *L'herméneutique formelle* (2nd ed., Paris: Klincksieck, 2013).

Léna Soler is associate professor at the Université de Lorraine, and member of the Laboratoire d'Histoire des Sciences et de Philosophie–Archives Henri Poincaré (Nancy, France). Her areas of specialization are philosophy of science and philosophy of physics. She wrote an *Introduction à l'épistémologie* (2nd ed., Ellipses, 2009), and has been the main editor of several recent collective publications, among which are *Characterizing the Robustness of Science* (Springer, 2012), *Science after the Practice Turn in the Philosophy, History, and Social Studies of Science* (Routledge, 2014), and a special issue on contingency in science (*Studies in History and Philosophy of Science*, 2008).

Emiliano Trizio (MPhil London School of Economics 2001, PhD Paris-X/ Ca' Foscari, Venice 2005) has taught at Lille III, and held a postdoctoral grant at Université de Nancy 2 with the group PratiScienS. He is currently a full-time instructor in the Philosophy Department of Seattle University, where he teaches philosophy of the human person, ethics, and introductory courses on ancient philosophy and culture. He is the author of several publications on phenomenology and on various aspects of philosophy of science. His main research focuses on the study of the epistemological, ontological, and ethical dimensions of Husserl's phenomenology, and its relations with the main contemporary philosophical currents.

Jean Paul Van Bendegem is professor at the Vrije Universiteit Brussel (Belgium) and director of the Center for Logic and Philosophy of Science (www.vub.ac.be/CLWF). He is also the editor of the logic journal *Logique & Analyse*. His research interests are in the domain of the philosophy of mathematics, more particularly, the study of strict finitism as a foundational theory and the study of mathematical practices. He was a cofounder of the Association for the Philosophy of Mathematical Practice.

INDEX